W9-BAV-291

Polymer Characterisation

POLYMER CHARACTERISATION

Edited by

B. J. HUNT
The Polymer Centre
Lancaster University

and

M. I. JAMES
Procter & Gamble
Egham

BLACKIE ACADEMIC & PROFESSIONAL
An Imprint of Chapman & Hall
London · Glasgow · New York · Tokyo · Melbourne · Madras

Published by
Blackie Academic & Professional, an imprint of Chapman & Hall,
Wester Cleddens Road,
Bishopbriggs, Glasgow G64 2NZ

Chapman & Hall, 2–6 Boundary Row, London SE1 8HN, UK

Blackie Academic & Professional, Wester Cleddens Road, Bishopbriggs, Glasgow G64 2NZ, UK

Chapman & Hall Inc., 29 West 35th Street, New York NY10001, USA

Chapman & Hall Japan, Thomson Publishing Japan, Hirakawacho Nemoto Building, 6F, 1–7–11 Hirakawa-cho, Chiyoda-ku, Tokyo 102, Japan

DA Book (Aust.) Pty Ltd., 648 Whitehorse Road, Mitcham 3132, Victoria, Australia

Chapman & Hall India, R. Seshadri, 32 Second Main Road, CIT East, Madras 600 035, India

First edition 1993

Typeset in 10/12 pt Times New Roman by Pure Tech Corporation, Pondicherry, India

Printed in Great Britain by St Edmundsbury Press Ltd, Bury St Edmunds, Suffolk

ISBN 0 7514 0082 3

A catalogue record for this book is available from the British Library

Preface

Polymers continue to play an ever increasing role in the modern world. In fact it is quite inconceivable to most people that we could ever have existed without them. As a result of the increased volume and variety of materials currently available, and the diversity of their application, characterisation has become an essential requirement of industrial and academic laboratories involved with polymeric materials. On the one hand requirements may come from polymer specialists involved in the design and synthesis of new materials who require a detailed understanding of the relationship between the precise molecular architecture and the properties of the polymer in order to improve its capabilities and range of applications. On the other hand, many analysts who are not polymer specialists are faced with the problems of analysing and testing a wide range of polymeric materials for quality control or material specification purposes.

We hope this book will be a useful reference for all scientists and technologists involved with polymers, whether in academic or industrial laboratories, and irrespective of their scientific discipline. We have attempted to include in one volume all of the most important techniques. Obviously it is not possible to do this in any great depth but we have encouraged the use of specific examples to illustrate the range of possibilities. In addition numerous references are given to more detailed texts on specific subjects, to direct the reader where appropriate.

The book is divided into 11 chapters. Following a short introductory chapter, chapter 2 describes, by way of specific examples, the chemical techniques which have been developed to 'deformulate' a commercial product and analyse the constituents. This is particularly important as most 'real' samples are not pure materials but contain a complex mixtures of additives. Chapters 3 and 4 deal with the invaluable spectroscopic techniques of NMR, IR and Raman. In chapter 5, methods for measuring the molar mass of polymers are discussed, except for the most commonly used technique of size exclusion chromatography. This is dealt with in chapter 6 along with the many other chromatographic techniques useful for analysing polymers such as HPLC, TLC and SFC. The techniques of thermal analysis are very important for characterising polymers and DSC, TGA, DMTA and DETA are described in chapter 7 along with some less familiar techinques. The theory and application of small angle neutron scattering and neutron reflectometry are described in chapter 8. Mechanical properties of polymers are extremely important as they govern the ultimate end use of these materials. The measurement of tensile and impact strength, creep and rheological properties are described in chapter 9. Measurement of these properties is not usually regarded as 'polymer characterisation' but we thought it useful and appropriate that they should be included. The final two chapters deal with light and electron microscopy (chapter 10) and the characterisation of polymer surfaces using XPS and SIMS (chapter 11).

Nomenclature and terminology in polymer science, as in all areas of science, change with time. For example 'gel permeation chromatography' is now more correctly referred to as 'size exclusion chromatography' and 'molecular weight' has become 'molar mass'. However, the old terms are still used and it has been left to the indivudual authors to decide on appropriate terminology.

Finally we would like to thank all the authors for their contributions and our colleagues for helpful advice and discussions. MIJ would also like to acknowledge the permission given by both his previous and current employers (Courtaulds Research and Procter and Gamble respectively) for his involvement in this project.

BJH

MIJ

Contents

Contributors

Mr J. M. Chalmers — ICI Wilton Materials Research Centre
PO Box 90
Middlesborough TS6 8JE, UK

Dr N. J. Everall — ICI Wilton Materials Research Centre
PO Box 90
Middlesborough TS6 8JE, UK

Mr M. Flanagan — Trowbridge College
Trowbridge
Wiltshire BA14 0ES, UK

Mr A. Handley — ICI Chemicals and Polymers Ltd
PO Box 8
Runcorn WA7 4QD, UK

Dr B. J. Hunt — The Polymer Centre
Lancaster University
Lancaster LA1 4YW, UK

Dr M. I. James — Procter & Gamble
(Health and Beauty Care Ltd)
Rusham Park
Whitehall Lane
Egham
Surrey TW20 9NW, UK

Dr K. Kamide — Asahi Chemical Industry Company Ltd
11-7 Hacchonawate, Takatsuki,
Osaka 569, Japan

Dr P. A. Mirau — A T & T Bell Laboratories
PO Box 636, Murray Hill
NJ 07974-0636, USA

Dr Y. Miyazaki — Asahi Chemical Industry Company Ltd
11-7 Hacchonawate, Takatsuki,
Osaka 569, Japan

Dr H. S. Munro — Courtaulds Research
Lockhurst Lane
Coventry CV6 5RH, UK

Mr I. D. Newton	Wilton Materials Research Centre PO Box 90 Middlesborough TS6 8JE, UK
Dr R. W. Richards	IRC in Polymer Science and Technology University of Durham Durham DH1 3LE, UK
Dr M. Saito	Asahi Chemical Industry Company Ltd 11-7 Hacchonawate, Takatsuki, Osaka 569, Japan
Dr S. Singh	Courtaulds Research Lockhurst Lane Coventry CV6 5RH, UK
Dr A. S. Vaughan	J J Thomson Physical Laboratory University of Reading Reading RG6 2AT, UK
Professor R. E. Wetton	Thermal Science Division Polymer Laboratories Loughborough LE11 0QE, UK

1 Introduction

B. J. HUNT and M. I. JAMES

1.1 Introduction

Polymeric materials now form an inextricable part of modern life. In fact it is difficult to imagine our everyday existence without these materials. Consider transport, energy production and transmission, agriculture, the building industry, clothing, consumer goods, packaging, food and the health and pharmaceutical industries; all these activities rely heavily on polymeric materials. The polymeric products that they use can take on many forms such as viscous liquids, fibres, films, mouldings, composites, powders and granules. Natural polymers are also very important. Apart from the commercial importance of cellulosic polymers, biological macromolecules are essential to life itself. The characterisation of these materials has been pursued with great vigour in recent years with the explosive increase in genetics and related research. As far as the scientific community is concerned polymers also impinge on many fields of research. As an example from the chemical disciplines, it has been estimated that one-third of all chemists will work with polymers at some stage in their careers.

Many of the high-tonnage polymers such as the polyolefins, poly (vinyl chloride) (PVC), polystyrene and natural and synthetic rubbers have been in use for more than half a century. Their methods of production have increased in sophistication and efficiency with a corresponding improvement in quality and reproducibility. One of the most important properties of a polymer is its molar mass and molar mass distribution. Indeed most of the important characteristics of polymeric materials are governed by these properties. Thus impact, tensile and adhesive strengths, brittleness, hardness, melt flow and solution properties are all determined by the molar mass and molar mass distribution. Some of these properties were originally monitored by rather crude techniques, some of which are still used today: for example, mechanical properties to test the strength of the product, viscosity (in the melt and solution) to indicate the degree of polymerisation (molar mass) and 'scorch time' to assess the degree of cure of rubbers. The rules and regulations associated with health and safety and product liability have resulted in much more stringent control of specifications for polymers (and other materials). Consequently more sophisticated techniques for characterising these materials, including any additives present, have been developed. In recent years high-value speciality polymers, such as those used in some engineering and aerospace applications, are being used in ever more demanding applications and require even higher specifications. Many of these materials are not pure homopolymers but may be co- or ter-polymers or blends of polymers. The requirement to characterise completely such materials has become considerable. Simple quality control tests may still be sufficient in some cases but not if the products are to be used in more demanding applications.

One of the limitations of polymers is their relatively poor thermal stability compared to inorganic and other engineering materials. The quest to synthesise polymers with increased thermal stability is still pursued vigorously. There are many applications for such materials in some of the previously mentioned industries. Although liquid crystals have been known for many years it is only in the last decade or so that polymers which exhibit liquid-crystal behaviour have been synthesised and characterised and are beginning to find commercial applications. The possibility of making a polymer that conducts electricity has been pursued by many scientists following the sucessful synthesis of polyacetylene. Again this would be a most desirable material. Unfortunately, the very nature of thermally stable polymers, liquid-crystal polymers and electrically conducting polymers make them intractable and insoluble and consequently extremely difficult to characterise. The ability to design polymers with specific properties for particular applications is an increasing reality as the polymerisation processes, mechanisms and structure–property relationships become better understood.

The aim of this book is to bring together in a single volume a survey of the more important, useful and common characterisation techniques, and to indicate by way of example and application what information can be obtained from the techniques. We hope that the readers of this book will be many and varied, but in particular analysts involved in the characterisation of polymeric materials, synthetic polymer chemists who need to know what they have made, and scientists and engineers in general who find that they are becoming increasingly involved with polymers and require information on the properties and characteristics of the materials they are dealing with.

1.2 Levels of characterisation

Polymers may be synthesised by many processes, very often the precise conditions used having an effect on the structure of the product. It is beyond the scope of this work to describe these processes in detail. In any case, this information is readily available in a variety of standard texts. In a similar way, the performance of polymers is influenced not only by the way in which the individual monomer units are joined together, but also by the way in which the polymer chains are arranged. For example, differences in the degree of crystallinity or extent of orientation can arise from processing the same polymer under different conditions. Such variations in structure may have enormous consequences in terms of the performance in use of a finished article. The relationship between polymer structure and properties is one that has received enormous attention, and is addressed in many publications. Examples of suitable sources of this type of information are given in the references [1–13].

The aim of this section is to give some indication of the different scales at which polymeric materials can be characterised. Which of these is appropriate in any given case is dictated by the the needs of the user. For example, in determining the applicability of a particular polymer for an engineering use, the most appropriate test will probably be some form of mechanical analysis to indicate its physical properties. If, subsequently, a batch of the selected polymer fails to perform to the required standard, then determination of its microscopic nature could reveal the nature of the problem. For example, the degree of crystallinity, or even molecular characteristics such as the molecular-weight distribution or comonomer composition might be appropriate.

It will become apparent that, in most instances, no single technique will be available to give totally unambiguous information. However, when two or more methods are used in combination even the most difficult problems can usually be resolved.

In the following sections, descriptions are given of four subdivisions within the overall context of characterisation. These are somewhat arbitrary, and obviously overlap. They do, however, give a framework around which to present examples of what is possible using the techniques described in individual chapters. It should be noted that the techniques listed here are simply selected as examples, and other methods are also applicable. The scope and limitations of different techniques will of course become apparent from the individual chapters.

1.2.1 Macroscopic properties

Macroscopic properties, alternatively referred to as bulk properties or simply 'performance', are of the utmost importance in material selection. For any application it is essential that the material provides the properties desired, under the conditions of use. In addition, it is wise to characterise the material more fully in order to understand what the effect might be, for example, of changing the temperature. Consideration should also be given to time-related phenomena, such as creep or stress relaxation. What are the consequences of dimensional instability? Techniques that can provide this type of information directly include mechanical testing, rheology and thermal analysis. In cases where knowledge of the relationship between structure and properties is desirable, then obviously the techniques described here must be used in combination with those which follow.

1.2.2 Microscopic properties

Here we consider those aspects of characterisation which fall between measurement of molecular structure and the bulk properties described above. A typical example might include the overall degree of crystallinity in a partially crystalline polymer, which could be determined by thermal analysis, scattering techniques or microscopy. The most appropriate method will of course be determined by the particular system of interest. Another example is taken from the area of polymer blends. In many cases the component materials are immiscible at the molecular level, and a phase-separated structure is formed. The morphology of this structure largely determines the way in which the blend will perform. Again, any of the above techniques could be used. Microscopy, in conjunction with preferential staining of one component, has proved particularly powerful in this area.

1.2.3 Molecular structure

'Molecular structure' is used to describe those attributes which are characteristic of individual polymer chains. Here we are concerned with the way in which the monomers are joined together. Within this heading are included tacticity, conformational isomerism, monomer sequence distribution, branching and the presence of minor structural defects. It is in studies of this kind that the spectroscopic techniques particularly excel. In recent years, the advent of the microcomputer for data collection and manipulation, in conjunction with modern Fourier-transform instruments, has allowed enormous advances in the quality of information that is now routinely available.

The average number of monomer units per chain, or degree of polymerisation, may also be included under this heading. This is usually converted to, and quoted as, the average molar mass or molecular weight. The importance of this parameter is reflected by the fact that a complete chapter is devoted to methods for its determination.

1.2.4 Other aspects of characterisation

Many commercial polymers contain not only the base material, but also a plethora of performance-enhancing additives such as anti-oxidants, stabilisers, plasticisers, fillers and pigments. Consequently, characterisation of such commercial materials can become extremely complex, requiring skills not only in analysis of the components, but also in their 'deformulation'. In chapter 2, several examples are given of systematic approaches to this type of problem in different systems. It is fascinating to observe the combination of traditional separation techniques, such as solvent extraction, with modern chromatographic and spectroscopic methods.

It is increasingly widely recognised that the performance of polymers in many applications is a function not only of the bulk properties, but particularly of the nature of the surface. The composition of the outermost few nanometres is critical in areas such as adhesion, and these surface layers can be very different from the bulk of the sample. A variety of modern spectroscopic techniques is now available with sufficient surface sensistivity to allow studies of this kind to take place. In chapter 11, two of the most powerful of these, XPS and SIMS, are described in some detail. From the results presented, the value of these techniques will be appreciated.

In addition to the levels of structural and performance characterisation described above, there is also some interest in measurement of the dynamics of local motions in polymer chains. Techniques such as NMR have the capability to generate this type of information through the measurement of relaxation times. The results can be useful in, for example, morphological studies of polymer blends.

1.3 Organisation of the book

The book has been divided into eleven chapters including this introductory chapter. It is difficult to divide a subject as complex and wide-ranging as this into a relatively small number of discrete topics but we hope this has been achieved in a reasonable way with a minimum of overlap between chapters. An attempt has been made to ensure that the terminology and general format is reasonably consistent. However, some old terms and phraseology are still used alongside more modern (and correct) wording. For example 'molar mass' is the more modern and correct term but 'molecular weight' is still commonly used. Similarly, 'size exclusion chromatography' (SEC) is the more correct description of the separation of polymer molecules according to their size in solution but the original terminology of 'gel permeation chromatography' (GPC) is widely used.

Chemical methods of analysis are simple and cheap but labour intensive. However, they are essential, coupled with spectroscopic techniques, for deformulation work and in the analysis of additives such as anti-oxidants and stabilisers. Specific examples of the use of these procedures for polypropylene, PVC, polytetrafluoroethylene (PTFE) and polyamides are detailed in chapter 2.

Spectroscopic techniques are of great importance in polymer science, as they are in all branches of the chemical sciences, and these are dealt with in chapters 3 and 4. NMR is of prime importance as a method for identification and elucidation of the structure of polymer chains, and a complete chapter has been devoted to this technique. In recent years it has become an increasingly sophisticated and correspondingly expensive technique, requiring considerable skill in operating the spectrometer and in interpreting spectra, but it is still available in a simple and routine form in many laboratories. Vibrational spectroscopy encompasses infrared (IR) and Raman. These techniques, IR in particular, are much more common. With the advent of Fourier-transform (FT) spectrometers, spectra are quickly acquired and data manipulation is sophisticated. FTIR spectrometers are now commonplace and FT–Raman spectrometers are becoming available. The latter are proving particularly useful to polymer scientists as the problems of fluorescence and sample degradation associated with Raman spectroscopy have been largely overcome. Ultraviolet (UV) spectroscopy has found only limited application to polymers and is only mentioned in passing (e.g. UV detectors in liquid chromatography).

Solution techniques are necessary to achieve separations of complex formulations and to determine molar mass and molar-mass distribution. These are primary characteristics of polymers and their measurement is dealt with in chapters 5 and 6. GPC is the most commonly used technique here, as it is the only simple way to obtain the molar-mass distribution, making it particularly useful for fingerprinting purposes. However, it is not an absolute technique but requires calibration, and consequently there is still a great interest in determining absolute molar masses as described in chapter 5.

Thermal methods of analysis are now routinely used to characterise polymers. Modern computerised instrumentation enables DSC, TGA, DMTA and associated analyses to be carried out quickly and simply. These techniques now rank alongside SEC and IR as routine procedures to be applied to any newly synthesised polymer. They are described in chapter 7.

Diffraction techniques are essential for obtaining information on the solid-state structure of materials. The use of X-rays, electrons, neutrons and light can be included here, the choice of radiation and technique depending on the dimensions of the species being investigated. This can range from a few angstroms to several thousand angstroms. The techniques discussed in chapter 8 are 'small-angle neutron scattering' and the more recent technique of 'neutron reflectometry'. The former has wide ranging applications in, for example, configuration studies, chain dimensions, copolymers and blends, while the latter is useful in the study of surfaces and interfaces. Space prevents the detailed discussion of the other techniques but the basic principles are similar. Unfortunately, equipment for neutron diffraction is hugely expensive and is available in only a few institutions around the world.

The ultimate use of any polymeric material is to fabricate it into a useful product with appropriate mechanical and structural properties such as strength, hardness and flexibility. Although the measurement of 'mechanical properties' is not usually regarded as 'polymer characterisation', it was thought relevant and worthy of inclusion as chapter 9.

One of the earliest scientific instruments invented was the light microscope but, apart from the biological sciences, its use in chemical laboratories has been sadly neglected. The simplicity and usefulness of this technique is highlighted in chapter 10 along with the more sophisticated techniques of scanning and transmission electron microscopy. These provide information on the physical structure of surfaces and very thin films. They are complemented by techniques such as XPS and SIMS which also yield information on the 'chemical' nature of surfaces and were chosen from a range of related techniques to illustrate the methods available for the characterisation of surfaces. They are described in some detail in chapter 11.

It is impossible to cover every analytical technique of use to the polymer scientist in a single volume; in fact several volumes would be required. We apologise in advance for any omissions which readers consider important to their particular field of work, but limitations on space have prevented the detailed discussion of many techniques. However, it is hoped that the extensive references included in some of the chapters will enable readers to find the specific information, examples and applications they require.

References

General

1. J. M. G. Cowie (1991). *Polymers: Chemistry and Physics of Modern Materials* (2nd edn), Blackie, Elasgow.
2. R. J. Young and P. A. Lovell (1991). *Introduction To Polymers* (2nd edn), Chapman & Hall, London.
3. F. W. Billmeyer (1984). *Textbook of Polymer Science* (3rd edn), Wiley.
4. R. B. Seymour and C. E. Carraher Jr. (1988). *Polymer Chemistry—An Introduction* (2nd edn), Marcel Decker.

Synthesis

5. G. Odian (1991). *Principles of Polymerization* (3rd edn), Wiley, New York.
6. J. R. Ebdon (ed.) (1991). *New Methods of Polymer Synthesis*, Blackie, Glasgow.

Structure/properties/applications

7. J. A. Brydson (1989). *Plastics Materials* (5th edn), Butterworth.
8. D. W. Van Krevelen (1990). *Properties of Polymers* (3rd edn), Elsevier, Amsterdam.

Characterisation

9. D. Campbell and J. R. White (1989). *Polymer Characterisation: Physical Techniques*, Chapman & Hall, London.
10. H. G. Barth and J. W. Mays (eds) (1991). *Modern Methods of Polymer Characterisation*, Wiley, New York.
11. J. I. Kroschwitz (ed.) (1990). *Polymers: Polymer Characterisation and Analysis*, Wiley–Interscience, New York.
12. H. G. Barth (ed.) (1990). *Polymer Analysis and Characterisation, II*, Applied Polymer Symposium No. 45, Wiley, New York.
13. C. D. Craver and T. Provder (eds) (1990). *Polymer Characterisation: Physical Property, Spectroscopic and Chromatographic Methods*, Advances in Chemistry series No. 227, American Chemical Society, Washington, DC.

2 The separation and analysis of additives in polymers

I. D. NEWTON

2.1 Introduction

This chapter describes the extraction, separation, identification and estimation of various organic and inorganic additives and polymeric materials in a range of plastic formulations such as poly(vinyl chloride), polypropylene, polyamide and polytetrafluoroethylene. It will concentrate primarily on their fractionation into, typically:

(a) a solvent-soluble organic fraction
(b) a solvent-insoluble inorganic or polymeric fraction
(c) the solvent-soluble base polymer.

The extraction techniques employed will include solvent dissolution and extraction, centrifugation, polymer precipitation, filtration and ashing. Details will be given of the identification of these fractions using IR spectroscopy, pyrolysis/GC, HPLC, mass spectrometry and NMR spectroscopy, often after further separation using either specific solvents or TLC [21]. All IR spectra are recorded over the region 4000–667 cm^{-1} (2.5–15.0 μm).

The estimation of individual components by gravimetric techniques, IR spectroscopy, UV–visible spectroscopy, HPLC or titrimetry will be described.

The basic principles of solvent extraction, etc., for the analysis of plastics formulations are well established and documented [1–4] and are still valid today. However, changes and improvements have been made, particularly in the fields of separation science and spectroscopy, to incorporate new ideas and instrumentation and to accommodate new additives. Despite such trends as robotics, the need for an experienced practical plastics analyst working at a bench using solvents, extraction flasks and beakers still exists.

One theme running through this chapter will be that of method validation and the measurement of precision by way of calculated estimated standard deviations (sd) and coefficients of variance (cv) using 'check samples' of known formulation and a collection of reference polymers and additives of known provenance. This approach not only will be necessary for any laboratories seeking accreditation such as BS5750 or NAMAS, particularly as a means of demonstrating to external auditors the competence of laboratory staff, but also can be used as a means of comparing the precision of analytical procedures. A selection of measured sd and cv values is given in Table 2.1.

Table 2.1 A selection of measured sd and cv values

Procedure	$\bar{\chi}$	sd	cv
Ethoxylated amine in polypropylene by colorimetry	1236 ppm	5 ppm	0.40%
Irganox 1010 in polypropylene by HPLC	784 ppm	5 ppm	0.62%
Talc in polypropylene by ashing	38.37%	0.04%	0.10%
Chimassorb 944 in polypropylene by UV spectroscopy	1425 ppm	36 ppm	2.50%
Nigrosine dye in nylon by visible spectroscopy	1.29%	0.02%	1.55%
Tinuvin P in acrylic by UV spectroscopy	201 ppm	4 ppm	1.99%
Rubber in acrylic by gravimetry	24.04%	0.55%	2.29%
Rubber in nylon by gravimetry	22.43%	0.18%	0.81%
Filler in PVC by gravimetry	14.85%	0.05%	0.36%
Plasticiser in PVC by gravimetry	28.61%	0.35%	1.23%
Sodium lauryl sulphate in PTFE dispersion by titration	1.536%	0.012%	0.75%
Total solids in PTFE dispersion by gravimetry	63.55%	0.02%	0.024%

2.2 Polypropylene

2.2.1 Introduction

The analysis of polypropylene-based materials involves testing a variety of materials over a wide range of applications from thin films for packaging through to large mouldings for car bumpers. While this spread of materials will invariably contain a wide range of additives imparting properties such as UV stability, clarity, processability, impact resistance or flame retardancy, not all will be present for any one application. A knowledge of the final use of the test material can be very helpful to the analyst both in the choice of test procedures and as an aid in any interpretation involved. The analysis can be readily split into three main fields:

(a) extractable organic additives
(b) fillers and pigments
(c) polymeric materials.

2.2.2 Extractable additives

This will be the most complex field of the three with sample extracts derived from up to six of the following groups:

(a) Heat stabilisers or primary anti-oxidants such as hindered phenols, e.g. Irganox 1010 (pentaerythrityl-tetrakis(3-[3,5-di-*tert*-butyl-4-hydroxyphenyl]-propionate), Irganox 1330 (1,3,5-tris(3′-5′-di-*tert*-butyl-4′-hydroxybenzyl)-2,4,6-trimethylbenzene) or BHT (2,6-di-*tert*-butyl-4-methylphenol)

(b) melt stabilisers or secondary anti-oxidants such as thioesters, e.g. DLTDP (dilaurylthiodipropionate) or DSTDP (distearylthiodipropionate), or phosphites, e.g. Irgafos 168 (tris[2,4-di-*tert*-butylphenyl] phosphite)
(c) UV absorbers such as substituted benzophenones or benzotriazoles, e.g. Chimassorb 81 (2-hydroxy-4-*n*-octoxybenzophenone), Tinuvin 326 (5-chloro-2-(2′-hydroxy-3′-*tert*-butyl-5′-methylphenyl)benzotriazole) or Tinuvin 327 (5-chloro-2-(2′-hydroxy-3′,5′-di-*tert*-butylphenyl)benzotriazole)
(d) hindered amine light stabilisers (HALS) such as Tinuvin 770 (bis(2,2,6,6-tetramethylpiperidin-4-yl) sebacate)
(e) antistatic additives such as glyceryl esters of stearic acid, e.g. Atmer 129, or ethoxylated tertiary amines of the general formula $R\text{-}N(CH_2CH_2OH)_n$ where R is typically C_{14}–C_{18}, e.g. Atmer 163
(f) slip additives such as long-chain aliphatic amides, e.g. Crodamide ER (erucamide) or Crodamide OR (oleamide).

Various other additives such as fire retardants, blowing agents or nucleating agents may also be extracted.

2.2.2.1 Extraction techniques. Various procedures for the extraction of additives from polypropylene have been reported [1–6], many directly coupled to techniques such as GC, HPLC or UV–visible spectroscopy. Some use Soxhlet extraction or extraction under reflux with solvents such as chloroform, hexane or diethyl ether. Others use a dissolution procedure with solvents such as toluene or xylene followed by precipitation of the polymer, either by the addition of methanol or ethanol or by cooling. A more recent development has been the use of a microwave oven for the rapid extraction in 3–6 min of additives from polypropylene [7].

The extraction procedure chosen, apart from being compatible with any associated chromatographic or spectroscopic technique used, has to extract the additives quantitatively. In order to quantify this, ‘check samples’ are used to measure the extraction of additives with different solvents against parameters such as time and sample form. These check samples are not certified standards and generally cannot easily be checked unless the additives contain elements such as phosphorus, nitrogen or sulphur that can be determined by an alternative technique. However, for comparing different extraction solvents it is usually adequate to work on the premise that once a plateau has been reached in the graph of ‘percentage extracted’ against time, all the extractable additive has been removed, and to relate this figure to the nominal level.

Extraction data for some common additives using chloroform are given in Table 2.2. From this set of results it can be seen that optimum extraction is obtained from a 1 h extraction on sieved freeze-ground material, although for Tinuvin 326 the results are marginal. Additionally, all the results for the freeze-ground materials are in agreement with their nominal values.

Table 2.2 Extraction data for some common additives using chloroform; all results are in ppm

Sample form and extraction time	Irganox 1010[a]	Irganox 1330[a]	Irgafos 168[a]	Irganox 3114[a]	Atmer 163[b]	Tinuvin 326[c]
Granules, 1 h	392	1045	1086	641	811	2560
Granules, 3 h	564	1379	1386	961	1147	2600
Granules, 6 h	643	1523	1480	—	—	—
Freeze-ground to <1.18 mm, 1 h	785	1556	1556	1103	1270	2730
Freeze-ground to <1.18 mm, 2 h	784	1546	1552	1109	1278	2740

[a] Determined by HPLC.
[b] Determined by colorimetry.
[c] Determined by UV spectroscopy.

For additives containing phosphorus, nitrogen or sulphur, their extractability can be monitored by measuring the reduction in the level of the appropriate element against extraction time.

Although the prime requirement must be the quantitative extraction of the additives care must also be taken to ensure that the extraction procedure used, as well as any sample preparation, does not have deleterious effects on the additives present. These effects could be the aerial oxidation of the additive during extraction, particularly for phosphite-type melt stabilisers, where Frietag [5] suggests the addition of triethyl phosphite as a reducing agent, or the loss of volatile additives such as BHT during any evaporation or concentration steps. These effects can readily be tested for by putting a known amount of the additive through the test procedure and determining the percentage recovery. Additionally, for techniques such as HPLC the chromatogram from an additive thus treated can be compared with that of an untreated additive of identical concentration, to check for any differences such as changes in peak shape or the appearance of extra peaks.

2.2.2.2 General scheme of analysis. The first stage during the analysis of polypropylene of an unknown formulation will be to establish the identity of the additives present. The scheme shown in Figure 2.1 combines the extraction, separation and identification of chloroform-extractable additives with quantitative techniques such as HPLC, UV spectroscopy and colorimetry on a single extract.

2.2.2.2.1 Preparation of test solution. Polypropylene (3 g) is extracted under reflux with 50.0 ml of chloroform using a hotplate or a heating mantle. If quantitative analysis by HPLC, UV spectroscopy or colorimetry is required, then the sample should be ground, preferably with liquid nitrogen, and the fraction passing a 14 mesh (1.18 mm aperture) sieve extracted for 1 h. If only granular material is available then 3 h extraction will be adequate for identification purposes. For film samples 3 h extraction will be adequate for quantitative results.

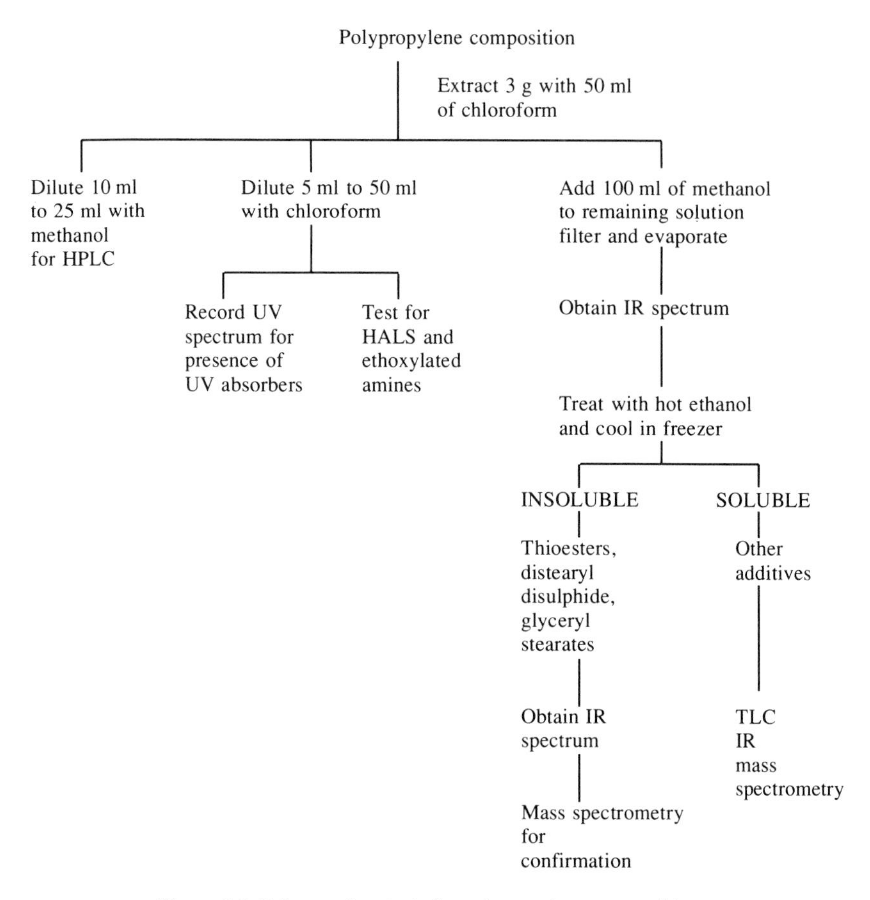

Figure 2.1 Scheme of analysis for polypropylene compositions.

2.2.2.2.2 HPLC examination. A 10.0 ml aliquot of the test solution is pipetted into a 25 ml volumetric flask and diluted to volume with methanol. This will precipitate out any extracted polymer. The mixture is shaken and allowed to stand for 20 min before filtering through a glass microfibre filter; 100 µl of clear filtrate are injected on to an HPLC system consisting of a 25 cm S50DS-2 analytical column together with a 5 cm S50DS-2 guard column, a manual injection valve fitted with a 10 µl sample loop, an HPLC pump delivering methanol at 1.0 ml per minute, a column oven at 50°C and a UV detector set at 275 nm. Typical retention times obtained for several common additives using this system are given in Table 2.3.

The retention times obtained in the sample chromatogram are compared with those in a current list of additive retention times. If any are within 0.2 min, then a reference solution of the appropriate standards is prepared and run. An agreement to within 0.01 min is deemed a positive match. Quantification can be carried out by the measurement of the relevant peak areas for the sample and for the standard(s).

Table 2.3 Retention times on HPLC for common additives

Additive	Retention time (min)
Hostanox 03	3.03
Topanol CA	3.39
BHT	4.02
Irganox 3114	4.67
Topanol 205	4.97
Irganox 1010	5.92
Irganox 1330	7.31
Tinuvin 326	8.55
Tinuvin 327	9.13
Irgafos 168 phosphate[a]	9.68
Irganox 1076	13.56
Irgafos 168	17.00

[a] Formed during processing as a result of the Irgafos 168 performing its role as a melt stabiliser; this is a benign material taking no further part in the stabilisation process.

2.2.2.2.3 Examination for UV absorbers. A 5.0 ml aliquot of the test solution is pipetted into a 50 ml volumetric flask and diluted to volume with chloroform. The UV absorption spectrum of this solution is recorded in a 5 m path-length quartz cell over the region 450–250 nm against chloroform in the reference cell. The presence of a UV absorber is indicated by a strong absorption maximum in the region 355–320 nm; other strong maxima occur at shorter wavelengths. The UV absorption spectra of some phenolic antioxidants and UV absorbers have been published [6]. The exact nature of any material thus detected will need to be confirmed by one of, or a combination of, HPLC, IR spectroscopy and mass spectrometry. Once identified it can be quantified either by HPLC or more simply by measurement of the UV absorbance at the wavelength of maximum absorbance (λ_{max}) which occurs in the region 355–320 nm, e.g. 351 nm for Tinuvin 326 and Tinuvin 327, or 325 nm for Chimassorb 81 (2-hydroxy-4-*n*-octoxybenzophenone). As a zero baseline measurement any absorbance at 420 nm should be subtracted from the measured absorbance at the λ_{max}. This corrected absorbance is converted into the amount of UV absorber present by referring to a calibration graph prepared from a reference sample.

2.2.2.2.4 Examination for HALS and ethoxylated amines. A 25.0 ml aliquot of the diluted test solution (see Figure 2.1) is pipetted into a 250 ml separating funnel together with 25.0 ml of chloroform. Also added are 20 ml of a pH 3.5 buffered indicator reagent (prepared by dissolving 14.7 g of citric acid and 8.3 g of anhydrous disodium hydrogen orthophosphate in 1000 ml of water and mixing with a solution of 0.1 g of bromophenol blue and 0.1 g of bromocresol green in 100 ml of methanol), 10 ml of water and 10 ml of methanol. The mixture is shaken for 15 s, the layers are allowed to separate, 50 ml of water is added and the mixture shaken again for 15 s. On separation, a yellow colour in the lower chloroform layer due to the formation of the amine salt of the sulphonphthalein dyestuffs [8] indicates the presence of either an ethoxylated tertiary amine antistatic agent or a HALS. This test cannot distinguish between the two types of additive, and confirmation by either IR spectroscopy (probably after preparative scale TLC) or mass spectrometry will be required. It is unlikely that they will be present together. HALS can be found in materials for outdoor uses, such as automative parts, or in agricultural and horticultural applications. Ethoxylated amine antistatic agents can be found in polypropylene packaging films or thin-wall containers.

The amount present can be determined by recording the absorption spectrum of the yellow chloroform phase in a 10 mm path-length quartz cell over the region 600–280 nm with chloroform in the reference cell. A baseline is drawn tangentially from 520 nm to 300 nm and the corrected absorbance measured at 415 nm. The corrected absorbance is converted into the amount of additive present by referring to a calibration graph prepared from a series of standard solutions of a reference material treated in the same manner as the test solution.

Mixtures of a UV absorber, such as Tinuvin 326, and a HALS, such as Tinuvin 770, may be encountered. In this case the HALS, being a non-UV absorbing material, will not interfere with the determination of the UV absorber (as in section 2.2.2.2.3). However, although the UV absorber does not react with the buffered indicator reagent to form a yellow product, as does the HALS, it nevertheless absorbs in the region of 350–300 nm, distorting the

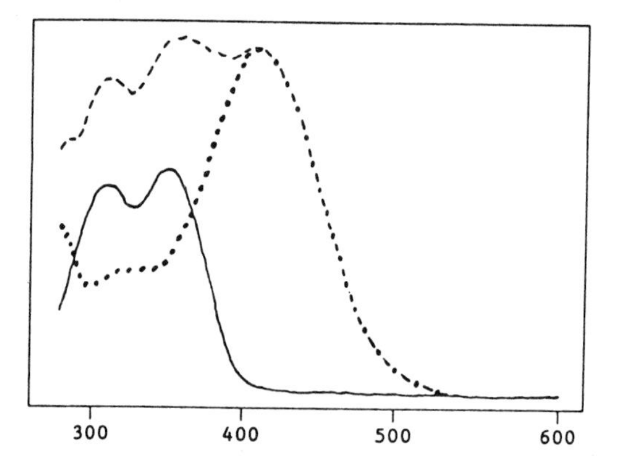

Figure 2.2 _ _ _ _ _ _ _ , 600 μg of Tinuvin 326 + 600 μg of Tinuvin 770 per 50 ml of chloroform. Mixture reacted with buffered indicator reagent and measured against chloroform.
........, 600 μg of Tinuvin 326 + 600 μg of Tinuvin 770 per 50 ml of chloroform. Mixture reacted with buffered indicator reagent and measured against unreacted mixture in chloroform.
———, 600 μg of Tinuvin 326 + 600 μg of Tinuvin 770 per 50 ml of chloroform. Measured against chloroform.

spectral curve and thus making direct measurements impossible. To overcome this, the yellow chloroform phase containing the two chromophores is run against an unreacted test solution containing the UV absorber and the HALS at identical concentrations to those present in the yellow chloroform phase. This will cancel out the chromophore due to the UV absorber, leaving an absorption spectrum due only to the HALS-buffered indicator reaction product. This is shown in Figure 2.2.

2.2.2.2.5 IR spectroscopic examination. Methanol (100 ml) is added to approximately 35 ml of remaining test solution to precipitate out any dissolved low-molecular-weight polymer. The precipitated polymer is allowed to stand for 1 h before filtering through a glass microfibre filter circle into a 150 ml extraction flask. The circle is washed with about 30 ml of methanol and the combined filtrate and washings are evaporated to dryness on a water bath contained in a fume cupboard. The flask and extract are dried in a drying oven for 10–15 min and allowed to cool in a desiccator.

N.B. Following the above procedure will almost certainly result in the loss of the volatile primary anti-oxidant BHT, and in order to obtain an extract containing this, or any other equally volatile additive, it is advisable to carry out a Soxhlet extract with diethyl ether, allowing the solvent to evaporate gently without the use of air as an aid to evaporation.

A capillary layer of the extract is prepared between two sodium chloride plates, either by transferring a small amount with a microspatula or, if the extract is small, by the use of 1,2-dichloroethane as a transfer solvent. Care is taken to remove all of the solvent. The plates are squeezed together to form the capillary layer. It may be necessary to apply gentle heat to melt the extract before squeezing. In all cases, the plates are allowed to cool to ambient temperature so that any additives which have a low melting point have solidified.

The IR spectrum is recorded and compared with standard reference spectra. These can comprise either a collection of spectra recorded locally on the same instrument or a published collection [2, 6]. The spectra obtained will invariably be complex, as most extracts will be mixtures, and their interpretation will require a basic knowledge of the wavelengths at which various functional groups absorb. This can be obtained from a reference book such as ref. 9, allied with considerable experience in the field and good memory recall.

It may also be helpful to have some preconceived ideas as to the likely nature of the extract, both from background knowledge of the sample and its application, and from information already gained from preceeding tests (sections 2.2.2.2.2–2.2.2.2.4). Information obtained from a further technique such as an XRF analysis of the original material is very useful, and may include examination for the presence or absence of phosphorus or sulphur, or, in the case of a fire-retardant material, the presence of large amounts of phosphorus, bromine or chlorine. Because of the complex nature of many of the extracts a precise identification of all of the components will be extremely difficult and confirmation by HPLC or by mass spectrometry will be required.

2.2.2.2.6 Treatment with ethanol. A further separation stage that has proved particularly useful as a rapid means of isolating thioesters from mixtures in a form pure enough to enable a positive IR identification, without the need for mass spectrometry, has been to add 5 ml of ethanol to the flask containing the extract remaining after the IR examination, and to reflux for 30 min. The solution is decanted while still hot into a glass vial leaving behind any atactic polypropylene that may have been soluble in the chloroform–methanol mixture. The vial is capped and the cooled contents are chilled either in a freezer or in ice. A white precipitate will indicate the presence of a thioester (DLTDP, DMTDP or DSTDP), dioctadecyl disulphide or glycerol esters of stearic acid. The precipitate is isolated by centrifugation and care must be taken to ensure that all ethanol used is kept chilled since the precipitate can rapidly redissolve if it warms up to ambient temperature. If a refrigerated centrifuge is used then it should be operated at 0°C and spun for 2–3 min. If a small bench-top centrifuge is used it is advisable to chill the centrifuge tubes before use and to spin for about 45 s. The precipitate should be thoroughly washed with chilled ethanol and recentrifuged three times. The centrifugate from the original spin should be retained. The washings may be discarded.

Some of the cold ethanol-insoluble material is transferred to a single sodium chloride plate and heated in a drying oven for 2–3 min in order to remove any residual ethanol and to melt the material. A second sodium chloride plate is placed on top and gently squeezed to form a capillary layer. The plates are allowed to cool to ambient temperature to solidify the additive, and the IR spectrum is recorded and compared with reference spectra. It should be possible to identify the additive positively without the need to examine it by mass spectrometry. The IR spectra of DLTDP, DMTDP and DSTDP can be distinguished one from another by carefully comparing the 'fingerprint' region (1100–900 cm^{-1}) and also by observing the wavelength of the weak band that occurs in the 770–740 cm^{-1} region.

The ethanol centrifugate can be used directly for the TLC identification of phenolic anti-oxidants and UV absorbers by following the BS method [10]. Although this procedure lacks the precision and speed of HPLC it is still a reliable and relatively cheap method of identifying phenolic anti-oxidants. Recent work by Everall *et al.* [11] has investigated the use of FT-Raman spectroscopy for the *in situ* analysis of fractions on silica TLC plates after the separation of polypropylene additives. In the most favourable cases good-quality spectra were obtained from sample loadings of about 3 $\mu g\ mm^{-2}$. However, since this technique samples only a small area of an eluted spot (*ca.* 1 mm^2) problems are encountered with diffuse spots.

Alternatively, the ethanol can be removed by evaporation and if required an IR spectrum of the residue can be obtained as described in section 2.2.2.2.5. This would be the case if the cold-ethanol treatment had removed a significant amount of material, such as a thioester thereby giving a more intense spectrum of the other components. If confirmation of the IR examination is required the residue can be examined by mass spectrometry.

In some instances it may be necessary to carry out preparative scale TLC on the residue in order further to separate complex extracts, thus improving the chances of identifying the individual fraction by either IR spectroscopy or mass spectrometry, or to separate out one specific component from a mixture for identification. One example of the latter would be the precise identification of an amide slip additive. Bands in the IR spectrum of a mixed extract in the region of 3360 and 3190 cm^{-1} due to N–H stretching vibrations and in the region of 1660 and 1630 cm^{-1} due to C=O stretching vibrations and N–H bending vibrations would clearly indicate the presence of a primary amide. Here, the 'finger-print' region of 1430–1110 cm^{-1} was masked by bands from other components.

TLC [21] is carried out using 20 × 20 cm PLC plates with a 2 mm layer of Silica Gel 60 containing an $F_{254+361}$ indicator. The extract is dissolved in 1 ml of 1,2-dichloroethane and, using a syringe or a disposable Pasteur pipette, applied as a narrow band about 30 mm from and parallel to the edge of the plate. The solvent is gently removed using a warm air blower, the plate placed into a prepared chromatography tank containing 95:5 toluene:ethyl acetate and eluted for a distance of about 170–180 mm. The plate is removed from the tank and the solvent gently evaporated in a fume cupboard using a warm air blower. Detection of the separated bands is carried out either by use of a UV lamp, for UV-absorbing materials, or by spraying in a fume cupboard with a 3% solution of iodine in methanol. This will detect all organic materials by turning them brown with absorbed iodine. The bands are marked with pencil and, after gently removing any remaining iodine with a warm air blower, removed from the plate with the careful use of a razor blade or scalpel. Each fraction is transferred to a cellulose extraction thimble, plugged with cotton wool and extracted for 2 h with diethyl ether using a Soxhlet extraction apparatus. After evaporation of the diethyl ether, the residue is examined by IR spectroscopy or mass spectrometry as appropriate.

2.2.2.3 Titrimetric and volumetric procedures. Procedures were described in the 1960s for the determination of ethoxylated tertiary amine antistatic agents and long-chain amide slip additives by solvent extraction followed by potentiometric non-aqueous titration against perchloric acid [2, 12]. More recent extraction studies [13] have shown that, for an ethoxylated tertiary amine, quantitative extraction is obtained after 1 h of extraction of 30 g of ground material with 100.0 ml of 1,2-dichloroethane. Only 85% is removed by extracting granular material for 3 h. For the slip additive oleamide, quantitative extraction with the same solvent is obtained both after 1 h for ground material and 3 h for granular material.

For polypropylene masterbatches where up to 20% of the additive is present, then a quick and effective procedure is to weigh 1–2 g of sample into an extraction flask, add 100.0 ml of xylene and gently reflux on a magnetic stirrer/hotplate until dissolved. This takes about 30 min. The mixture is allowed to cool to ambient temperature whilst still attached to the condenser, removed, stoppered and cooled under running water to precipitate the polymer. It is then allowed to return to ambient temperature and an aliquot taken and titrated potentiometrically against 0.02 M perchloric acid in 1:1 acetic acid : acetic anhydride. This procedure can be applied to samples containing either the ethoxylated amine or the amide previously mentioned, or a mixture of the two, or to a single HALS such as Tinuvin 770, Tinuvin 622 (an oligomer of *N*-(2-hydroxyethyl)-2,2,6,6-tetramethyl-4-hydroxypiperidine and succinic acid) or Chimassorb 944 (poly(*N*-1,1,3,3-tetramethylbutyl-*N*′, *N*″-di (2,2,6,6-tetramethylpiperidinyl)-*N*′, *N*″-melaminoditrimethylene). For a sample of polypropylene that had been shown by nitrogen content to contain 7.9% of Tinuvin 622, this procedure gave a level of 7.6%. Extraction of granular material with 1,2-dichloroethane for 3 h gave a level of 4.4%.

Mixed mono- and di-glycerides of stearic acid are used as antistatic agents in polypropylene, and their determination is based on the periodate oxidation of adjacent hydroxyl groups in the α-monoglyceride [14]. The ground material (10 g), or proportionately less if more than 2% of additive is expected (so that the amount in the extract does not exceed 200 mg), is extracted in a 500 ml conical flask with 50 ml of chloroform by gently refluxing on a hotplate for 2 h. To the cooled mixture are added 5 ml of 1:1 pyridine:glacial acetic acid and, by pipette, 50.0 ml of potassium periodate solution (prepared by dissolving 2.8 g of the reagent in 50 ml of 25% v/v sulphuric acid with gentle warming, adding 975 ml of glacial acetic acid and mixing). The flask is stoppered, swirled gently and allowed to stand in the dark for 1 h with occasional agitation by swirling; 100 ml of water and 20 ml of 10% potassium iodide solution are added and the liberated iodine is titrated with 0.1 M sodium thiosulphate; 2 ml of a 5% aqueous solution of the starch–urea complex 'Thyodene' are added at the endpoint, and the titration is continued until the iodine disappears from the chloroform layer and the blue colour of the iodine–starch–urea complex disappears from the aqueous phase. Vigorous agitation is necessary for the complete removal of iodine from the chloroform layer. A blank determination is carried out on all the reagents. This should be done in duplicate. The titre due to the additive is found by subtracting the sample titre from the blank titre.

Standardisation is carried out by accurately weighing about 250 mg of additive, dissolving it in 50 ml of chloroform and proceeding as for the sample. This should also be carried out in duplicate. Film samples are cut into one-inch squares and extracted for 3 h with 100.0 ml of chloroform. On cooling, a 50.0 ml aliquot is pipetted into a second flask and treated as for a ground-polymer sample. Additives such as thioesters, BHT, and Irgafos 168 interfere, and a correction, established by titrating a known weight of a reference standard to obtain a correction factor relating millilitres titre to milligrams titrated, and adjusted for the actual amount of interfering additive present, must be applied.

Metal stearates such as those of calcium or zinc are used as antacids in polypropylene and can react with acidic species to form 'free' stearic acid. To determine the amount of 'free' acid present, 20 g of ground material should be weighed into a pre-extracted cellulose extraction thimble, sealed with a small plug of pre-extracted cotton wool, placed in a Soxhlet apparatus and extracted with diethyl ether for 18 h on a water bath. The diethyl ether is removed by evaporation on a water bath in a fume cupboard and the residue re-extracted by gently refluxing on a hotplate with 20.0 ml of ethanol for 1 h. On cooling, the stearic acid is titrated against freshly prepared aqueous 0.01 M sodium hydroxide to a phenolphthalein endpoint. A blank on all the reagents used must be carried out. This procedure is obviously not specific to stearic acid and will give a measure of the total extractable acidity present. Depending on the amount of metal stearate present, levels up to about 500 ppm of free acid may be present. It has been established that a similar extraction procedure on granules gives only a 50% recovery [15].

2.2.2.4 The extraction of additive concentrates. These are mixed additive packages (which may also contain polypropylene) containing up to six additives at individual levels of typically 2–20%. They do not need an exhaustive extraction procedure but do require a sampling technique that ensures a representative sample, and in some cases a dilution step will be necessary. The extraction procedure generally used is a direct extraction with a fixed volume of solvent at ambient temperature using an ultrasonic bath. Extraction periods of up to 30 min are used. The solvents generally used are carbon tetrachloride for IR measurements, 1,2-dichloroethane for non-aqueous potentiometric measurements, and chloroform with subsequent dilution with excess methanol for HPLC measurements [16].

2.2.3 Polymers

The identification and estimation of polypropylene and its copolymers by IR and NMR spectroscopy is well established and documented, as are the difficulties encountered during the application of pyrolysis/gas chromatography [2, 3]. IR spectroscopy is a rapid and precise means of characterisation, but the detection of copolymers can be difficult in filled materials and separation of the polymer is advantageous.

This is carried out by weighing 0.6–0.8 g of sample into a 25 × 80 mm double-thickness cellulose extraction thimble lined with a folded Whatman 542 filter paper. A second folded 542 paper is inserted into the thimble on top of the sample and the thimble is sealed with just enough cotton wool to prevent any 'boiling-over' of filler/pigment during the subsequent extraction. The thimble is extracted in a Soxhlet assembly for 18 h with 100 ml of xylene contained in a 250 ml extraction flask, using either an electric hotplate or a heating mantle. After extraction, the bulk of the xylene in the flask is reduced by evaporation to about 50 ml, cooled, and the polymer is precipitated by the slow addition from a burette of 100 ml of ethanol, whilst stirring with a magnetic stirrer. The polymer is allowed to settle for 30 min and then filtered under suction, washing with about 200 ml of ethanol. It is sucked dry and heated in a vacuum oven at 100°C for a least 4 h. This will produce polymer free of fillers/pigment such as glass, talc, chalk, barium sulphate and titanium dioxide. Carbon black can give problems and usually black samples yield, at the very best, grey extracted polymer. The polymer is hot pressed into a 0.2 mm film, its IR spectrum recorded and compared with reference spectra to classify the polymer type. Propylene-ethylene block and random copolymers, propylene-ethylene-butylene terpolymers, polypropylene-polyethylene blends and blends of polypropylene with ethylene-propropylene rubber (EPR) can readily be identified, as can low levels of an ethylene-vinyl acetate copolymer (EVA) used as a carrier for a pigment masterbatch. The film can be used for the quantification of the copolymers by IR and the isolated polymer can also be examined by other techniques such as ^{13}C NMR or DSC.

2.2.4 *Fillers and pigments*

The xylene-insoluble material remaining in the extraction thimble following the previous procedure can be dried and examined by IR spectroscopy to identify the fillers/pigments present. Examination by other techniques such as X-ray fluorescence spectroscopy (XRF) or X-ray diffraction (XRD) may be helpful. Alternatively, the polymer can be removed by ashing, leaving the filler/pigment. This is generally satisfactory as the common fillers (chalk, talc and glass) are stable at 500°C, a suitable ashing temperature for polypropylene. Some hydrous silicates such as china-clay lose water on ashing, with a correspondingly dramatic change in their IR spectrum.

Ashing is also the simplest means of determining filler contents. The sample (1 g) is accurately weighed into a tared porcelain crucible and placed in a muffle furnace situated in a fume cupboard at 600°C for 30 min (except for chalk, where it is heated at 500°C for 45 min). The crucible is allowed to cool in a desiccator and reweighed, and the weight of ash is calculated. This ash content will equate to the total inorganic material present, and if a low level of a pigment such as titanium dioxide were present would only give a total figure. Most neat fillers will lose 1–2% on ashing and if a precise filler content is required then a sample of the actual filler used must be obtained, its ash content determined and a correction factor applied to the sample ash level to give a true filler level.

A two-stage ashing procedure has been developed that can be applied to a mixture of chalk with a second filler such as talc. A known weight of sample is ashed at 500°C to remove the polymer. The residual chalk and talc are weighed and the percentage ash calculated. From this figure a total filler level is derived. The residue is then heated at 800°C, at which temperature the chalk quantitatively decomposes according to the equation $CaCO_3 \rightarrow CaO + CO_2$. This second residue of talc and calcium oxide is weighed and the percentage ash calculated. Using formulae derived by solving simultaneous equations involving 'total filler' and 'ash at 800°C' the amounts of chalk and of talc can be calculated.

In order to establish the conditions for ashing, approximately 0.5 g of current sources of talc and chalk, accurately weighed, were each ashed in a muffle furnace for 30 min at 500°C and then for 60 min at 800°C. The talc was found to lose 1% by weight at 500°C and a total of 3% after further heating at 800°C. The chalk lost 2% by weight at 500°C and a total of 42% by further heating at 800°C. Neither filler lost any further weight on heating at 800°C in excess of 60 min. These actual losses are incorporated in the method of calculation. If fillers from different sources are used then this exercise should be repeated to obtain current data.

For a nominal 1:1 mix of chalk and talc, the % total filler = % ash at 500°C × 1.015. Two equations can be written:

(a) % Total filler = % chalk + % talc

(b) % Total filler − % ash at 800°C = [% chalk × 0.42] + [% talc × 0.03]

Solving these two equations simultaneously gives:

$$\%\ \text{chalk} = \frac{[0.97 \times \%\ \text{total filler}] - \%\ \text{ash at } 800^\circ\text{C}}{0.39}$$

$$\%\ \text{talc} = \%\ \text{total filler} - \%\ \text{chalk}$$

Using this procedure on ten portions of a check sample gave a mean chalk level of 19.50% with a sd of 0.08% and a cv of 0.42%, and a mean talc level of 18.05% with a sd of 0.07% and a cv of 0.39%.

2.2.5 'Non-extractable' additives

The use of moulded films for the direct measurement of additives by IR and UV spectroscopy has been described [1, 5]. While problems can exist with interferences from other additives present, for certain applications it is a quick method of analysis. It can be particularly useful in analysing for additives that are difficult to extract, although in such cases the calibration of standards may present a problem.

The determination of the oligomeric HALS Tinuvin 622 by its saponification with tetrabutylammonium hydroxide (during the precipitation of the polymer) to form a diol and its subsequent quantification by HPLC has been reported [17].

A method for the determination of the polymeric HALS Chimassorb 944 has also been reported [18]. Here about 1 g of polymer is dissolved by refluxing with 100 ml of decalin. In order to prevent autoxidation giving rise to interfering UV-absorbing species, 100 mg of Irganox 1010 are added. The cooled suspension is quantitatively transferred to a separating funnel and the Chimassorb 944 extracted by shaking with 100.0 ml of 0.5 M sulphuric acid, which contains 0.05% of diethanolamine to prevent absorption of the Chimassorb 944 on to glassware. On separation, the UV spectrum of the lower acid layer is recorded over the region 400–200 nm. The absorbance due to Chimassorb 944 is measured at 245 nm. The sulphuric acid extraction of the decalin is repeated until no more Chimassorb 944 is extracted. Among the additives currently in commercial use this method is specific for Chimassorb 944.

Another 'non-extractable' additive in current use is Irganox 1425WL (calcium (3,5-di-*tert*-butyl-4-hydroxybenzyl monoethyl phosphonate) on a 50% wax base), which contains 4.47% phosphorus and 2.88% calcium. This is generally found in the presence of Irgafos 168, which contains 4.79% phosphorus, and calcium stearate, which contains 6.60% calcium. The total phosphorus level, from Irgafos 168 and Irganox 1425WL, and the total calcium level, from calcium stearate and Irganox 1425WL, are determined by an inorganic technique such as XRF. The Irgafos 168 and its phosphate are determined by HPLC as described in section 2.2.2.2. The total phosphorus level is corrected for the phosphorus present as Irgafos 168 and its phosphate, as determined by HPLC. This gives the phosphorus due to Irganox 1425WL from which its level is calculated. This level of Irganox 1425WL is then used to calculate the calcium due to Irganox 1425WL, and the total calcium level is corrected for this amount. This gives the calcium due to calcium stearate, from which its level is calculated.

2.3 Poly(vinyl chloride)

2.3.1 Introduction

PVC formulations can be, by virtue of both the range of functional additives used to cover a wide spread of applications and the variety of each type, the most complex, difficult yet challenging plastics to analyse. Nevertheless most additive levels are macro-scale, yielding adequate separated materials for a variety of tests. Many procedures exist for the determination of various additives in PVC by solvent extraction followed by an instrumental technique such as GC or IR spectroscopy, and these are comprehensively listed in ref. 3. Schemes of analysis for the separation of PVC formulations into solvent-soluble, solvent-insoluble and polymeric fractions have also been published [4, 20].

2.3.2 General scheme of analysis

The scheme shown in Figure 2.3 is a simplified version of those earlier ones. Previously, the sample was extracted successively with diethyl ether and methanol, dissolved in THF, and centrifuged at two speeds, the polymer was precipitated with methanol, and the evaporated THF/methanol was examined for non-extracted material. This yielded the base polymer and five other fractions to examine. The two sets of centrifuged material, one at 3000 rpm and supposedly inorganic in nature and the other at 17 000 rpm and supposedly a polymeric modifier, were often found to be cross-contaminated.

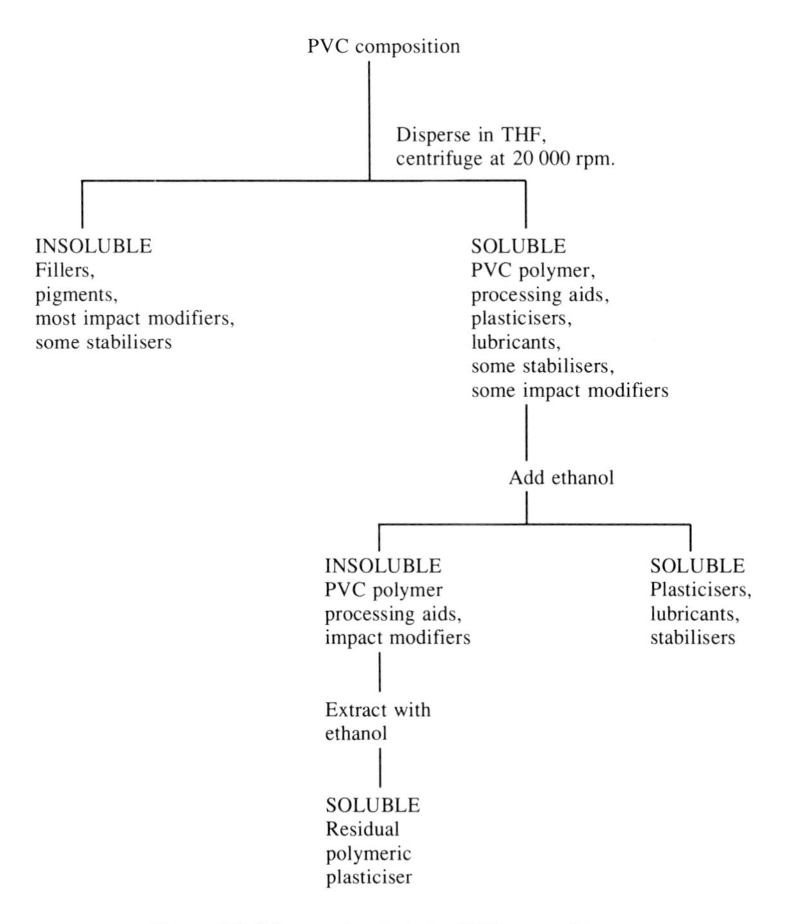

Figure 2.3 Scheme of analysis for PVC compositions.

In this scheme the sample is dissolved in THF and centrifuged at 20 000 rpm to isolate inorganic materials and polymeric impact modifiers together. The polymer is precipitated from the THF centrifugate with ethanol and isolated by filtration. The THF–ethanol filtrate is evaporated to dryness to yield the soluble organic additives. This yields a total of three fractions to examine. When a polymeric plasticiser is either known to be present or found to be present in the THF–ethanol-soluble fraction, then the isolated polymer is Soxhlet extracted with hot ethanol to remove residual non-extracted plasticiser. All fractions are examined by IR spectroscopy and use is made of reference spectra.

Validation studies carried out by analysing mixtures of DOP (di-2-ethylhexyl phthalate), calcium carbonate and 'blank' PVC and of polypropylene adipate, calcium carbonate and 'blank' PVC have given quantitative recoveries, as shown in Table 2.4.

Table 2.4 Validation studies on PVC analysis method

DOP added	0.6154 g	PPA added	0.6748 g
DOP found[a]	0.6102 g	PPA found[a]	0.6343
% recovery	99.2	% recovery	94.0
DOP found[b]	< 0.002 g	PPA found[b]	0.0477
% recovery	< 0.3	% recovery	7.1
Chalk added	0.2913 g	Chalk added	0.2849
Chalk found	0.2918 g	Chalk found	0.2844
% recovery	100.2	% recovery	99.8

[a]As THF–ethanol-solubles on original.
[b]As ethanol-extractables on polymer.

2.3.2.1 Preliminary examination. For rigid or uPVC materials 50–60 mg of material are hot-pressed at a temperature of 180°C to obtain a film of about 0.05 mm thickness, and the IR spectrum is recorded. This can indicate, depending on the total additives present, the presence of inorganic fillers, pigments and stabilisers and polymeric impact modifiers. There is less virtue in examining flexible materials since the resulting spectrum will be dominated by the plasticiser, unless a quick check of the generic plasticiser type is required. Examination by an elemental technique such as XRF for the levels of stabilisers, fillers or pigments is also necessary.

2.3.2.2 Separation of fractions. Sample (2 g) is weighed into a 150 ml conical flask and 'dissolved' either by standing overnight or by the gentle application of heat. The use of an ultrasonic bath also aids dissolution. Unless the solution is crystal-clear (in which case it is quantitatively transferred to a 600 ml beaker using THF as a transfer solvent), it is transferred quantitatively using THF to a 50 ml polypropylene centrifuge tube, capped and centrifuged at 20 000 rpm for 45 min or until the supernatant THF is crystal-clear and free of suspended matter. The clear supernatant is transferred to a 600 ml beaker. If all the sample solution had not been transferred to the tube it is now added and centrifuged as before. The THF-insoluble material in the tube is washed with 25 ml of THF using an ultrasonic bath to break it up. It is then centrifuged at 20 000 rpm for 20 min. These washings are added to the THF supernatant in the 600 ml beaker. This is repeated twice more.

Using a small volume of THF as the transfer solvent, the washed THF-insoluble material is transferred quantitatively to a tared glass basin. The solvent is removed by evaporation and the residue dried in an 80°C vacuum oven for 45 min. After cooling the dish and contents are reweighed and the percentage of THF-insoluble material in the original sample is calculated.

The combined THF washings, or the original sample solution if it was crystal-clear, are reduced in bulk by evaporation to about 80 ml. Whilst stirring, the polymer is precipitated by the dropwise addition of about 150 ml of ethanol from a burette or dropping funnel. The beaker and contents are allowed to stand for 1 h before filtering under suction through a porosity-3 sintered-glass crucible, washing the beaker and the polymer in the sinter with about 300 ml of hot ethanol. The polymer is sucked dry and placed in a vacuum oven at 80°C to remove the last traces of ethanol, while the THF–ethanol filtrate is quantitatively retained and evaporated to dryness in a 600 ml beaker. The residue is then transferred with hot ethanol to a tared 150 ml extraction flask, filtering to remove any low-molecular-weight polymer. The beaker and filter paper are washed well with hot ethanol, the ethanol is removed by evaporation and the residue dried in an oven at 110°C for 30 min. Finally the flask and extract are reweighed and the percentage of THF–ethanol-soluble material in the original sample is calculated.

As shown in Table 2.4 this procedure will quantitatively extract monomeric plasticisers such as DOP, but for polymerics such as PPA (polypropylene adipate) only 94% was found to be extracted and therefore the dried polymer is Soxhlet extracted with ethanol for 18 h to remove the rest. The validation was carried out by weighing the plasticiser and filler into a 600 ml beaker together with sufficient 'blank' PVC to give a total of 2 g, dissolving in 40 ml of THF and proceeding as described.

A sample of flexible PVC nominally containing 28.5% of a mixture of DIOP and a chloroparaffin and 15.1% of fillers and inorganic lead stabiliser was also analysed. The THF–ethanol-soluble fraction was 28.6% and the THF-insoluble fraction was 14.9%.

2.3.2.3 Examination of the THF-insoluble fraction. This fraction consists of: inorganic pigments and fillers such as chalk, titanium dioxide, alumina trihydrate, clay, etc.; fire retardants such as antimony trioxide; lead-based stabilisers such as TBLS or DBLP; polymeric impact modifiers that are totally insoluble in THF, such as MBS and various acrylic rubbers; and polymeric impact modifiers that are partially soluble in THF, such as nitrile rubber and some grades of ABS which, while mainly insoluble in THF, will yield a fraction consisting of SAN which is soluble in THF and will be found with the precipitated polymer.

The IR spectrum of the fraction is recorded by using the KBr-disc technique. Care must be taken to ensure that a representative sample is taken, particularly if impact modifier and filler are present together. If impact modifier is present the addition of a few drops of THF to the KBr will aid grinding. Frequently this fraction will contain major amounts of chalk, making the identification of impact modifiers difficult especially at low levels. In such cases the chalk is removed using dilute hydrochloric acid, the residue washed and dried and its IR spectrum recorded (see Figure 2.4).

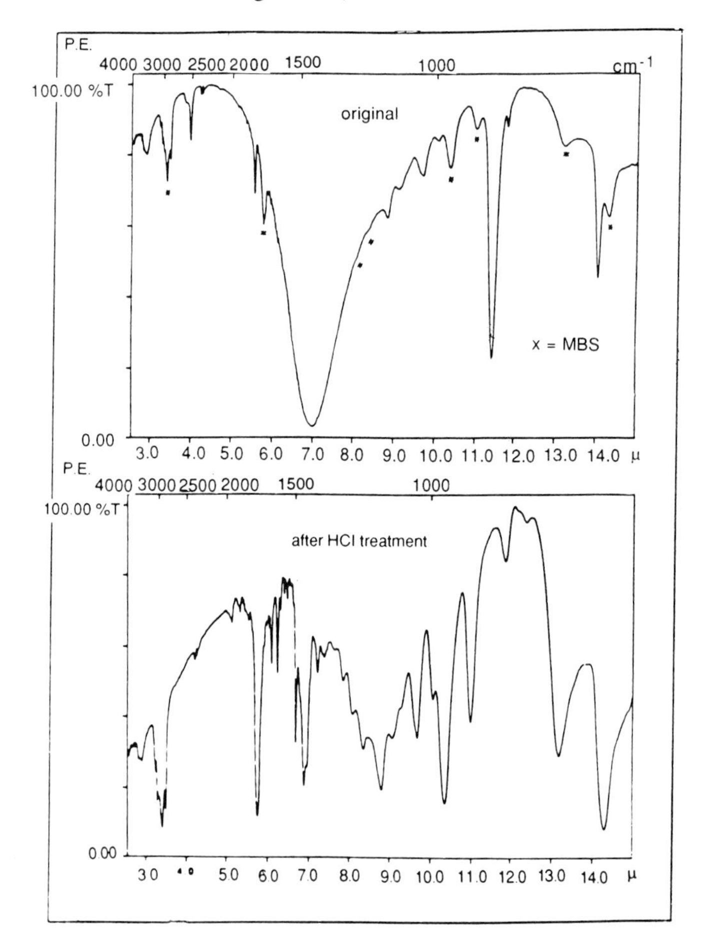

Figure 2.4 IR spectra of THF-insoluble fraction from PVC analysis.

Knowledge of the elements present, obtained either from direct examination or by reference to those found in the original material, will help in the identification of the inorganics present. Confirmation of any impact modifier thought to be present by IR is readily achieved by pyrolysis/GC.

The total percentage of this fraction will have already been determined gravimetrically, and if only impact modifier is present can be reported as the impact modifier level. However, as in many cases it will be a mixture of impact modifier and inorganics, two relatively simple routes can be followed in order to determine the proportion of impact modifier present:

(a) The THF-insoluble material is transferred quantitatively to a tared porcelain crucible and ashed in a muffle furnace at 500°C for the minimum length of time necessary to remove all the impact modifier. Prolonged heating should be avoided if lead compounds are present. The cooled crucible and residue are reweighed to give the loss in weight on ashing, which is equivalent to the amount of impact modifier present in the fraction. Hence the percentage of impact modifier in the original material can be calculated.
(b) If the fraction is a relatively simple mixture of an impact modifier with one or two inorganics such as chalk or titanium dioxide, then their levels as determined on the original material (by, for example, XRF) can be summated and subtracted from the percentage of THF-insoluble material to give the percentage of impact modifier in the original material.

2.3.2.4 Examination of the THF–ethanol-insoluble fraction. This will consist of the base polymer, which will usually be a PVC homopolymer but could be a copolymer of vinyl chloride (VC) with another monomer such as vinyl acetate (VA). Also present may be polymeric processing aids such as Paraloid K120N (90 : 10 methylmethacrylate : ethyl acrylate copolymer), impact modifiers that are totally soluble in THF such as EVA or CPE, and impact modifiers that are partially soluble in THF such as nitrile rubber and SAN fractions from ABS.

A sample (50–60 mg) is hot-pressed at a temperature of 180°C to obtain a film of about 0.05 mm thickness, and the IR spectrum is recorded and compared with standard reference spectra. Difficulties will be encountered in distinguishing between a mixture of PVC and EVA and VC/VA copolymer. The main difference is seen in changes to the C–H stretching vibration bands of PVC in the region of 3030–2900 cm^{-1} due to the presence of $(CH_2)_n$ units in the polymerised ethylene portion of EVA. Equally difficult to detect is CPE, where the same effect is observed. Pyrolysis/GC and NMR spectroscopy are used as confirmatory techniques, with the latter also being used to quantify some polymeric compounds.

A quick way of confirming the presence of CPE, EVA or nitrile rubber is to add about 20 ml of toluene to the polymer mixture and leave it to stand at ambient temperature for 2–3 h to dissolve the modifier. The toluene is separated by filtration and concentrated to low bulk, and sufficient is evaporated on to a sodium chloride plate to produce a thin film. The last traces of toluene are removed in a vacuum oven and the IR spectrum is recorded.

2.3.2.5 Examination of the THF–ethanol soluble fraction. This fraction may be complex and will consist of the soluble organic additives such as plasticisers, lubricants and some stabilisers. Comprehensive listings of plasticisers as found in flexible PVC and lubricants as found in rigid PVC are given in ref. 6.

The IR spectrum of the fraction is recorded as a capillary layer between two sodium chloride plates. When the extract is waxy, as will be the case with most rigid PVC, then it is melted to form a capillary layer and allowed to resolidify by cooling before its spectrum is recorded.

For flexible PVC, where the extract will be 25–40%, plasticisers such as *o*-phthalates, trimellitates, adipates and phosphates are easily recognisable generically. Of the *o*-phthalates in use, only DOP, DBP and the alkyl benzyl *o*-phthalates such as BBP can be identified with any degree of certainty. Di-isoalkyl *o*-phthalates such as DIOP, DINP, DIDP, etc., can be identified as a group by virtue of a weak band at 1365 cm^{-1} due to a C–H bending vibration of either a *gem*-dimethylalkyl or a *tert*-butyl group. Similarly with trimellitates, only TOTM (tri-2-ethylhexyl trimellitate) can be identified with any degree of certainty. For alkyl phosphates such as the isopropylated phosphates and aryl/alkyl phosphates such as octyl diphenyl phosphate, precise identification is again not easy.

For all the above groups, as long as reference materials are available, HPLC or capillary GC are excellent techniques for their identification. This is particularly true for *o*-phthalates and trimellitates derived from mixed chain-length alcohols such as C_6–C_{10}. These techniques can also be used to quantify components in mixed plasticiser systems.

Aliphatic dicarboxylic acid ester plasticisers such as adipates are readily identifiable when present on their own. Significant levels of DOA (di-2-ethylhexyl adipate) can be seen in the presence of DOP and levels of 20%, 30% and 40% DOA in DOP can be distinguished one from another. ESBO (epoxidised soya bean oil) cannot be detected in DOP at the 10% level, a typical amount.

Chloroparaffin can be detected in DOP at the 40% level, a typical amount. Polymeric plasticisers (e.g. polymeric adipate), which are very much more viscous than monomeric plasticisers, are at the worst generically identifiable.

Mass spectrometry and NMR, especially for polymerics, are used to identify plasticisers that cannot be identified by IR spectroscopy, HPLC or GC.

For rigid PVC the extract will be much smaller, typically 1–5%, and will consist of lubricants such as long-chain aliphatic esters, e.g. glyceryl monoricinoleate, long-chain aliphatic acids and alcohols, and hydrocarbon and amide waxes. Metal-based stabilisers such as metal carboxylates or organo-tin compounds may be present. Also present may be nonmetal stabilisers such as ESBO, organic phosphites, bisphenol A, and polyols such as trimethylolpropane or pentaerythritol.

Although, as with flexibles, the extracts will frequently be complex mixtures, at least a partial identification into generic groups such as long-chain glyceryl ester or long-chain aliphatic acid can be made. To achieve this best, a comparison is made between the sample spectrum and reference spectra as to the presence or absence and relative intensities of bands in the following regions:

(a) –OH stretching vibration in the region of 3450–3225 cm^{-1} due to a glycerol mono- or di-ester, a long-chain alcohol or a polyol.
(b) –C=O stretching vibration in the region of 1750–1740 cm^{-1} together with a $-CH_2COOR$ stretching vibration in the region of 1180 cm^{-1} due to an aliphatic ester.
(c) –C=O stretching vibration in the region of 1700 cm^{-1} due to a saturated aliphatic acid.
(d) –C–O stretching vibration and –OH bending vibration in the region of 1050 cm^{-1} due to the $-CH_2OH$ group in either a primary alcohol or a glycerol mono- or di-ester.
(e) $(CH_2)_n$ stretching vibration in the region of 720 cm^{-1} from long-chain aliphatic groups; particular attention should be paid to whether the band is a singlet or a doublet.

Following the TLC procedure described in section 2.2.2.2.6 using 60–80 mg of extract will separate plasticisers extracted from flexible PVC and lubricants extracted from rigid PVC. For plasticisers, mineral oils and chloroparaffins travel with the solvent front whilst polymeric plasticisers and ESBO remain on, or near to, the application line. *o*-Phthalates, trimellitates, phosphates and adipate elute to the top half of the plate. DOP and DOA are found in close proximity, usually occurring as two adjoining bands, but in the hands of an experienced chromatographer can be separated. This is achieved by drying the plate after one elution and returning it to the tank for a second elution. By using this technique dialkyl *o*-phthalates such as C_8 and C_{10} can, with skill, be separated from each other.

For lubricants, although not ideal, the same solvent system is used with the components separating very much according to polarity. The more polar materials such as ESBO, long-chain acids, alcohols and polyols remain at the application line, whilst long-chain esters such as stearyl stearate and distearyl phthalate (another common lubricant) elute to the top half of the plate. As with plasticisers, a second elution can be used to separate the more polar components further. Under these conditions ESBO can separate into four or five fractions of differing degrees of epoxidation, with the most highly epoxidised fraction having the lowest R_f value.

2.3.2.6 Examination of the ethanol-extractables from the THF–ethanol-insoluble fraction. The IR spectrum is recorded as a capillary layer between two sodium chloride plates and compared with the spectrum obtained from the THF–ethanol-soluble fraction. As well as polymeric plasticiser, this fraction may contain a small amount of long-chain metal carboxylate stabiliser.

2.4 Polytetrafluoroethylene

Since PTFE is an inert polymer which is virtually non-extractable once processed it is of limited interest to a 'wet chemist', and only two areas will be considered here. These are the determination of mixed filler systems in granular PTFE and the separation, identification and determination of non-ionic and anionic surfactants in PTFE aqueous dispersions.

2.4.1 Mixed fillers in PTFE

Haslam *et al.* [2] described the determination of the total filler content of PTFE by pyrolysing the sample at 700°C in a stream of nitrogen and weighing the residue. This procedure has been extended to cover the determination of mixtures of carbon-based fillers such as graphite or coke with an inorganic filler such as glass.

Sample (1 g) is weighed into a tared combustion boat and placed in an aluminous porcelain combustion tube contained in a tube-furnace, ensuring that the boat and sample are situated in the 'hot-spot' of the furnace. A tightly fitting rubber bung, carrying an inlet tube for an oxygen-free nitrogen supply, is inserted into the mouth of the combustion tube. The flow is adjusted to 2 l min^{-1} and left for 5 min to flush all the air from the system. Since certain pyrolysis products of PTFE are toxic, the fumes evolved from the combustion tube must be removed into an efficient extraction system. The temperature of the furnace is raised at the rate of 15°C min^{-1} to 700°C and maintained at this temperature for a further 30 min. The tube and contents are then allowed to cool to 100°C with the nitrogen still flowing, and the cooled boat and contents are reweighed to give the weight of the residue of glass and carbon. The boat containing this residue is then placed in a muffle furnace at 700°C for 1 h to remove the carbon, cooled, and reweighed to give the weight of glass and hence the glass content. The carbon content is calculated from the loss in weight on ashing in air. Repeatability studies on six portions of a 'check sample' gave a mean glass level of 27.32% with a sd of 0.12% and a cv of 0.45%, and a mean graphite level of 4.17% with a sd of 0.10% and a cv of 2.45%.

2.4.2 Surfactants in PTFE dispersions

2.4.2.1 Separation and identification of nonionic and anionic surfactants. This procedure is based on that described previously [2]. The polymer is precipitated by adding 5 g of the dispersion dropwise to 50 ml of stirred ethanol in a 250 ml beaker, warming the mixture if necessary to coagulate the polymer. The polymer is removed by filtration, and the solvent is collected in a beaker and evaporated to dryness to yield the total surfactant present. This will generally be nonionic, such as an alkylphenol ethoxylate, or anionic, such as a sodium alkyl sulphate or a sodium alkyl benzenesulphonate, or a mixture of the two types.

To separate these two types, about 20 ml of diethyl ether (dried over anhydrous sodium sulphate) are added to the surfactant extract and allowed to stand for 1 h with occasional swirling, to dissolve any nonionic material. The ether is decanted through a filter paper into a glass basin. The residue in the beaker, if any, is washed twice more with dry ether and the total washings are evaporated to dryness, yielding nonionic surfactant. The ether-insoluble residue is treated with 10 ml of cold water, which is decanted through a filter paper into a glass basin and evaporated to dryness to yield anionic surfactant.

The IR spectrum of each of these fractions is recorded and compared with reference spectra as found in ref. 23, where useful aids to interpretation precede each section. If confirmation of the IR findings or more detailed information is required then NMR spectroscopy is used.

2.4.2.2 The determination of anionic surfactants. This procedure is based on the method described by Epton [24]. About 1 g of the aqueous dispersion is weighed into a 100 ml volumetric flask and diluted to volume with water. A 25.0 ml aliquot is transferred to a 100 ml stoppered measuring cylinder together with 15 ml of chloroform and 25 ml of methylene blue indicator solution (prepared by dissolving 0.03 g of methylene blue and 50 g of anhydrous sodium sulphate in 200 ml of water, carefully adding 6.5 ml of concentrated sulphuric acid and diluting to 1000 ml). The measuring cylinder is stoppered and the mixture gently shaken, avoiding the formation of an emulsion. In the presence of anionic material the lower chloroform phase will assume a deep blue colour. The anionic surfactant is titrated by the addition from a 10 ml microburette of an aqueous 0.004 M solution of cetyltrimethylammonium bromide (CTAB), gently shaking the stoppered measuring cylinder between additions. During the course of the titration the blue colour will gradually transfer to the upper aqueous phase, the end-point being reached when the intensity of colour in the two phases matches. Standardisation is carried out by titrating about 10 mg, by way of a stock solution, of the anionic surfactant present. One millilitre of 0.004 M CTAB is equivalent to about 1.2 mg of sodium lauryl sulphate. Repeatability studies on six portions of a 'check sample' gave a mean sodium lauryl sulphate level of 1.536% with an sd of 0.012% and a cv of 0.75%.

2.4.2.3 The determination of an alkylphenol ethoxylate by UV spectroscopy. This can be carried out on the same test solution as used for the determination of anionic surfactant. The spectrum is recorded in a 10 mm pathlength cell against water over the region 350–230 nm, and the absorbance, due to the alkylphenol chromophore, is measured at the maximum which occurs at about

275 nm with a baseline drawn tangentially between 295 and 245 nm. Calibration is carried out by using a reference sample of the alkylphenol ethoxylate present, and should cover the approximate range 0–40 mg per 100 ml. Anionic surfactants such as sodium alkyl sulphates and sodium alkyl sulphonates do not interfere, but sodium alkyl benzenesulphonates, which absorb in the region 250–275 nm, do.

2.5 Polyamides

2.5.1 Introduction

Polyamides are to a large extent used as high-temperature, abrasion resistant, high-impact-strength moulding compounds. Their analysis is concentrated on the following areas:

(a) Lubricants, such as metal stearates, stearic acid, aliphatic esters, aliphatic secondary amides and EBS (ethylene bis-stearamide).
(b) Impact modifiers (or rubbers), such as EPDM or EPR, particularly when grafted with materials such as acrylic acid, maleic acid or maleic anhydride.
(c) Fillers, particularly glass fibre.
(d) Pigments such as carbon black and organic dyes such as nigrosine.
(e) Heat stabilisers such as phenolic anti-oxidants and flame retardants such as melamine or melamine derivatives.

2.5.2 Extraction techniques

Most procedures for the isolation of additives from polyamides involve solution/precipitation techniques such as solution in formic acid with precipitation by a combination of methanol and diethyl ether [2], or more recently by solution in either hexafluoroisopropanol or trifluoroethane with precipitation by either methanol or acetonitrile [5]. Recent work [19] has shown that direct methanol extraction of a polyamide, which had been freeze-ground to pass a 1.18 mm sieve, removed phenolic anti-oxidants (present as heat stabilisers) prior to their determination by HPLC. The ease of solubility of polyamides 6 and 66 in both hydrochloric acid and formic acid is used in several gravimetric and spectroscopic procedures which will be described later.

2.5.3 Lubricants

The determination of metal stearate lubricants by means of a hydrochloric acid/diethyl ether hydrolysis procedure has been described [2] and can be used if, for example, calcium or zinc stearate is known to be present. However, for unknown lubricant systems this procedure cannot distinguish between hydrolysis-produced and added stearic acid, and analysis of metal stearates by way of the metal is preferred. A scheme was developed by Palmer and Slowther [22] to isolate lubricants other than metal stearates from formic acid-soluble polyamides without decomposition. This is shown in Figure 2.5 using the apparatus shown in Figure 2.6.

Polyamide composition

Dissolve 5 g in 30 ml of formic acid and liquid/liquid extract with 150 ml of toluene for 18 hrs.

SOLUBLE IN TOLUENE
Lubricants such as
aliphatic secondary amides,
stearic acid,
aliphatic esters and EBS.
Polyamide oligomers

Evaporate toluene to dryness, treat with boiling ethanol and filter to remove fillers and impact modifiers. Evaporate ethanol to dryness. Extract residue with boiling petroleum ether (40–60) and filter.

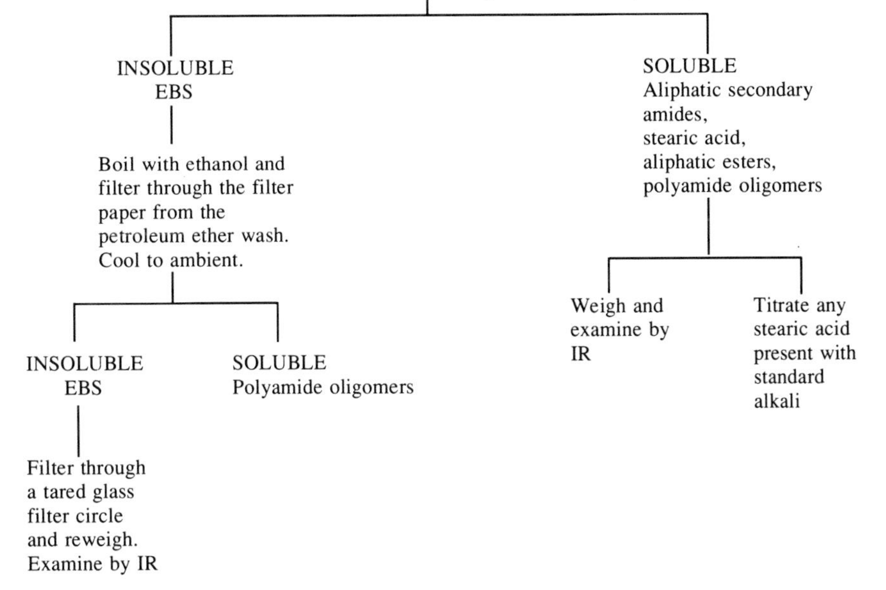

Figure 2.5 Scheme of analysis for lubricants in polyamide.

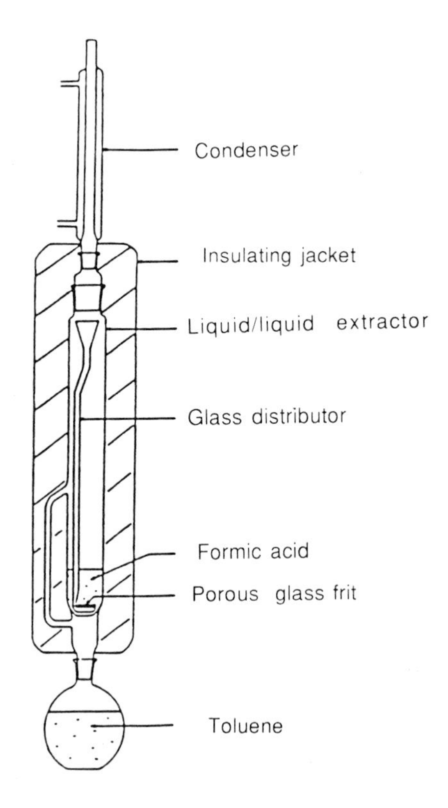

Figure 2.6 Apparatus of Palmer and Slowther [22] for the isolation of lubricants from polyamides.

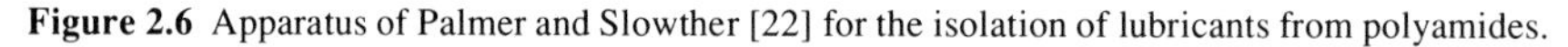

The recovery of EBS is consistently 85% and a correction is applied to the EBS level found. This is partly due to a 5% solubility of EBS in cold ethanol. The recoveries of aliphatic secondary amides and stearic acid are still being investigated.

2.5.4 The determination of 50% hydrochloric acid-insoluble material

Refluxing 2 g of polyamide 6 or 66 with 50 ml of 50% hydrochloric acid for 2 h readily dissolves the polyamide. Any impact modifier, any carbon black, together with any polymer such as EVA used as a pigment carrier, and the bulk of any glass filler are insoluble.

The insoluble material is quantitatively separated by filtering, under suction, through a tared glass filter circle, and washing with 50% hydrochloric acid, water and finally ethanol. The oven-dried filter circle and insolubles are reweighed and the total percentage of insoluble material is calculated.

If this material is only carbon black or only impact modifier, then the percentage of insolubles is taken as the percentage of that additive. A mixture of the two is reported as the total organic hydrochloric-acid insolubles. If glass is present then 2 g of the original sample is ashed in a muffle furnace at 600°C for 30 min to determine the glass level. The glass from ashing is treated with 50% hydrochloric acid exactly following the procedure previously described and the percentage of glass insoluble in acid is obtained. For a particular grade of glass currently available this has consistently been found to be 63%. The glass figure found by ashing is corrected accordingly to give the percentage of glass that will be in the hydrochloric acid-insoluble fraction. This is subtracted from the total hydrochloric acid-insoluble figure to give the percentage of impact modifier and/or carbon black present. Carrying out this procedure on weighed amounts of 'blank' polyamide 6, glass and polyolefin rubber mixtures gave, in the presence of a typical level of glass, recoveries of the rubber of better than 99.5%, for a range of levels down to a few per cent.

2.5.5 The determination of melamine-based flame retardants

Melamine and melamine derivatives are widely used as flame retardants in polyamides. They can be determined by dissolving 0.3 g of sample in 20.0 ml of conc. hydrochloric acid and diluting 2.0 ml of this solution to 100 ml with water. This precipitates out the polymer, which is removed by filtration. The UV spectrum of the filtrate is recorded in 5 mm pathlength quartz cells over the region 400–200 nm. The absorbance of the peak due to either melamine or a melamine derivative, which occurs at about 235 nm, is measured and a baseline correction applied by subtracting the absorbance at 280 nm. Calibration is carried out by preparing a series of standards, covering the range 0–40 mg per 20.0 ml of conc. hydrochloric acid in the presence of 0.3 g of 'blank' polyamide, and proceeding as above. A small 'blank' absorbance is obtained for polyamide. This method cannot distinguish between the additives.

2.5.6 The determination of nigrosine

Nigrosine is a black dye which can be used as an alternative to carbon black for colouring polyamides. It is usually added at levels of up to 1%, and is determined by dissolving 0.3 g of sample in 100 ml of formic acid and recording the visible spectrum of the solution in 10 mm pathlength cells against formic acid over the region 750–350 nm. The absorbance due to nigrosine is measured at 635 nm. Calibration is carried out by preparing, by dilution of a stock solution, a series of standards covering the range 0–2.5 mg of nigrosine per 100 ml of formic acid. Despite the small sample weight taken repeatability studies on six portions of a 'check' sample gave a mean nigrosine content of 1.29% with a sd of 0.02% and a cv of 1.55%.

References

1. T. R. Crompton (1971). *Chemical Analysis of Additives in Plastics*, Pergamon Press, Oxford.
2. J. Haslam, H. A. Willis and D. C. M. Squirrell (1977). *Identification and Analysis of Plastics*, (2nd edn), Iliffe Books, London.
3. T. R. Crompton (1984). *The Analysis of Plastics*, Pergamon Press, Oxford.
4. D. C. M. Squirrell (1981). *Analyst*, **106**, 1042–56.
5. W. Freitag (1990). In *Plastics Additives Handbook*, eds. R. Gachter, H. Muller and P. P. Klemchuk, Hanser, Munich, pp. 909–946.
6. F. Scholl (1981). *Atlas of Polymer and Plastics Analysis*, Vol. 3, Hanser, Munich.
7. W. Freitag and O. John (1990). *Angewandte Makromolekulare Chemie*, **175**, 181–5.
8. A. Mukerjee and P. Mukerjee (1962). *J. Appl. Chem.*, **12**, 127–9.
9. G. Socrates (1980). *Infrared Characteristic Group Frequencies*, John Wiley, Chichester.
10. BS 2782 (1975), Part 4, Method 434A.
11. N. Everall, J. M. Chalmers and I. D. Newton (1992). *Applied Spectroscopy*, in the press.
12. D. C. M. Squirrell (1964). *Automatic Methods in Volumetric Analysis*, Hilger and Watts, London.
13. I. D. Newton, I.C.I. plc, unpublished.
14. W. D. Pohle and V. C. Mehlenbacher (1950). *J. Assoc. Offic. Anal. Chem.*, **27**, 54–56.
15. J. Nelson, I.C.I. plc, unpublished.
16. C. Rawson *et al.*, I.C.I. plc, unpublished.
17. W. Freitag, R. F. Wurster and N. Mady (1988). *J. Chromatogr.*, **455**, 426.
18. W. Freitag (1983). *Fresenius Z. Anal. Chem.*, **316**, 495.
19. C. J. Slowther, I.C.I. plc, unpublished.
20. L. H. Ruddle, S. D. Swift, J. Udris and P. E. Arnold (1965). In *Proceedings of the SAC Conference*, Nottingham, pp. 224–31.
21. E. Stahl (1969). *Thin Layer Chromatography, A Laboratory Handbook*, Springer-Verlag, Berlin.
22. R. J. Palmer and C. J. Slowther, I.C.I. plc, unpublished.
23. *The Infrared Spectra Atlas of Surface Active Agents* (1982). Heyden, London.
24. S. R. Epton (1948). *Trans. Faraday Soc.*, **44**, 226–30.

3 NMR characterisation of polymers

P. A. MIRAU

3.1 Fundamental concepts

3.1.1 Introduction

Nuclear magnetic resonance (NMR) spectroscopy is an important method for materials characterisation and for the study of polymer structure–property relationships. The importance of NMR as a technique arises in part because the signals can be assigned to specific atoms along the polymer backbone and side chains [1, 2]. The properties of the NMR signals depend on the magnetic environment of the NMR active nuclei, and the local fields that they experience. Since the NMR spectrum is determined by local forces, this method provides valuable and unique information about polymers on an atomic-length scale.

The observation of NMR signals is possible because the nuclei have angular momenta [1], that are characterised from quantum mechanics by the spin quantum number I. Those nuclei that have $I \neq 0$ have magnetic moments and $2I + 1$ spin states. The degeneracies in the spin states are lifted in the magnetic field and the transitions between the $2I$ levels can be observed, as shown in Figure 3.1 for $I = \frac{1}{2}$ and $I = 1$.

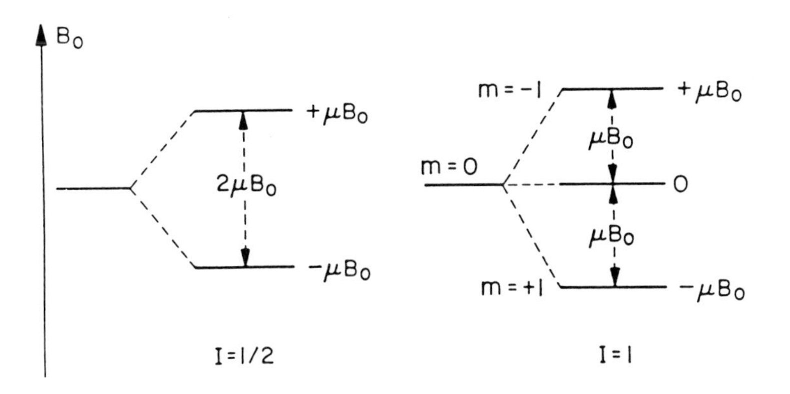

Figure 3.1 The magnetic energy levels for nuclei of spin $\frac{1}{2}$ and spin 1.

The signal frequencies depend on the magnetic moment and the strength of the magnetic field. The magnitude of the nuclear magnetic moment is specified as the ratio of the magnetic moment to the angular momentum, or the magnetogyric ratio γ which is defined as:

$$\gamma = \frac{2\pi\mu}{Ih} \tag{3.1}$$

where μ is the nuclear moment and h is Planck's constant. The observation frequency ω_0 depends on the magnetic field strength B_0 by the relationship

$$\omega_0 = \gamma B_0 \tag{3.2}$$

NMR spectrometers with fields strengths as high as 14.1 T (600 MHz for protons) are currently available, but the highest possible field strengths are not required for most applications.

The NMR sensitivity depends on the magnetogyric ratio and the natural abundance of the NMR active nuclei. Table 3.1 lists the properties of some of the nuclei that are of particular interest to polymer chemists. The sensitivity is highest for protons and fluorines, but nuclei such as carbon, silicon and phosphorus can be routinely studied. Deuterium and nitrogen have poor sensitivities, but they provide valuable insight into the structure and dynamics of polymers when isotopically enriched samples are prepared.

Table 3.1. The nuclear properties of isotopes of interest to polymer chemists

Isotope	Abundance (%)	Spin[a]	$\gamma^b \times 10^{-4}$	Relative sensitivity[c]	Frequency at 2.35 T[d] (MHz)
^1H	99.98	$\frac{1}{2}$	2.67	1.00	100.0
^2H	0.0156	1	0.410	0.009	15.35
^{13}C	1.108	$\frac{1}{2}$	0.672	0.016	25.11
^{15}N	0.365	$\frac{1}{2}$	−0.27	0.001	10.13
^{19}F	100.0	$\frac{1}{2}$	2.516	0.834	94.07
^{29}Si	4.70	$\frac{1}{2}$	−0.53	0.78	19.86
^{31}P	100.0	$\frac{1}{2}$	1.082	0.066	40.47

[a] The spin quantum number I.
[b] The magnetogyric ratio.
[c] The sensitivity relative to protons.
[d] The signal frequency in a field strength of 2.35 T.

The NMR spectral parameters provide detailed information for polymer characterisation. Of primary importance are the frequency (or chemical shift), intensity, line width, J coupling constants, and relaxation rates (see Figure 3.2) [2]. High-resolution spectra are obtained because the frequencies for the backbone and side-chain atoms are extremely sensitive to the local magnetic environment. Not only are the signals from the methyl, methylene, methine, and other substituted carbons resolved from each other, but atoms in asymmetric stereochemical environments can also be resolved. Polymers with a complex microstructure may give extremely complex NMR spectra that can be quantitatively analysed because the line intensities depend on the number of nuclei. The through-bond or J couplings lead to line splittings that depend on the number and type of neighboring nuclei, and sometimes on chain conformation [2]. These parameters provide the critical information needed for spectral assignments, the first step in materials characterisation.

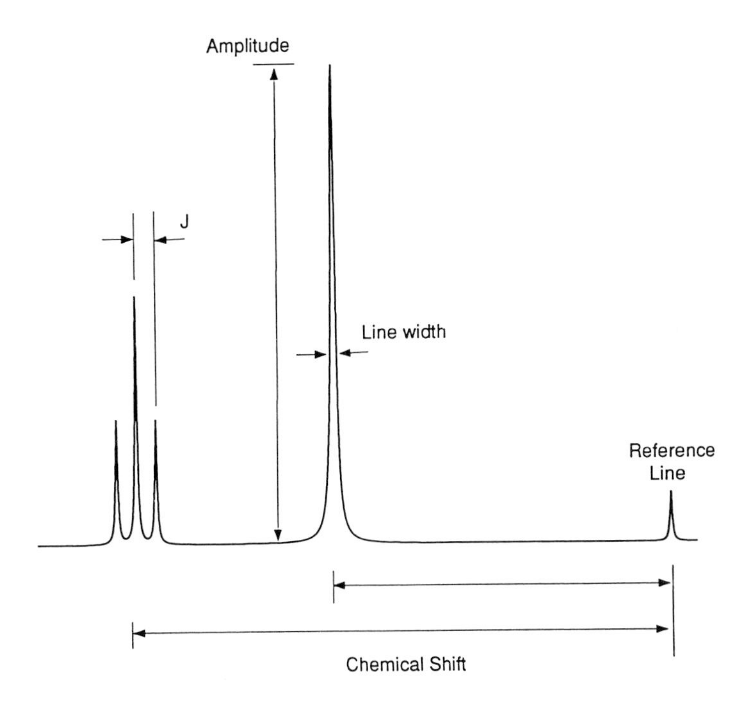

Figure 3.2 The NMR parameters used in polymer characterisation. The line width, amplitude, and J-coupling constants are measured directly from the spectra, and the integrated intensity of the signals is related to the number of nuclei.

The line widths and relaxation times are determined by the local atomic fluctuations in the polymer main- and side-chain atoms. The line widths can be directly determined from the spectrum, while the relaxation times are measured by monitoring the return of the spin system to equilibrium following some perturbation. The relaxation times are typically of the order of milliseconds to seconds in high polymers, but depend on the molecular motions that occur on the picosecond to microsecond time scale.

3.1.2 Pulsed NMR spectroscopy

The NMR signals in virtually all modern NMR spectrometers are obtained by a combination of pulsed NMR and Fourier transformation, as illustrated in

Figure 3.3. In the absence of radio frequency (rf) pulses, the nuclear magnetic moments tend to align with the magnetic field. The application of rf pulses rotates the magnetisation into the xy plane, where it rotates, or precesses, at a rate that depends on the magnetic environment. This rotation is detected as the oscillating signal known as a free induction decay (fid) and converted into a frequency spectrum by Fourier transformation [1].

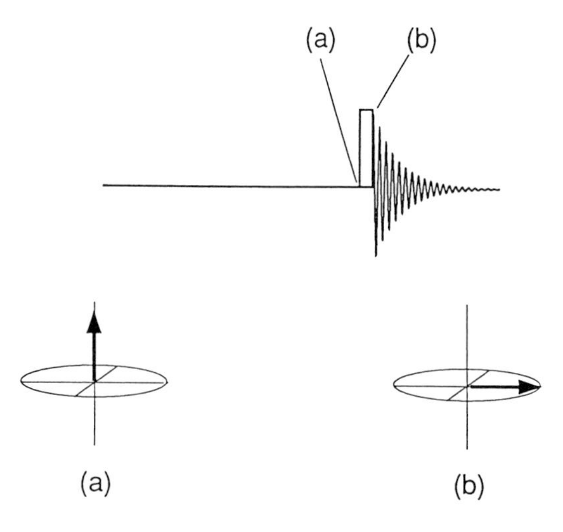

Figure 3.3 The principles of pulsed NMR spectroscopy. A 90° pulse rotates the magnetisation into the xy plane, where it precesses at a frequency that depends on the local environment. This oscillating signal (fid) is Fourier-transformed to give a frequency spectrum.

It is the use of pulsed NMR that allows us to probe in detail the structure and dynamics of polymers. The polymer spin system can be perturbed with rf pulses and its return to equilibrium can be measured. The degree to which the spins are rotated by the rf pulses depends on the nuclei, the strength of the rf field, and the pulse length. The tip angle (θ) is given by:

$$\theta = \gamma H_1 t \tag{3.3}$$

where γ is the magnetogyric ratio, H_1 is the rf field strength, and t is the pulse length (typically microseconds). The experiments used to measure spin–lattice relaxation rates are one example of measuring the return to equilibrium following perturbation of the spin system. Figure 3.4 shows the 180°–delay–90° pulse sequence and vector diagram that illustrate the measurement of the spin-lattice relaxation time (T_1). The magnetisation is inverted with the initial pulse and returns exponentially to equilibrium during the variable relaxation delay period. The final pulse is used to measure the state of the spin system following the delay, and the relaxation time constant is obtained from a plot of the intensity versus time. Information about the molecular dynamics is

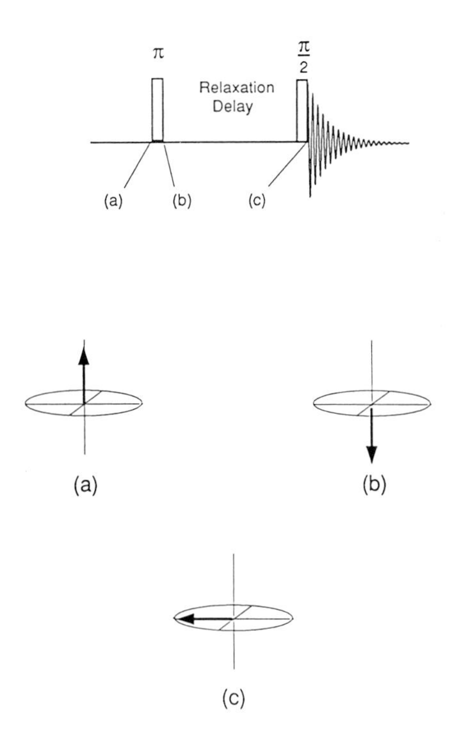

Figure 3.4 The pulse sequence and vector diagram for spin–lattice relaxation-rate measurements by inversion recovery. The magnetisation is inverted with a 180° pulse (b) and returns toward equilibrium during the relaxation delay. The magnetisation is sampled with a 90° pulse and T_1 is calculated from a plot of the signal intensity versus relaxation delay.

obtained because the relaxation rates depend on the rotational correlation time τ_c. For the special case of very rapid isotropic molecular motion, the spin–lattice relaxation rate for a proton pair is given by:

$$\frac{1}{T_1} = \frac{3}{2}\frac{\gamma^4 h^2 \tau_c}{r_{HH}^6} \tag{3.4}$$

Note that such methods are also powerful tools to study the molecular structure, since the relaxation rates depend on the inverse sixth power of the internuclear distance (r_{HH}). Because of this dependence, these methods are useful for structural measurements in the range of 2–5 Å.

3.1.3 Instrumentation

NMR spectrometers are currently used in many areas of polymer science. Common to all instruments are a high field magnet, a source of rf energy, and a computer. The modern superconducting magnets have fields in the range of 4–12 T, such that the observation frequency for protons is typically between 200 and 600 MHz. Inside the magnet is a probe tuned to deliver rf pulses and to observe the signals for the nuclei of interest. There are many types of probes designed for the wide variety of NMR experiments. Among the capabilities of probes are the ability to irradiate more than one type of nucleus simultaneously (such as carbon and protons) and the ability to spin the samples rapidly for the observation of signals in solids.

The rf source is controlled by the computer, which is also used to record, transform, plot, and store the data. Many types of relaxation and two-dimensional NMR experiments are possible because of the precise computer control over the application of the rf pulses.

3.2 NMR characterisation of polymers in solution

3.2.1 Proton NMR studies

NMR in solution, primarily proton and carbon, is one of the most extensively used tools for materials characterisation. The importance of these methods is due in part to the extreme sensitivity of the NMR signals to small changes in the chemical structure of polymers, such as those that arise from polymer stereochemistry, defects and other types of microstructure.

The proton NMR sensitivity to polymer stereochemistry is illustrated in the 500 MHz NMR spectra of poly(methyl methacrylate) shown in Figure 3.5 [3]. The stereochemistry of poly(methyl methacrylate) is determined by the conditions used in the polymer synthesis. A relatively simple spectrum is observed for the predominantly isotactic (>95%) poly(methyl methacrylate) obtained from anionic polymerisation in toluene at low temperature [3]. Three major signals are observed in Figure 3.5 that can be assigned to the methyl resonance and the non-equivalent main-chain methylene protons labelled *e* and *t* (for *erythro* and *threo*). The methoxyl protons are not plotted, but appear as a sharp resonance at 3.4 ppm. Among the features to note about the proton NMR spectrum of isotactic poly(methyl methacrylate) are that sharp lines are observed, and that the resonances are well separated based on the resonance type (methyl, methylene, or methoxyl). In addition, the non-equivalent methylene protons are separated by 0.7 ppm. This chemical shift difference can be understood from molecular modelling studies that show that the environments for the methylene protons are different from each other, since one is in closer proximity to the carbonyl group and may be shifted from its normal position. This effect is observed only in *m* centred dyads, and such splittings are one basis for making stereochemical assignments in polymers. The other feature to note about Figure 3.5 is that the methylene proton signals are split into doublets due to the 15 Hz through-bond coupling between the geminal methylene protons.

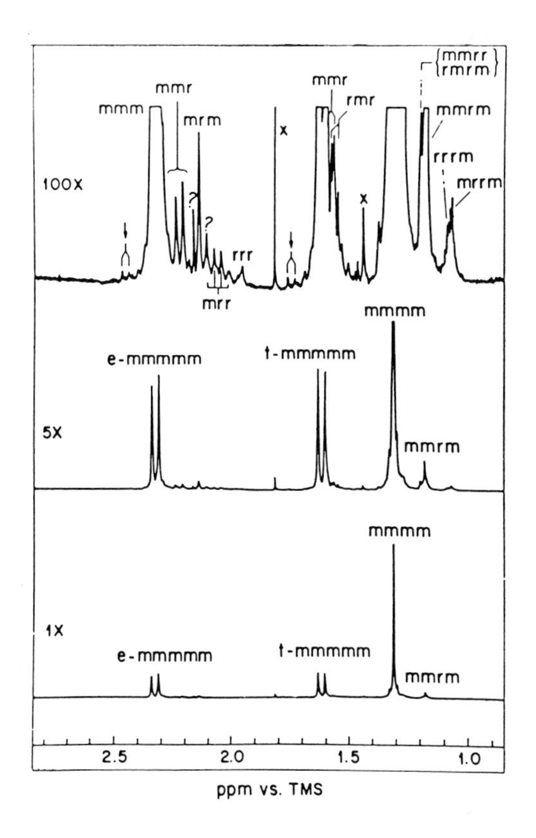

Figure 3.5 The 500 MHz proton NMR spectra of isotactic (>95%) poly(methyl methacrylate) plotted at different gain levels as a 10% w/v solution in chlorobenzene-d_5 at 100°C. Resonances marked by arrows are due to oligomers or end groups and impurities are marked x. Reprinted with permission from ref. 3.

The chain irregularities in isotactic poly(methyl methacrylate) become apparent as the gain is increased in Figure 3.5. These results show that polymer stereochemistry can be characterised in detail, since many of the peaks can be assigned and the amount of each stereosequence can be determined from the peak intensity. Some of the peaks are assigned to stereosequences that have probabilities below the 1% level.

Figure 3.6 shows the spectrum of free-radical polymerised poly(methyl methacrylate), a polymer in which the fractions of *m* and *r* dyads are approximately equal [3]. The spectrum differs considerably from that of the isotactic polymer, and comparison of Figures 3.5 and 3.6 illustrates the power of solution NMR for the study of polymer microstructure. The results can provide information about reaction mechanism if the stereosequence peaks can be assigned. The traditional assignment methods rely on spectral comparisons with model compounds

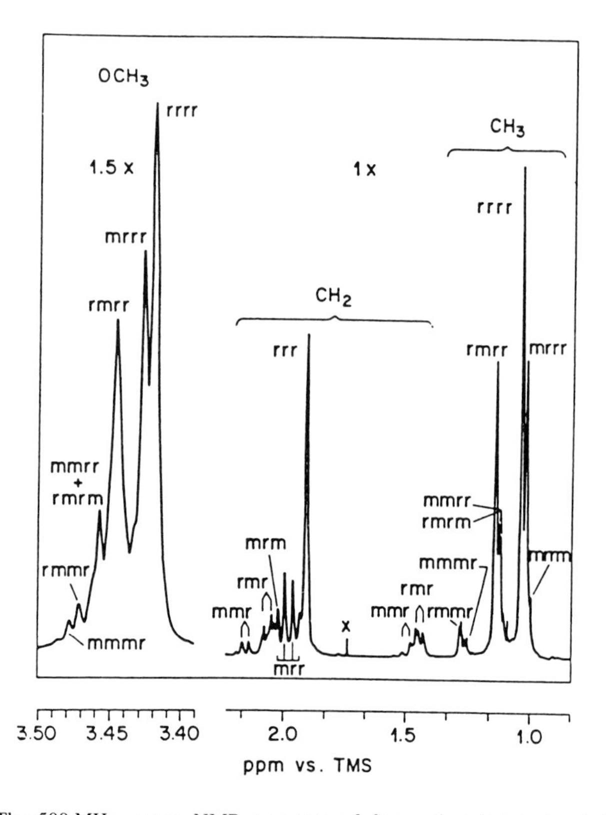

Figure 3.6 The 500 MHz proton NMR spectrum of free-radical (atactic) poly(methyl methacrylate) (10 wt% in chlorobenzene-d_5). The stereosequence assignments are listed. Reprinted with permission from ref. 3.

or polymers. With the advent of two-dimensional (2D) NMR, many of the assignments can now be established without the synthesis of new materials.

The spectrum of poly(methyl methacrylate) is among the most highly resolved. Many polymers do not have such a large chemical shift dispersion, so the lines appear as broadened signals from a distribution of unresolved lines. This behavior is illustrated in the proton NMR spectrum of poly(vinyl chloride) of molecular weight 83 000 shown in Figure 3.7 [4]. Again the methine and methylene protons are resolved from each other, but simple Lorentzian lines are not observed. The signals in the methylene region (1.5–2.5 ppm) arise from the overlap of many stereosequences, as well as from the non-equivalent methylene protons in *m* centred dyads. Even with such extensive overlap, the signals can frequently be assigned and analysed using 2D NMR methods [5].

The power of proton NMR has been illustrated with stereochemical examples. The spectrum is also sensitive to other types of defects, including branches,

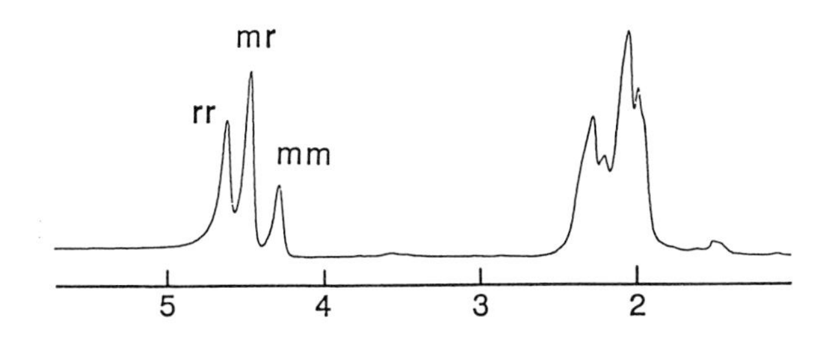

Figure 3.7 The 500 MHz proton NMR spectrum of a 10 wt% solution of 83 000 molecular-weight poly(vinyl chloride) in deuterated benzene at 45°C. Reprinted with permission from ref. 4.

crosslinks, and geometric as well as head-to-head and tail-to-tail regeo-isomers. In addition, the signals from chain ends are sometimes visible for polymers with a degree of polymerisation less than 500. The next section shows that ^{13}C-NMR is a powerful tool for materials characterisation because of its wider chemical-shift range. Among the advantages of proton NMR are its high sensitivity and the ease of acquiring a spectrum, making it an important tool for routine characterisation of materials.

3.2.2 Carbon NMR studies

Carbon NMR in solution is the method of choice for many types of material characterisation. While the sensitivity is much lower than for protons (Table 3.1), synthetic polymers are often soluble at concentrations as high as 5–20 wt%, so ^{13}C-NMR spectra with a high signal-to-noise ratio can be acquired in a few hours. One major advantage of carbon NMR is that the chemical shifts are dispersed over 200 ppm rather than the 10 ppm typically observed for protons [1]. In addition, ^{13}C-NMR is frequently used to study polymer molecular dynamics because the relaxation is predominantly due to carbon–hydrogen dipolar interactions with directly bonded protons. The C–H distances are fixed by the bond lengths, so the relaxation times can be used to measure the rotational correlation times directly (eqn (3.4)).

As with protons, the carbon spectrum is observed by pulsed NMR and Fourier transformation. However, the carbon resonances are frequently split by the through-bond couplings to directly bonded and nearest-neighbour protons. While these couplings provide information about the number of directly bonded protons and the carbon type, they are usually removed by proton irradiation during carbon signal acquisition to simplify the spectra. Figure 3.8 shows the 125 MHz ^{13}C-NMR spectra of a 10 wt% solution of poly(vinyl chloride) in deuterated benzene acquired at 45°C [4]. Note the increase in resolution compared with the 500 MHz proton NMR spectrum of the same material shown in Figure 3.7. Comparison of the proton and carbon spectra of poly(vinyl chloride) demonstrates the higher resolution of ^{13}C-NMR due to the extreme sensitivity of the chemical shifts to polymer microstructure. While only triads are resolved in the proton spectrum, tetrads can be resolved in the carbon spectrum of poly(vinyl chloride), and even higher resolution has been reported in other polymer systems [6]. As in proton NMR, a variety of methods are used to establish the resonance assignments. These methods include comparisons with the NMR spectra of model compounds and polymers, calculation of the chemical shifts from rotational isomeric state models [7], and 2D NMR [5].

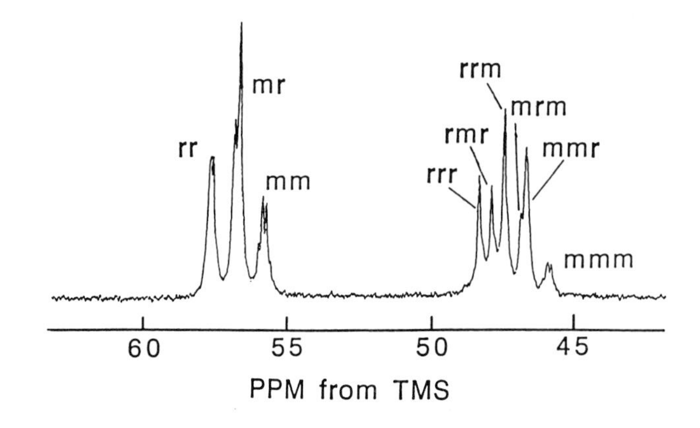

Figure 3.8 The 125 MHz ^{13}C-NMR spectrum of poly(vinyl chloride). See Figure 3.7 for details. Reprinted with permission from ref. 4.

In addition to polymer stereochemistry, the ^{13}C spectra are sensitive to other types of defects, such as branching, isomerism, head-to-head and tail-to-tail additions, and to chain ends. The sensitivity of the ^{13}C-NMR spectra to the defects and isomers is illustrated in the spectrum of free-radical polymerised polybutadiene shown in Figure 3.9 [2]. The many resonances are observed because of the statistical incorporation of *cis*- and *trans*-1,4-butadiene, and the random stereochemistry for the addition of 1,2-butadiene. The inset to Figure 3.9 shows a simulation of the olefinic region calculated from a random distribution of *cis*- and *trans*-1,4 units and 1,2 units. When quantitative spectra are acquired, the molecular weights can be determined from the ratio of the end groups to main-chain resonances.

The carbon chemical shifts depend not only on the stereochemistry and the chain defects, but also on chain conformation through the γ-gauche effect (7). The γ-gauche effect is an empirical correlation between the chemical shift of a particular carbon and the orientation of its γ neighbouring carbons. The γ-gauche effect can shift resonances by as much as 5 ppm, depending on the chain conformation and the microstructure. In many cases the chemical shifts can be assigned by comparison of the calculated and observed line intensities [7].

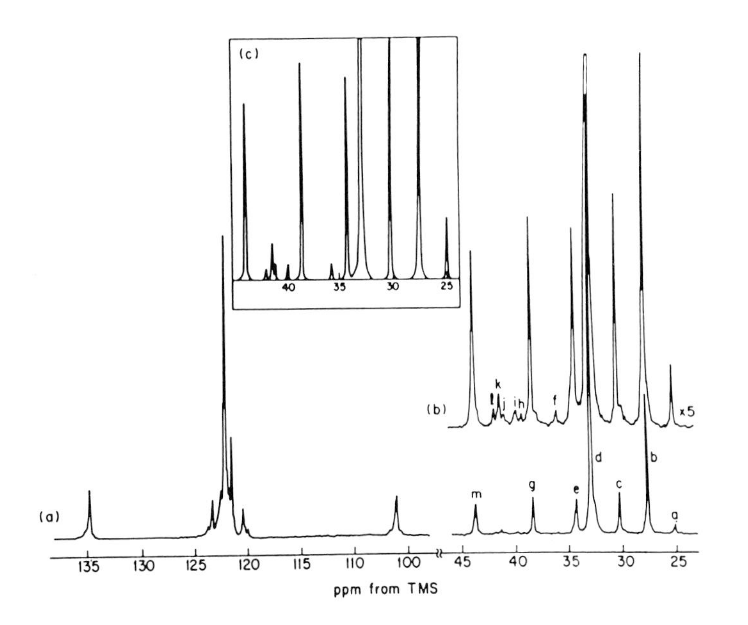

Figure 3.9 The 50 MHz ^{13}C-NMR spectrum of free-radical polymerised polybutadiene observed at 50°C as a 20 wt% solution in $CDCl_3$. The inset (c) is a computer simulation of the olefinic region (b) based on a random distribution of *cis, trans*, and 1,2 units in a 23:58:19 ratio. Reprinted with permission from ref. 2.

Table 3.2 The spin–lattice relaxation and correlation times for polymers in solution

Polymer	Solvent[a]	Temp. (°C)	NT_1^b (s)	τ_c^c (ns)
Polyethylene	ODCB	30	1.24	0.040
Poly(ethylene oxide)	C_6D_6	30	2.80	0.018
Polypropylene	ODCB	30	0.30	0.130
Poly(propylene oxide)	$CDCl_3$	30	1.00	0.049
1,4-Polybutadiene (*cis*)	$CDCl_3$	54	3.00	0.016
1,4-Polybutadiene (*trans*)	$CDCl_3$	54	2.38	0.021
Poly(vinyl chloride)	TCB	107	0.32	0.15
Polystyrene	$CDCl_3$	38	0.12	0.39
Poly(butene-1-sulphone)	$CDCl_3$	40	0.09	23.0

[a] The solvents are *ortho*-dichlorobenzene (ODCB), benzene (C_6D_6), chloroform ($CDCl_3$), and 1,2,4-trichlorobenzene (TCB).
[b] The spin–lattice relaxation times are listed as the relaxation time multiplied by the number of directly bonded protons.
[c] The correlation times are calculated assuming isotropic motion and are listed in nanoseconds.

Carbon NMR relaxation is one of the most important methods for the study of polymer dynamics, since the relaxation times are directly related to the rotational correlation times [1]. The correlation times for many polymers have been measured, and a few are shown in Table 3.2 [8]. Such results have made important contributions to our understanding of polymer structure–property relationships, since the entries in Table 3.2 illustrate the relationships between the chemical structure of polymers and the local molecular dynamics. Note, for example, how the introduction of oxygen into the polymer main chain (polypropylene versus poly(propylene oxide)) greatly increases the mobility. The molecular dynamics of a wide variety of polymers have been measured as a function of temperature, molecular weight, and concentration [9].

These NMR spectra are useful for materials characterisation in part because the line intensities are proportional to the number of carbons. However, care must be taken to ensure that the carbon spectra can be quantitatively analysed. The protons are usually irradiated to suppress the through-bond J couplings, and this irradiation can also affect the line intensities through the nuclear Overhauser effect (NOE). Figure 3.10 shows a plot of the effect of the rotational correlation time on the signal intensity for NOE-enhanced spectra [1]. The maximum intensity is observed when the molecular dynamics are fast compared to the inverse of the observation frequency (0.1–0.5 ns). The signals can be stronger by a factor of three because of the NOE, and this technique is frequently used to improve the signal-to-noise ratio. However, for those cases where the dynamics vary along the chain, the signals may be differentially enhanced by the NOE. This leads to a spectrum in which the more mobile carbons have greater intensities than the more restricted ones, or the quater-

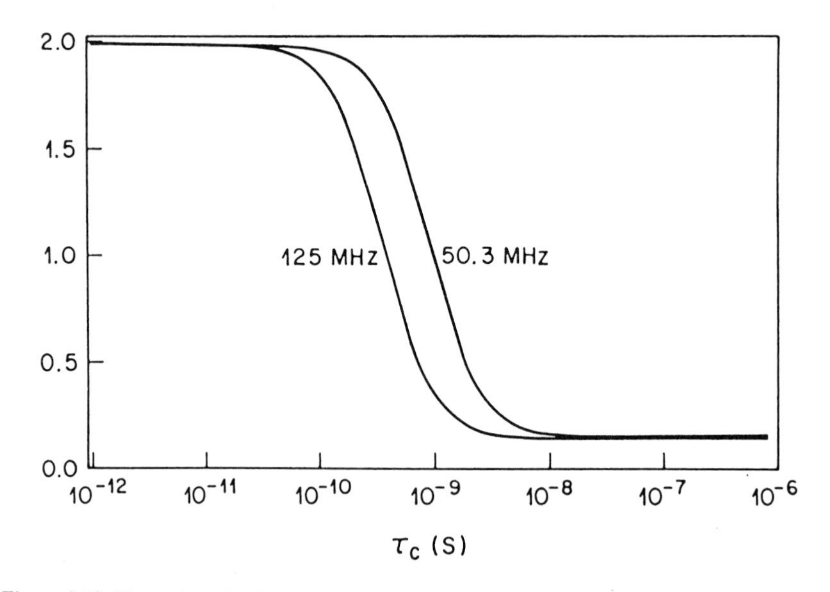

Figure 3.10 The nuclear Overhauser enhancement factor η for carbons as a function of correlation time at 125 and 50 Mhz. The intensity measured with the NOE is $1 + \eta$. Reprinted with permission from ref. 2.

nary or carbonyl carbons that do not have directly bonded protons. For quantitative comparisons, the spectra must be acquired by turning on the proton decoupling only during the acquisition period to suppress the NOE. It is also likely that the spin–lattice relaxation times will vary depending on the number of directly bonded protons and the correlation times, which can differ by orders of magnitude between the main-chain and side-chain atoms. In such cases, long delay times between acquisitions are used to ensure that all carbons have returned to equilibrium between experiments.

3.2.3 NMR studies of other nuclei

While carbon and proton NMR studies are most commonly reported, several of the other nuclei listed in Table 3.1 (fluorine, phosphorus, nitrogen, and silicon) have made important contributions to materials characterisation. Fluorine NMR ranks only below proton NMR in sensitivity and provides a valuable probe of fluorine-containing polymers, such as poly(viny fluoride) [10]. The fluorine chemical shifts are much more sensitive to structural defects than protons, as shown in Figure 3.11, which compares the ^{19}F spectra for

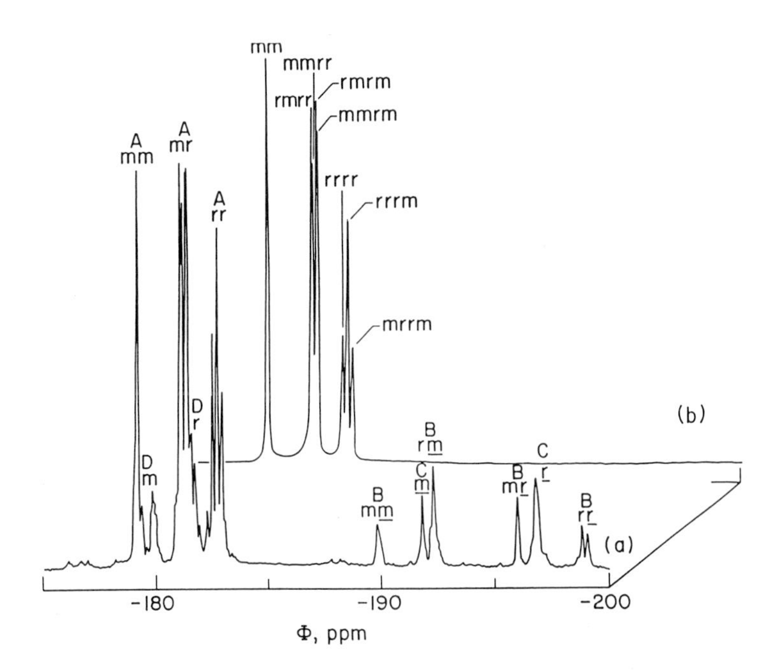

Figure 3.11 The ^{19}F spectra of (A) commercial poly(vinyl fluoride) and (B) poly(vinyl fluoride) prepared without head-to-head and tail-to-tail defects. The spectra were acquired as an 8% solution in N,N-dimethylformamide-d$_7$. The resonances marked A correspond to head-to-tail sequences while B and C represent the head-to-head and tail-to-tail defects. Reprinted with permission from ref. 10.

commercial poly(vinyl fluoride) and a sample prepared without regeo defects [10]. The lines are split both from stereochemistry and from the presence of head-to-head and tail-to-tail isomeric defects in the commercial material (Figure 3.11(a)). The higher field portion of the spectra (– 175 to – 185 ppm) is assigned to the stereosequences in the head-to-tail monomers. This assignment is established by comparing the spectrum of the commercial material with that of poly(vinyl fluoride) prepared without monomer reversals (shown in Figure 3.11(b)), and by 2D NMR.

^{29}Si-NMR has been primarily used to study silicon-containing main-chain polymers such as polysiloxanes and polysilanes. The sensitivity of ^{29}Si is greater than for carbon and the resonances are observed over a large chemical-shift range. Figure 3.12 shows the 59.6 MHz ^{29}Si-NMR spectrum of a 20 wt% solution of a tetramethyl-*p*-silphenylene siloxane and dimethyl siloxane block copolymer in chloroform [11]. Signals are observed for the blocks of the two different monomers, and for the resonances at the interface between the blocks, as shown by the inset to Figure 3.12. The small signal (e) has been assigned to dimethyl siloxane monomers between two tetramethyl-*p*-silphenylene siloxane monomers. It is clear that this type of analysis provides detailed information about the sequence distribution and the reaction mechan-

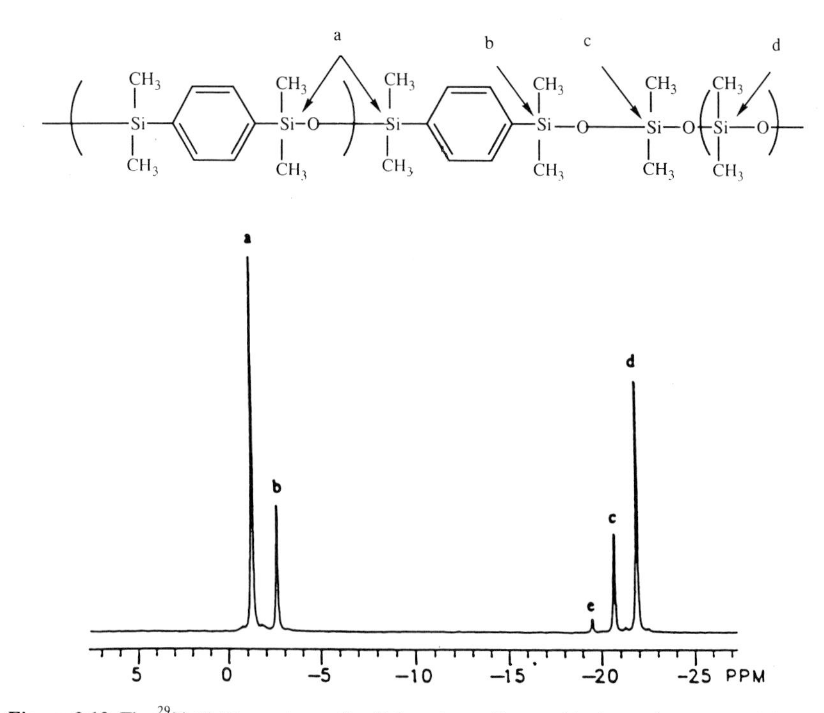

Figure 3.12 The ^{29}Si-NMR spectrum of a silphenylene–siloxane block copolymer containing 67% dimethylsiloxane with an average silphenylene block length of four and an average dimethylsiloxane block length of seven. Reprinted with permission from ref. 11.

ism. ^{29}Si-NMR has also been used to study polysilanes, unique polymers containing only silicon in the backbone [12]. As in carbon NMR, the ^{29}Si signals in polysilanes are sensitive to stereochemistry.

Phosphorus is an abundant spin-$\frac{1}{2}$ nucleus that occurs more frequently in biopolymers than in synthetic polymers. In biopolymers, the ^{31}P chemical shifts are related to the phosphate conformation and the ionisation state. Phosphorus NMR has been used to characterise synthetic polymers such as poly(phosphazines) [13]. The ^{31}P chemical shifts provide a probe of the polymer backbone in such polymers.

Nitrogen is a commonly occurring element in many polymers, such as nylons, polyimides, and biopolymers. The use of ^{15}N-NMR to characterise these materials is limited in sensitivity by the low magnetogyric ratio and natural abundance. These materials can be studied in concentrated solutions or by the synthesis of materials enriched with ^{15}N. The results show that the ^{15}N chemical shifts are extremely sensitive to the polymer sequence and environment. The chemical shifts can be used to study the conformation of biopolymers in important materials such as silk and wool.

3.3 Solid-state NMR studies of polymers

3.3.1 Introduction

The intense interest in solid-state polymers has driven development of new methods for materials characterisation by solid-state NMR. Although solid-state NMR spectra are more difficult to acquire than solution spectra, solid-state NMR has become a routine technique in many laboratories. These spectra can be acquired in the high-resolution or wide-line mode, methods that yield different but complementary information.

High-resolution solid-state NMR spectra are acquired by using a combination of high-power proton irradiation and rapid sample spinning to remove the line broadening from dipolar couplings and chemical-shift anisotropy [14]. The local magnetic field B_{loc} of a ^{13}C nucleus in a solid polymer due to dipolar interactions with protons is given by:

$$B_{\text{loc}} = \pm \frac{h}{4\pi} \gamma_H \frac{(3\cos^2\theta_{CH} - 1)}{r_{CH}^3} \tag{3.5}$$

where θ is the angle between the C–H bond vector and the magnetic field and r_{CH} is the carbon–hydrogen bond distance. The dipolar broadening in samples not oriented with respect to the magnetic field are much larger (50 kHz) than the chemical-shift range (20 kHz), so broad featureless signals are observed. This dipolar broadening can be removed by high-power proton irradiation during carbon signal acquisition. The carbon signals are also broadened by chemical-shift anisotropy (CSA). This broadening arises because the chemical shift depends on the orientation of the molecular axis (defined by the electron density around the nuclei of interest) with respect to the magnetic field. The chemical shift is given by:

$$\sigma_i = \sigma_{11}\lambda_{11}^2 + \sigma_{22}\lambda_{22}^2 + \sigma_{33}\lambda_{33}^2 \tag{3.6}$$

where λ_{ii} are the direction of cosines relating the molecular axis to the laboratory axis and σ_{ii} are the principle values of the chemical-shift tensor [14]. If the sample is spun rapidly (a few kilohertz) the chemical shift is given by:

$$\sigma = \sin^2\beta(\sigma_{11} + \sigma_{22} + \sigma_{33}) + \tfrac{1}{2}(3\cos^2\beta - 1) \tag{3.7}$$

where β is the angle between the rotation axis and the magnetic field. For rapid spinning at the magic angle ($\beta = 54.6°$) the first term is equal to one-third of the trace of the CSA tensor (i.e. the isotropic chemical shift) and the second term is zero, so high-resolution carbon spectra are observed. The effect of high-power decoupling and magic-angle spinning is illustrated in Figure 3.13 for bisphenol A, the repeat unit in polycarbonate [2]. Broad, featureless signals are observed in the absence of high-power decoupling or magic-angle spinning, and these are resolved into the aromatic and methyl signals with the application of either magic-angle spinning or decoupling. High-resolution spectra are obtained with the simultaneous application of magic-angle spinning and high-power proton decoupling. Separate signals are obtained for the C_1, C_4, $C_{3,5}$, and $C_{2,6}$, resonances under these conditions and the resolution approaches that observed in the liquid-state spectrum shown in the bottom of Figure 3.13, although the line widths in solid polymers (50–100 Hz) are typically greater than those in liquids (1–5 Hz). High-power decoupling and magic-angle spinning are not required in solution because the carbon–hydrogen dipolar couplings and carbon chemical-shift anisotropies are averaged by rapid molecular motion. The resolution in solid-state NMR, while not as good as for solutions, is sufficient for many types of materials characterisation. In addition, the solid-state NMR spectra provide information about the conformation of crystalline polymers through the γ-gauche effect [7] described in section 3.2.2. Polymer dynamics can be investigated via the relaxation rates as in solution or by measuring the effects of molecular motions on the averaging of the chemical-shift anisotropies or dipolar interactions [14].

The molecular dynamics of polymers depend on the chemical structure and the local environment. The dynamics consist of vibrations, *gauche–trans* isomerization, ring flips, and segmental motions of several monomers involving combinations of these motions. The local dynamics have a large effect on some NMR parameters, and provide a method to discriminate between polymers in different environments. Large differences in dynamics are observed, for example, between crystalline, amorphous, and rubbery polymer chains. Cross-polarisation (Figure 3.14) is one method to distinguish between

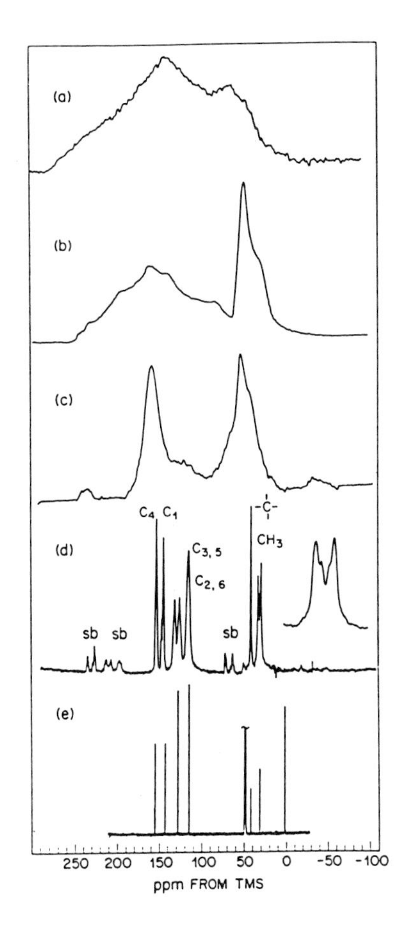

Figure 3.13 The effect of magic-angle spinning and high-power decoupling on the solid-state ^{13}C-NMR spectrum of crystalline bisphenol A: (a) nonspinning with no proton decoupling; (b) non-spinning with low-power proton decoupling; (c) nonspinning with high power decoupling; (c) magic angle spinning without decoupling; (d) magic-angle spinning with proton decoupling; (e) the spectrum in solution. Reprinted with permission from ref. 1.

these different environments. The pulse sequence is initiated with a 90° pulse that rotates the proton magnetisation into the *xy* plane. Long pulses are simultaneously applied to the carbons and protons (called the contact, cross-polarisation, or spin-lock time) and the signals are observed with high-power proton decoupling and magic-angle spinning. The key part of this sequence is the spin-lock period, during which magnetisation is equilibrated between the protons and carbons along the *y* axis, provided that the field strengths are exactly matched. The match of the carbon and proton rf fields is known as the 'Hartman–Hahn' condition [14]. For materials where molecular motions are

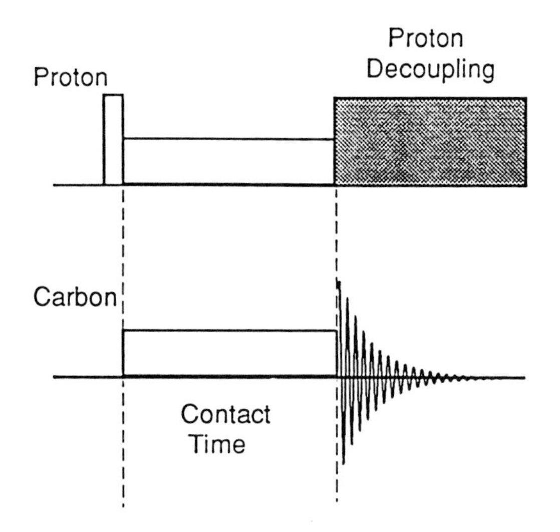

Figure 3.14 The pulse sequence for cross-polarisation. The sequence is initiated with a 90° proton pulse followed by spin locking (the contact time) of the carbon and proton magnetisations. This leads to the efficient polarisation of the carbons via the protons when the carbon and proton field strengths are matched. The signals are then observed with high-power proton decoupling.

restricted by the environment, such as crystalline materials or amorphous polymers below their glass transition temperatures, the carbon–proton transfer is very efficient. The time dependence of the cross-polarised intensity is one method to distinguish carbons in different dynamic environments, as illustrated in the simulated intensity versus contact time plot shown in Figure 3.15. Signals from the crystalline, glassy or rubbery material can be emphasised with the proper choice of contact time. The signal intensities can also be enhanced by a factor of four through cross-polarisation, since the carbon and proton signals are equilibrated, and the protons have a higher sensitivity. The polymer dynamics also affect the NMR spectra through the spin–lattice relaxation time, which can be very long for crystalline polymers. Such long T_1 values limit the rate at which spectra can be acquired, since the signals must return to equilibrium between acquisitions. In cross-polarisation, however, the repetition time is determined by the *proton* T_1 values that are much shorter than the carbon T_1 values. Thus, the advantages of the combination of cross-polarisation with magic-angle spinning (CPMAS) are that the signals are enhanced, the spectra can be acquired more rapidly, and polymers in different environments can be distinguished.

3.3.2 Carbon CPMAS studies of polymers

Carbon CPMAS is the solid-state method most commonly used to study polymers. These spectra are now routinely acquired and can provide information

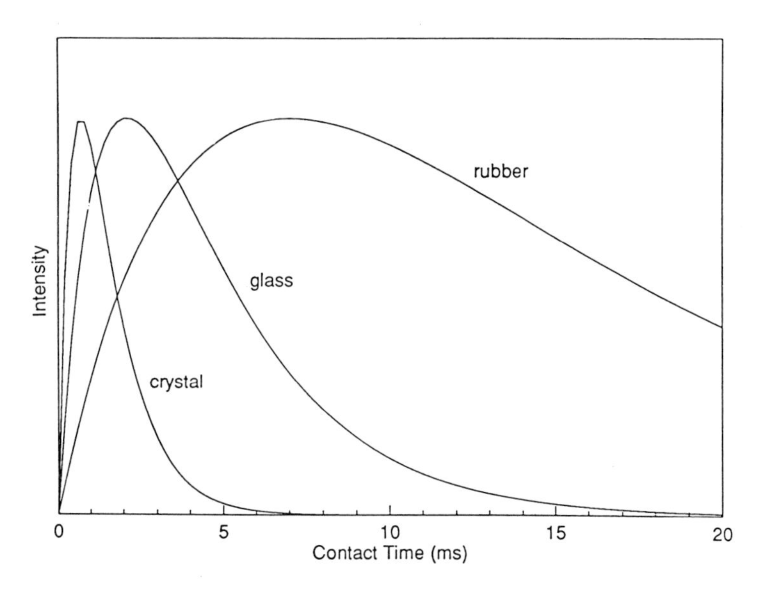

Figure 3.15 The time-dependence of cross-polarised signal intensity for crystalline, glassy and rubbery polymers.

about polymer microstructure as shown above for solution-state NMR. In addition, the solid-state spectra and relaxation times depend on the polymer conformation, dynamics and phase structure. In crystalline polymers the γ-gauche effect [7] leads to chemical-shift changes that depend on the crystalline conformation, and different regions of the sample can be visualised, since the molecular dynamics vary greatly between the crystalline, interfacial, amorphous, and rubbery regions. Proton spin diffusion detected by CPMAS can be used to measure the phase structure in semicrystalline polymers, polymer blends, and block copolymers over the length scale of 20–200 Å [15].

The conformational sensitivity of the solid-state ^{13}C chemical shifts is illustrated in Figure 3.16, which shows the solid-state spectra of isotactic and syndiotactic polypropylene [16]. While both polymers are crystalline, the isotactic polymer crystallises in a 2_1 . . . ggttggtt . . . conformation. One consequence of this conformational difference is that the methylene carbons are in a symmetric environment in the isotactic polymer and are non-equivalent in the syndiotactic polymer. The methylene carbons are separated by 8.7 ppm in the CPMAS spectrum of syndiotactic polypropylene.

Cross-polarisation can be used to distinguish between areas that differ in molecular mobility, since material in a crystalline environment cross-polarises much more efficiently than material in glassy or rubbery environments. Figure 3.17 shows the solid-state NMR spectra of a triblock copolymer of polycaprolactone–poly(dimethyl siloxane)–polycaprolactone with block molecular weights of 2000 for the polycaprolactone and 3000 for the poly(dimethyl

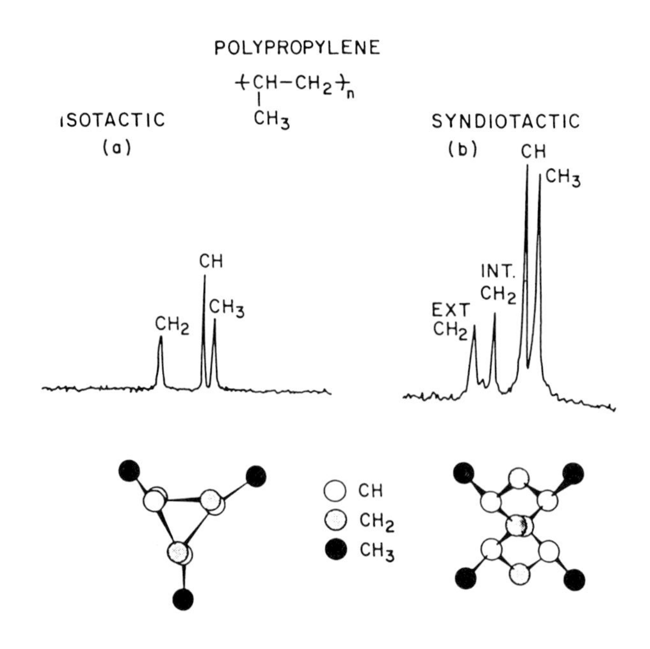

Figure 3.16 The solid-state ^{13}C-NMR spectra of (a) isotactic and (b) syndiotactic polypropylene. The inset plot illustrates the environment in the crystalline phase. Reprinted with permission from ref. 16.

siloxane) [17]. The polycaprolactone forms crystalline domains while the poly(dimethyl siloxane) is an amorphous material far above its T_g. The bottom trace shows the spectrum obtained with cross-polarisation, in which only the crystalline polycaprolactone carbons are observed. The top spectrum was observed without cross-polarisation, and shows both the crystalline and the amorphous material. Polymers below their T_g can frequently be cross-polarised, but at a different rate from the crystalline domains.

Proton spin diffusion between polymer chains can be used to characterise the phase structure of polymers. Spin diffusion is the process by which magnetisation is rapidly transferred between nearby protons in the solid state. If the materials are phase-separated on a length scale longer than the spin-diffusion length scale, then the relaxation rates are identical to those observed for the homopolymers. If the polymers are mixed, then the relaxation rates are a weighted average of the values for the two homopolymers. The length scale L for spin diffusion is approximately given by [18]:

$$L = (6D/k)^{\frac{1}{2}} \tag{3.8}$$

where D is the spin diffusion coefficient and k is the proton $T_{1\rho}$ or T_1 relaxation rate that is usually detected by CPMAS [15]. The diffusion coefficient for

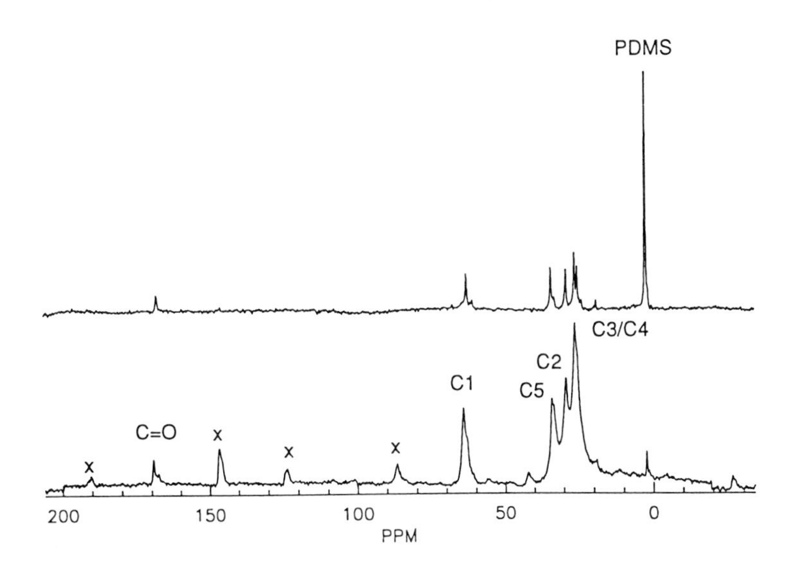

Figure 3.17 The 100 MHz solid-state ^{13}C-NMR spectra of the polycaprolactone–poly(dimethyl siloxane)–polycaprolactone triblock copolymer obtained with (bottom) and without (top) cross polarisation. The spinning side bands are marked x. Reprinted with permission from ref. 17.

polymers is typically of the order of 10^{-12}cm^{-1}s^{-1}, so the length scales measured by the T_1 and $T_{1\rho}$ relaxation rates are of the order of 20–200 Å. To evaluate the average domain size more quantitatively, however, the morphology must be characterised [19]. Few other measurements are able to provide information about the phase structure on this length scale.

Figure 3.18 shows the proton $T_{1\rho}$ relaxation for a mixture of a poly(ether sulphone) and a polyimide [20]. This relaxation rate is measured from the CPMAS signal intensity as a function of the spin-lock time after the initial build-up. The relaxation times for the homopolymers are 4.5 and 17 ms for polyimide and the poly(ether sulphone). Figure 3.18 shows that identical relaxation rates are observed for both polymers in the blend, and that the rate is a composition weighted average of the homopolymer relaxation rates. This demonstrates that the polymers are mixed on the length scale of 20 Å.

Molecular-level mixing can also be measured using mixtures of protonated and deuterated polymers [15]. In deuterated polymers, carbons are not in close proximity with protons and cannot be cross-polarised. However, if the polymers are mixed on the molecular level, then the protonated polymer can be *intermolecularly* cross-polarised from the protonated chain. Intermolecular cross-polarisation is demonstrated in Figure 3.19 for phase-separated and mixed samples of deuterated polystyrene and protonated poly(vinyl methyl ether) [21]. Rapid cross-polarisation of the poly(vinyl methyl ether) carbons

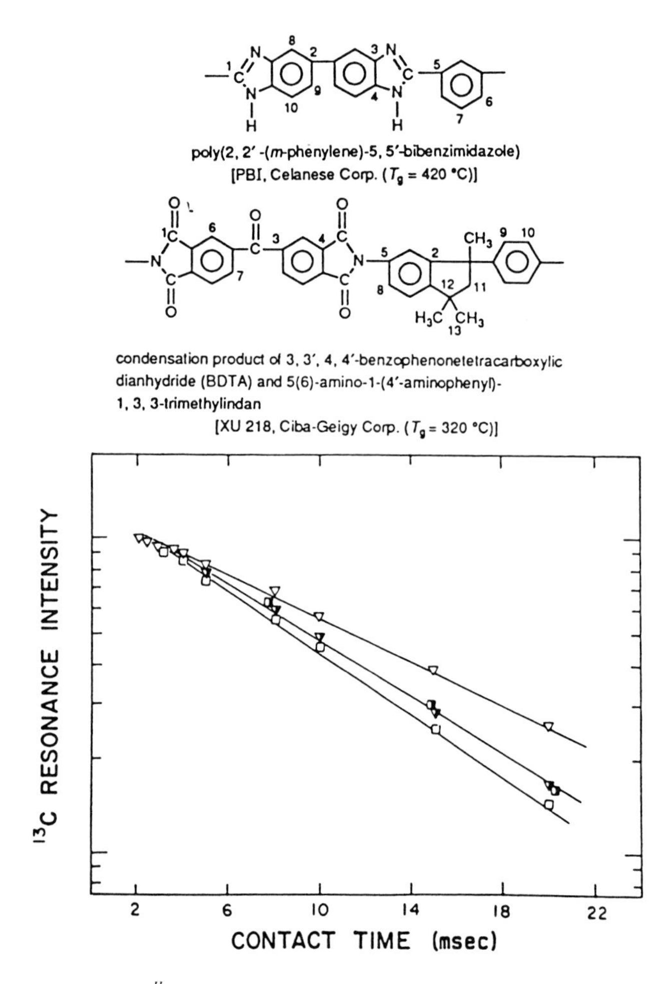

Figure 3.18 Plots of the $T_{1\rho}^{H}$ relaxation for PBI (□), XU218 (▽), and the 50/50 wt% PBI–XU218 blend (filled symbols). Reprinted with permission from ref. 20.

is observed in the mechanically mixed sample, and only small signals from the polystyrene are observed at the longest spin-lock time. In the annealed sample, the signals for the protonated and deuterated chains are cross-polarised at the same rate, demonstrating that the polymers are mixed on the $T_{1\rho}$ length scale (20 Å).

3.3.3 CPMAS studies of other nuclei

Many of the nuclei studied by solution NMR can also be studied by CPMAS in the solid state. The chemical shifts for ^{29}Si, ^{31}P, and ^{15}N are extremely sensitive to the local environment and the crystalline conformation. This effect is

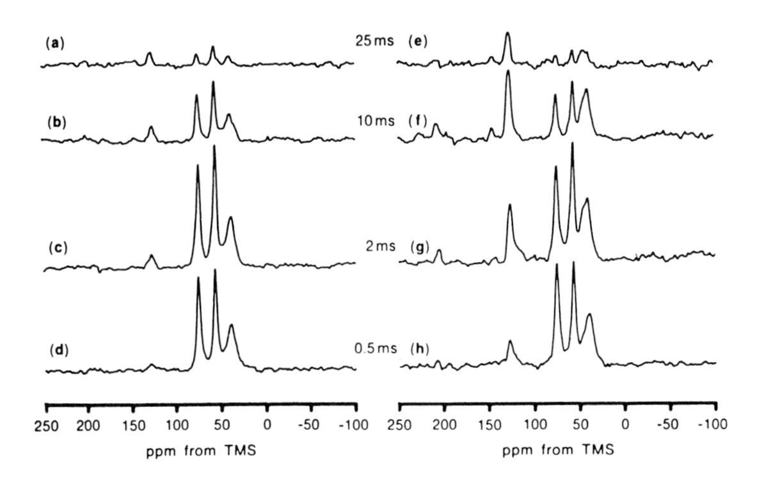

Figure 3.19 Intermolecular cross-polarisation in polymer blends. The cross-polarisation intensity as a function of contact time is shown for (a–d) a mechanical mixture and (e–h) a miscible blend prepared by annealing. Reprinted with permission from ref. 21.

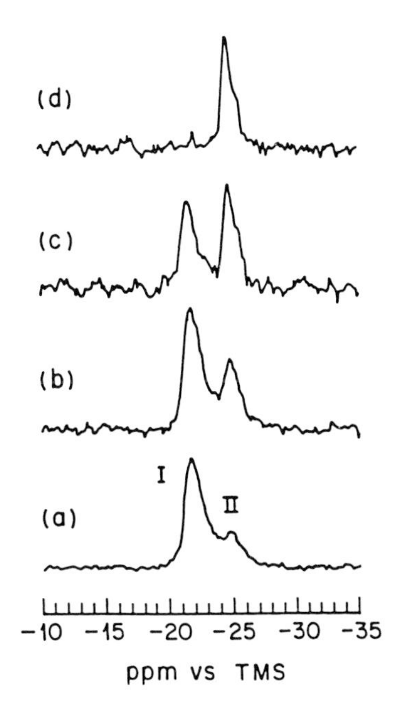

Figure 3.20 The solid-state 39.75 MHz ^{29}Si-CPMAS spectra of poly(dihexylsilane) at (a) 25, (b) 39.5, (c) 41.5, and (d) 44.3°C, showing the solid–solid phase transition. Reprinted with permission from ref. 12.

illustrated in the 39 MHz ^{29}Si-NMR spectrum of poly(dihexyl silane) shown in Figure 3.20 [12]. Optical and X-ray studies show that the low-temperature crystal form has an all-*trans* conformation that undergoes a transition to a conformationally disordered phase at a temperature that depends strongly on the length of the side chain. Figure 3.20 shows that the chemical shifts for the all-*trans* and conformationally disordered forms are separated by 4 ppm, and that the solid–solid phase transition can be clearly followed using CPMAS.

The ^{15}N spectra of nylons are also sensitive to the crystalline conformation and the extent of hydrogen bonding. Figure 3.21 shows the ^{15}N-CPMAS spectra of nylon-6 prepared in the γ and α crystalline forms, and a mixture of the two conformations [22]. Clearly, the ^{15}N chemical shifts are a sensitive probe of the crystalline chain conformation in nitrogen-containing polymers.

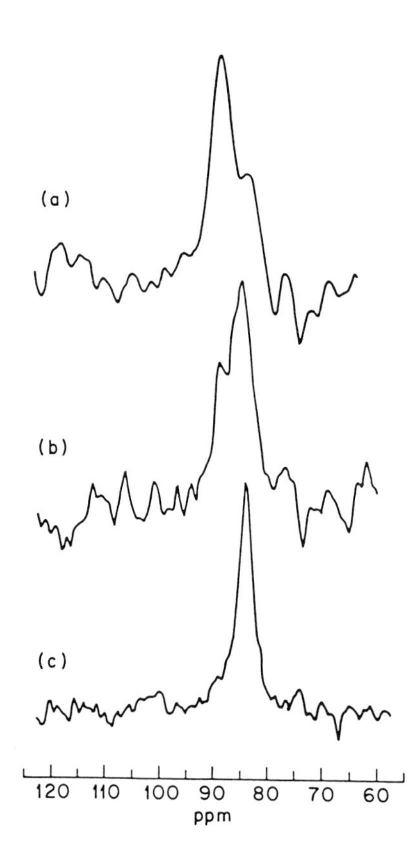

Figure 3.21 The ^{15}N-CPMAS NMR spectra of nylon-6 prepared in (a) predominantly γ-nylon-6, (b) a mixture of α- and γ-nylon-6, and (c) predominantly α-nylon-6. Reprinted with permission from ref. 22.

3.3.4 Deuterium NMR studies of polymers

Wide-line deuterium NMR is one of the most powerful methods used to study polymer dynamics, since changes in the line shape and the relaxation times are sensitive to motions on the time scale from 10^{-8} to 10^2. Wide lines are observed for deuterium in the solid state since the relaxation is dominated by the quadrupolar interactions with the electric field gradient tensor at the deuterium nuclei [23], as illustrated in Figure 3.22. Deuterium is a spin $I = 1$ nucleus, but the spacing of the Zeeman energy levels is perturbed by the quadrupolar coupling, leading to two observable transitions. The magnitude of the quadrupolar coupling is given by:

$$\Delta\nu_q = \frac{3}{4}\frac{e^2qQ}{h}(3\cos^2\theta - 1) \tag{3.9}$$

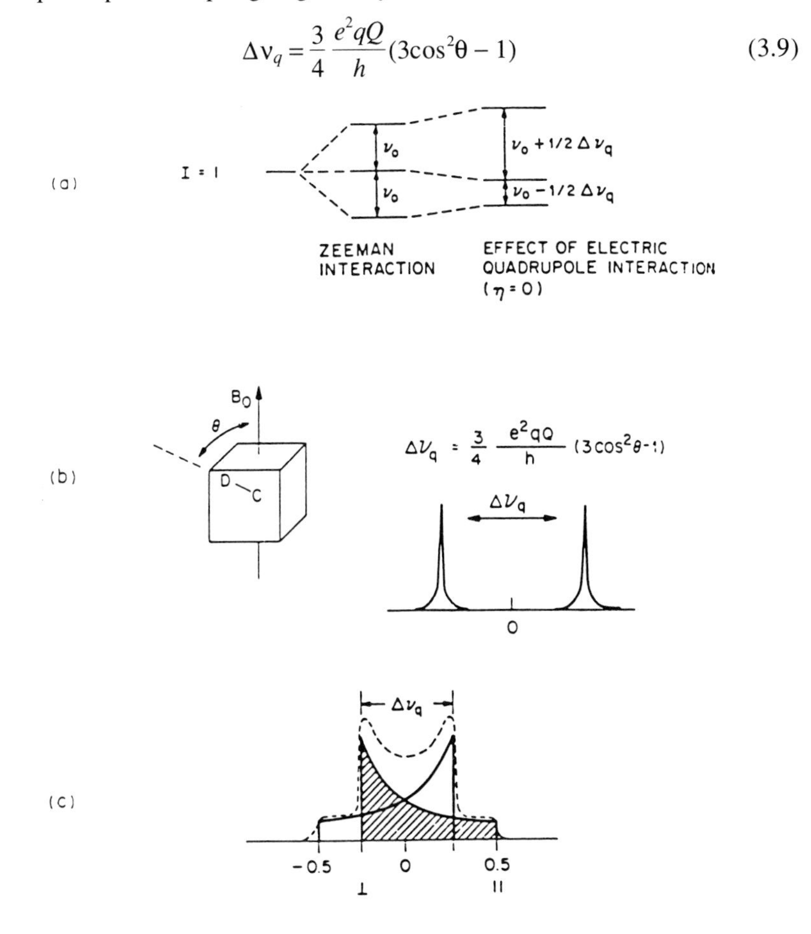

Figure 3.22 Deuterium NMR spectroscopy. ^{2}H is a spin 1 nucleus (a), but two transitions are observed because of the quadrupolar interactions. The quadrupolar coupling (b) depends on the orientation of the C–D bond vector with respect to the magnetic field. For samples not oriented with respect to the field (c) there is a distribution of angles that result in a powder pattern. Reprinted with permission from ref. 2.

where e^2qQ/h is the quadrupolar coupling constant and θ is the angle between the carbon–deuterium bond vector and the magnetic field (Figure 3.22(b)). For samples not oriented with respect to the magnetic field, the C–D bond vector will assume all possible orientations, resulting in a powder pattern (or Pake doublet) (Figure 3.22(c)). Selective deuterium labelling is usually employed in these studies, since the breadth of the powder pattern is typically much larger (128 kHz) than the deuterium chemical-shift range (1 kHz) and the sensitivity is low.

The acquisition of such wide-line signals is more demanding than for high-resolution spectra because the receiver must be turned on immediately following the pulse and the data must be acquired very quickly. For this reason, deuterium spectra are typically acquired using the ($90°–\tau_e–90°–\tau_e$–acquire) quadrupolar echo-pulse sequence in which the data are acquired as an echo so the receiver is not turned on immediately after the pulse.

Deuterium NMR has been extremely useful for the study of polymer dynamics in part because the line shape is sensitive to the *geometry* of molecular motion [24]. For example, the line shape resulting from a deuteron on a phenyl ring undergoing a 180° ring flip can be distinguished from a phenyl ring undergoing free diffusion. For a deuterium experiencing a jump between two orientations, the line shape is sensitive to the jump angle. The strategy for interpreting such spectra is to simulate the deuterium line shape [23] for a particular model of molecular motion and compare it with the observed spectra. While such comparisons are not unambiguous, they provide an important and unique insight into the molecular dynamics of polymers. The sensitivity of the deuterium line shape to the molecular dynamics is illustrated in Figure 3.23, which shows the spectra for a series of nylon-6,6 derivatives that are deuterium-labelled at different sites along the chain [25]. The differences in the line shapes are directly related to the differences in the molecular dynamics, and detailed models for the molecular motion can be constructed from such data [26]. The line shape is most sensitive to the molecular dynamics that are fast compared to the width of the spectra ($\tau_c < 10^{-5}$s). Combined with relaxation rate measurements, deuterium NMR can be used to characterise the molecular dynamics over a wide dynamic range [23].

3.4 Two-dimensional NMR

Solution and solid-state NMR are important tools to study the structure and dynamics of polymers. However, this analysis is frequently limited by spectral overlap, as shown in Figure 3.7 for poly(vinyl chloride) [4]. This overlap hampers signal assignment and measurement of the dipolar and scalar couplings used to determine structure–property relationships in polymers. In many cases these limitations can be overcome using two-dimensional (2D) NMR, a

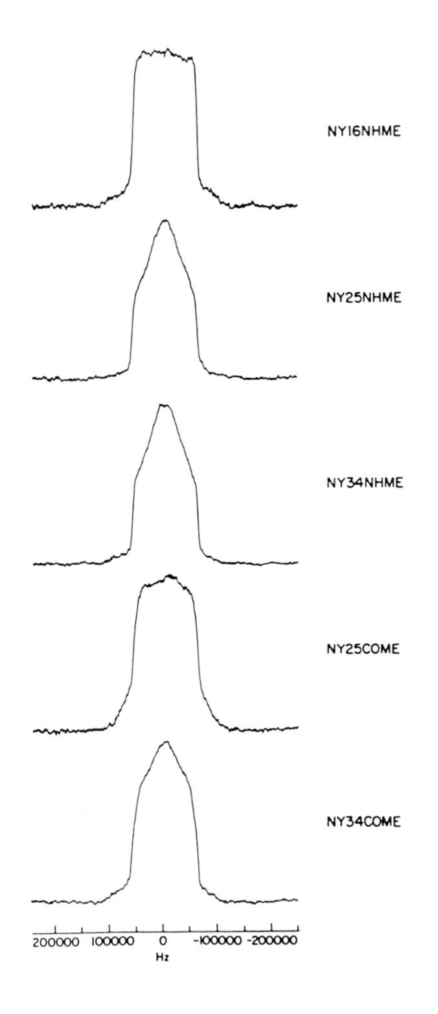

Figure 3.23 The ^{2}H-NMR spectra at 97°C of nylon-6,6 polymers that have been selectively deuterated at the diamine C_1 and C_6 carbons (NY16NHME), the C_2 and C_5 carbons (NY25NHME), and the C_3 and C_4 carbons (NY34NHME) and in the adipoyl C_2 and C_5 (NY25COME) and C_3 and C_4 (NY34COME) carbons. Reprinted with permission from ref. 25.

technique that leads to improved resolution since the interactions are spread out into two frequency dimensions. Since its introduction in 1976 [27] many experiments have been proposed, and these may be categorised as correlated and resolved 2D NMR experiments. The resolved experiments are useful for separating different types of interactions, such as in 2D *J*-resolved spectroscopy [27], where the chemical shifts appear in one frequency dimension and the *J* couplings appear in the other. In correlated experiments the frequencies in the two dimensions are correlated. In *J*-correlated experiments (COSY), for example, J coupling leads to exchanges of magnetisation via the through-bond couplings, and the appearance of peaks with different frequencies in the first and second dimensions (off-diagonal or cross peaks) means that these peaks are J coupled to each other.

Figure 3.24 illustrates the basic principles of 2D NMR. A 2D NMR pulse sequence consists of a series of pulses and delays that may be divided into the preparation, evolution, mixing and detection periods. In common with one-dimensional NMR, all 2D NMR experiments have a preparation period, during which the spin system is allowed to return to equilibrium, and the detection period, during which the signals are recorded. The second frequency is introduced by systematically varying the delay between the first and second pulses (the t_1 or evolution period). A series of spectra are recorded for many (128–512) equally spaced t_1 increments, leading to frequency-labelling of the spins. The mixing period is the time during which magnetisation may be exchanged between spins of different frequency. In this three-pulse sequence, known as 2D nuclear Overhauser effect spectroscopy (NOESY) [28], magnetisation may be exchanged via through-space dipolar interactions during the mixing period. If two protons are in close proximity (less than 5 Å) magnetisation will be exchanged, and magnetisation that was labelled at one frequency during the t_1 period appears at a different frequency during the detection period. After collecting the spectra at many values of t_1, the data are Fourier-transformed along both dimensions, giving a spectrum with two frequency dimensions. The diagonal spectrum contains signals for those protons that did not exchange magnetisation during the mixing time. The through-space connectivities and assignments are obtained from the smaller off-diagonal peaks.

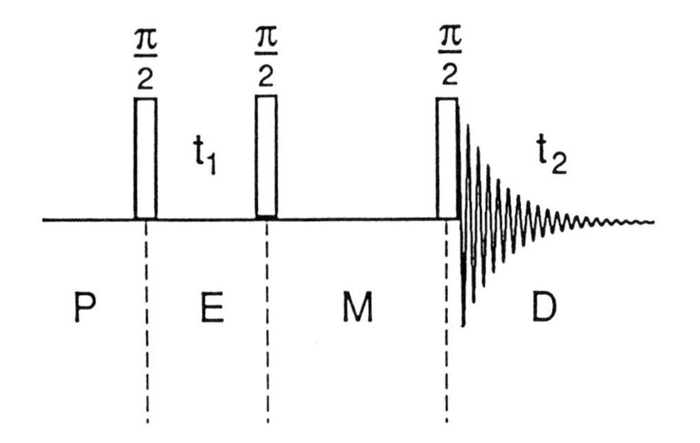

Figure 3.24 The pulse sequence for two-dimensional nuclear Overhauser effect spectroscopy (NOESY). The pulse sequence is divided into the preparation (P), evolution or t_1 (E), mixing (M), and detection or t_2 (D) periods. The data are recorded in the detection period for many equally spaced values of t_1 and double Fourier-transformed to give the two-dimensional frequency spectrum.

Two-dimensional NMR spectroscopy has been used for materials characterisation and for analysis of the structure and dynamics of polymers. Figure 3.25 shows the 2D correlation of the carbon and proton chemical shifts for a 10 wt% poly(vinyl chloride) solution in benzene-d_6 [4]. In carbon–proton correlations the proton signals frequency-labelled during the t_1 period are correlated with the carbon chemical shifts detected in the t_2 period. The pulse sequence and delays are chosen so that magnetisation is exchanged only for those protons that are directly coupled to carbons. A close examination of Figure 3.25 shows that the overlapping proton spectrum of Figure 3.7 has now been resolved via the carbon chemical shifts. Since the carbon spectra are more easily assigned than the protons, inspection of Figure 3.25 leads to the proton assignments if the carbons are assigned. In addition, such spectra contain information about the stereochemical assignments. Note in the inset to Figure 3.25 that some of the methylene carbon signals are correlated with two proton peaks. This can only occur for protons in asymmetric stereochemical sequences, such as *m* dyads. Thus, the observation of such pairs of peaks is important for stereochemical assignments.

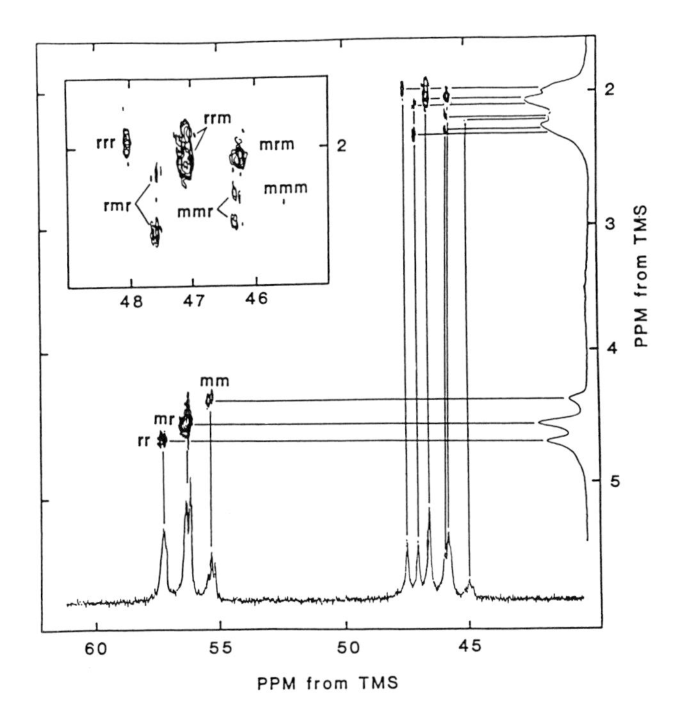

Figure 3.25 The two-dimensional ^{13}C–^{1}H correlation spectrum for 10 wt% poly(vinyl chloride). The inset plot shows an expansion of the correlations for the methylene protons and carbons. Reprinted with permission from ref. 4.

These 2D spectra can also provide detailed structural information about the solution structure of polymers. Figure 3.26 shows the 2D NOESY spectra of a 50:50 mixture of poly(vinyl chloride) and poly(methyl methacrylate) at 38 wt% concentration [29], a polymer pair reported to form a compatible blend in films cast from tetrahydrofuran (THF). Most polymer mixtures do not form such miscible blends because of the unfavourable entropy associated with molecular-level mixing, and for those pairs that do form compatible blends there is usually some specific interaction (ionic, hydrogen bonding, etc.) between the groups on the different chains [30]. From examination of the chemical structure of the two polymers, however, it is not obvious which interchain interactions lead to the formation of a compatible blend. It has been suggested that hydrogen-bond formation between the poly(methyl methacrylate) carbonyl and the poly(vinyl chloride) electron-deficient methine proton is responsible for this interaction [31]. The top spectrum shows several intra-

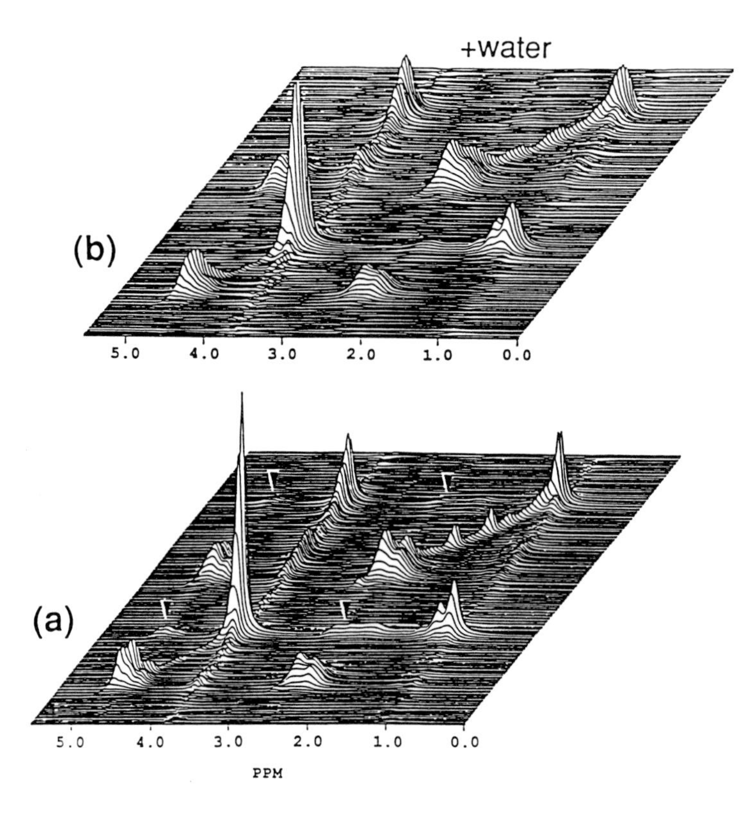

Figure 3.26 The two-dimensional NOESY spectra for 38 wt% 50:50 mixtures of poly(vinyl chloride) and poly(methyl methacrylate) in deuterated THF. Intermolecular cross-peaks are marked in (a). A small amount of water is added in (b), leading to a disappearance of the intermolecular cross-peaks. Reprinted with permission from ref. 29.

and inter-molecular cross peaks. Note, for example, that strong cross peaks are observed between the poly(vinyl chloride) methine and methylene peaks. Such cross peaks are expected since the distances between these protons are less than the 5 Å detected in this experiment. Similarly, intrachain cross peaks are observed between the methyl, methoxyl and methylene protons of the poly(methyl methacrylate). At concentrations above 25 wt%, interchain cross peaks are observed, providing detailed information about the structure of the intermolecular complex. Information about the nature of the intermolecular interactions is obtained by measuring these cross peaks in different solvents. Figure 3.26(b) shows the NOESY spectrum at the same concentration but with a trace of water added. Water is expected to be a much more efficient hydrogen-bond donor than the electron-deficient methine proton of poly(vinyl chloride), and these data show that the addition of water leads to a complete loss of the intermolecular cross peaks. These data strongly support the proposal that intermolecular hydrogen-bond formation is responsible for miscibility in the poly(vinyl chloride)–poly(methyl methacrylate) blends.

Two-dimensional NMR experiments can also be used in the solid state to study the structure and dynamics of polymers, and 2D solid-state NOESY has been used to provide a molecular-level assignment for the polyethylene α transition observed by dielectric and dynamic mechanical spectroscopy [32]. One proposal is that this transition can be assigned to chain diffusion between the crystalline and amorphous regions [33]. Two peaks are observed in the ^{13}C-CPMAS spectra of polyethylene that can be assigned to chains in crystalline and amorphous environments. Figure 3.27 shows the 2D spin exchange

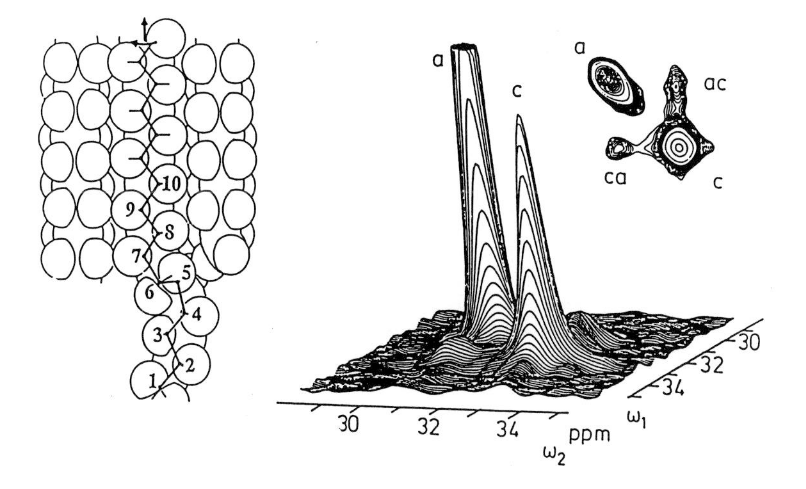

Figure 3.27 The two-dimensional spin exchange spectrum for semicrystalline polyethylene. The signals along the diagonal are labelled for the crystalline (c) and amorphous (a) peaks. The cross-peaks ac and ca result from exchange between the crystalline and amorphous fractions. Reprinted with permission from ref. 33.

spectrum of semicrystalline polyethylene with a mixing time of 1 s [33]. The spin exchange experiment in the solid uses a pulse sequence similar to NOESY, but cross peaks are observed for carbons exchanging between crystalline and amorphous environments. The peak intensities are quantitatively related to the exchange rates, and the rate of chain transfer can be directly correlated with the peak maxima and the temperature dependence of the dielectric and dynamic mechanical spectra.

References

1. F. Bovey (1988). *Nuclear Magnetic Resonance Spectroscopy* (2nd edn), Academic Press, New York, p. 653.
2. F. Bovey (1982). *Chain Structure and Conformation of Macromolecules*, Academic Press, New York, p. 259.
3. F. Schilling, F. Bovey, M. Bruch and S. Kozlowski (1985). *Macromolecules*, **18**, 1418.
4. P. Mirau and F. Bovey (1986). *Macromolecules*, **19**, 210.
5. F. Bovey and P. Mirau (1988). *Acc. Chem. Res.*, **21**, 37.
6. A. Tonelli and F. Schilling (1981). *Acc. Chem. Res.*, **14**, 233.
7. A. Tonelli (1989). *NMR Spectroscopy and Polymer Microstructure: The Conformational Connection*, VCH Publishers, New York, p. 252.
8. F. Bovey and L. Jelinski (1985). *J. Phys. Chem.*, **89**, 571.
9. F. Heatley (1979). *Prog. NMR Spectroscopy*, **13**, 47.
10. R. Cais and J. Kometani (1984). *NMR and Macromolecules*, ACS Symposium series no. 247, American Chemical Society, Washington, DC.
11. E. Williams, J. Wengrovius, V. Van Valkenburgh and J. Smith (1991). *Macromolecules*, **24**, 1445.
12. F. Schilling, F. Bovey, A. Lovinger and J. Zeigler (1986). *Macromolecules*, **19**, 2660.
13. H. Tanaka, M. Gomez, A. Tonelli, S. V. Chichester-Hicks and R. C. Haddon (1989). *Macromolecules*, **22**, 1031.
14. M. Mehring (1983). *High Resolution NMR in Solids*, Springer-Verlag, Berlin.
15. E. Stejskal, J. Schaefer, M. Sefcik and R. McKay (1981). *Macromolecules*, **14**, 275.
16. A. Bunn, E. Cudby, R. Harris, K. Packer and B. Say (1981). *J. Chem. Soc., Chem. Commun.*, **15**, 15.
17. A. Lovinger, B. Han, F. Padden and P. Mirau (1992). *J. Polym. Sci.*, in the press.
18. V. McBrierty, D. Douglass and T. Kwei (1978). *Macromolecules*, **11**, 1265.
19. T. Cheung and B. Gerstein (1981). *J. Appl. Phys.*, **52**, 5517.
20. J. Grobelny, D. Rice, F. Karasz and W. MacKnight (1990). *Macromolecules*, **23**, 2139.
21. G. Gobbi, R. Silvestri, R. Thomas, J. Lyerla, W. Flemming and T. Nishi (1987). *J. Polym. Sci. C, Polym. Lett.*, **25**, 61.
22. D. Powell, A. Sikes and L. Mathias (1988). *Macromolecules*, **21**, 1533.
23. H. Spiess (1985). *Adv. Polym. Sci.*, **66**, 23.
24. H. Spiess (1991). *Chem. Rev.*, **91**, 1321.
25. H. Miura and A. English (1988). *Macromolecules*, **21**, 1543.
26. J. Wendoloski, K. Gardner, J. Hirschinger, H. Miura and A. English (1990). *Science USA*, **247**, 431.
27. W. Aue, E. Bartnoldi and R. Ernst (1976). *J. Chem. Phys.*, **64**, 2229.
28. J. Jeneer, B. Meier, P. Bachmann and R. Ernst (1979). *J. Chem. Phys.*, **71**, 4546.
29. P. Mirau, S. Heffner, G. Koegler and F. Bovey (1991). *Polym. Int.*, **26**, 29.
30. D. Walsh and S. Rostami (1985). *Adv. Polym. Sci.*, **70**, 119.
31. D. Varnell, E. Moskala, P. Painter and M. Coleman (1983). *Polym. Eng. Sci.*, **23**, 658.
32. J. Hoffmann, G. Williams and E. Passaglia (1966). *J. Polym. Sci.*, **C14**, 173.
33. K. Schmidt-Rohr and H. Spiess (1991). *Macromolecules*, **24**, 5288.

4 Vibrational spectroscopy

J. M. CHALMERS and N. J. EVERALL

4.1 Introduction

The intention in this chapter is to provide the reader with a balanced picture of the complementary roles of Raman and infrared spectroscopy in polymer characterisation. The emphasis chosen is towards practical applications, particularly those of significant industrial or analytical relevance. An attempt has been made to target as wide a range as possible, and to give examples of the many chemical and physical characterisations regularly undertaken. In doing this the intention is to provide as well exampled and referenced a text as could reasonably be accommodated, and to have provided those interested in more detail with a short route to important texts. Unfortunately, space constraints have inevitably necessitated the omission of certain important classes of materials such as biomolecules and conducting polymers.

4.2 General considerations

As the principles of vibrational spectroscopy have been covered extensively in a number of texts [1–5], only a few basic points are summarised here. An isolated (nonlinear) molecule containing N atoms can undergo $3N - 6$ normal modes of vibration, each often consisting of a complex mixture of bond stretches and deformations. The frequencies and intensities of the modes are sensitive to chemical and physical structure, and so provide a useful means of characterising the composition and geometry of molecules and crystals. Normal modes are often localised to discrete chemical groupings, so that observation of bands in well-defined ranges can indicate the presence of particular groups (hence the principle of group frequency correlations [6, 7]).

The two techniques most commonly used to observe vibrational spectra are infrared (IR) and Raman spectroscopy, although other techniques such as neutron scattering [8] can also be employed. Methods for obtaining IR and Raman data are considered in section 4.3, and comprehensive reviews have been given elsewhere [9, 10]. The IR spectrum arises from the absorption of radiation the frequency of which is resonant with a vibrational transition, while the Raman effect results from inelastic scattering of photons to leave a molecule or crystal in a vibrationally excited state. The shift in frequency of the scattered photon corresponds to the frequency of the normal mode that has been excited.

IR absorption occurs from direct coupling of radiation with a vibrating dipole, with the result that modes that produce strong oscillating dipole moments (e.g. O–H stretch) have strong absorption bands. In contrast, Raman scattering arises from vibrational modulation of sample polarisability, with the result that highly symmetric or nonpolar vibrations (e.g. ring breathing) tend to give strong Raman bands. Although a group-theoretical analysis is needed to predict mode activities [5], a reasonable 'rule of thumb' is that vibrations that are strongly IR active tend to be weak or inactive to Raman, and vice versa, with the extreme situation that molecules with an inversion centre of symmetry have no vibrations that are simultaneously Raman and IR active. The application of both techniques is therefore highly desirable in order to extract the maximum amount of information from a sample. The complementarity of the two techniques is illustrated by Figure 4.1, which shows the IR and Raman spectra of nylon-6,6.

4.2.1 Polymer vibrations

At this stage it may be helpful to note some aspects of polymer vibrational spectroscopy that differ from those of small molecules. Two comprehensive works provide an excellent and thorough review of the subject [11, 12].

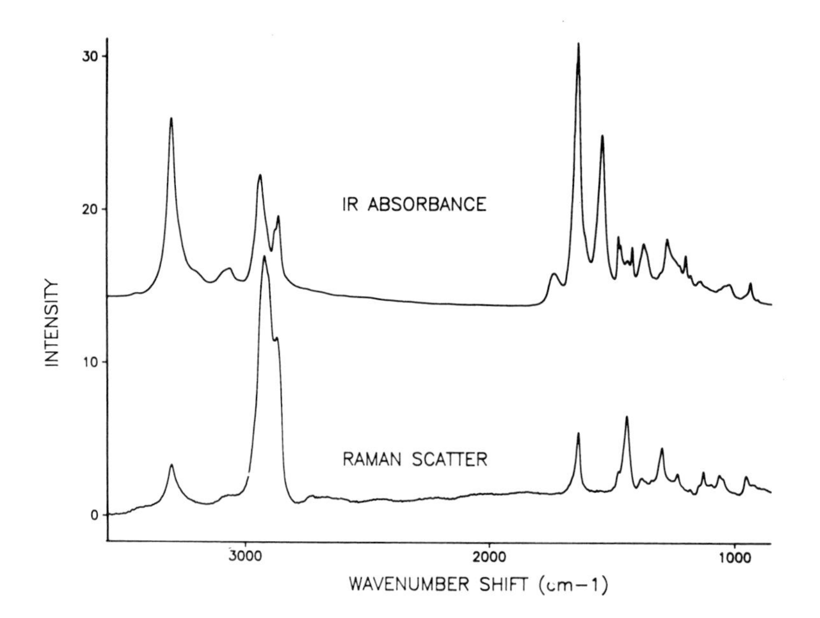

Figure 4.1 Comparison of the IR (upper) and Raman (lower) spectra of nylon-6,6. The IR spectrum was obtained by transmission through a film cast from formic acid, while the Raman spectrum was obtained from a chip of polymer. Note the domination of bands due to polar (amide) groups in the IR data, while the Raman spectrum is dominated by backbone and C–H modes. In fact, below 1650 cm^{-1} the Raman spectrum is reminiscent of that of polyethylene.

The foremost difference is that although a polymer molecule often consists of many thousands of atoms, one does not observe (in optical spectra) the $3N - 6$ discrete vibrations expected for a small molecule. Instead, for a regular polymer chain, only vibrations in which the repeat units vibrate with specific phases give IR or Raman fundamental bands, greatly reducing the spectral complexity. The IR and Raman activity of vibrations can be predicted from a 'line-group' symmetry analysis of a proposed structure [11–13], and this has proved extremely useful in determining factors such as polymer tacticity, helix versus extended-chain conformations and even helix pitch [13]. The vibrations that are not optically active can contribute to physical properties such as heat capacity, and can be observed by using neutron scattering [8].

In general, the interpretation of a polymer spectrum can be made on several levels. The most simple approach treats the polymer as a collection of individual functional groups and attempts to assign observed bands by using group frequency correlations. This can aid polymer identification (fingerprinting) or quantification of composition. The differentiation of single number nylons provides a good example of this approach [14]. The second level correlates spectral features with specific conformations or phases [15], and is valuable in morphological studies (see section 4.9). The third approach attempts to assign the individual bands to their symmetry species by measuring their response to polarised radiation. This both aids band assignments and helps to distinguish possible chain structures [13].

The final degree of complexity involves calculation of vibrational frequencies by using normal coordinate analysis (NCA). NCA has the advantages of confirming band assignments, of illustrating the effects of changes in polymer conformation or the introduction of defects, and of clarifying the nature of modes that cannot simply be assigned to individual groups but which involve cooperative vibrations over many species or repeat units. A very readable introduction to NCA is given in ref. 4. Although few practising polymer spectroscopists will normally carry out a full NCA, a particularly useful introduction to, and examples of, the analysis of polymers is given elsewhere [11, 12].

4.2.2 Qualitative vibrational spectroscopy (fingerprinting)

It was noted above that certain groupings usually vibrate within well defined frequency ranges. Thus, fundamental vibrations involving hydrogen stretching fall in the range 4000–2200 cm^{-1}, triple bonds (C≡C, C≡N) between 2300 and 2000 cm^{-1}, double bonds (C=O, C=C) between 1850 and 1550 cm^{-1}, and single bonds (e.g. C–S, C–S) below 1550 cm^{-1}. Combinations and overtones can occur up to *c.* 7500 cm^{-1}, while the region below 650 cm^{-1} is sensitive to lattice vibrations and librations. The region below 1400 cm^{-1} contains many bands that arise from other fundamental stretching vibrations (e.g. S=O, C–F, C–Cl) and from coupling of stretches and deformations of adjacent units; this gives rise to complex spectra that cannot normally be fully assigned to individual groups. However, this region is characteristic for an individual molecule, since small changes in chemical or geometrical structure can grossly perturb vibrational coupling. This provides a spectral 'fingerprint' which can be used to identify unknown materials by comparison with spectra from reference libraries. This can be facilitated by the use of difference spectroscopy, whereby reference spectra of individual compounds are subtracted from the spectrum of a mixture, thus simplifying identification of the remaining components. This is also termed 'spectral stripping', examples of which appear later in this chapter.

4.2.3 Quantitative vibrational spectroscopy

IR and Raman spectroscopy are both quantitative techniques, in that both the IR absorbance and the Raman scattered intensity are, to a first approximation, linearly proportional to the number density of vibrating species. Although quantitative information can be extracted from a single absorbance or scattered intensity (after calibration against absolute standards), it is more usual to measure the ratio of the analyte band of interest against an independent internal standard, in order to allow for variations in pathlength, deviations from ideality, detector nonlinearity, etc. The reader is referred elsewhere for a full discussion of quantitative methods (including multivariate techniques) [9, 10].

4.3 Instrumentation, sampling and data treatment

4.3.1 Instrumentation

Nowadays, worldwide the dominant infrared detection system for the characterisation of polymers is undoubtedly Fourier transform infrared (FTIR), which has over the last fifteen years rapidly replaced the more traditional dispersive spectrometers. The modus operandi and advantages (throughput, multiplex and precision) of FTIR are well covered elsewhere [e.g. 16–19] and are not a prime topic of this chapter.

Modern conventional laser–Raman systems for recording spontaneous Raman spectra comprise essentially an appropriate source, some dispersing and collection optics and a sensitive detector. The two most common approaches are either single-channel detection using a photomultiplier in combination with a multi (double or triple) monochromator set-up, or a multi channel option with an array detector coupled to a spectrograph [12, 20, 21]. The recent commercial development of FT-Raman systems, either as standalone spectrometers or as FTIR adjuncts, has added to the armoury of techniques, and without doubt offers a cost-effective, rapid, readily applicable technique for a wide range of polymer studies [22–24].

4.3.2 Sampling

The comparative simplicity of sampling that Raman measurement offers to many investigations, e.g. minimal preparation or direct analysis, or use of glass containers, now coupled with the capability FT-Raman brings of mostly avoiding intense fluorescence emissions (so often the drawback of polymer Raman studies in the real world), will help to popularise Raman measurements in many more industrial, research and investigative laboratories. Samples are usually mounted in a suitable holder or container and positioned directly in the laser beam for 180° back-scatter or 90° excitation collection [12, 22], with the former configuration being that favoured for routine FT-Raman applications comparable with IR operations [22]. As with IR, specialised cells and apparatus are available for variable temperature and pressure, and for dynamic molecular property measurements [12, 22].

Since the most commonly sought information from an IR experiment is the absorption characteristics, a wide variety of sampling methods and procedures are employed to obtain this in a useful form. For polymers, the standard is to measure the transmission spectrum of a thin film (particularly for mid-infrared qualitative purposes) which typically needs to be between 20 and 200 μm thickness. Generally applicable methods for a large number of materials are melt casting or hot pressing. Commonly used alternatives include solvent or latex casting and microtoming. However, traditional solid dispersion methods (KBr discs, mineral-oil mulls) are less used in polymer analysis, because of the requirement to reduce the powder particle size preferably to well below the wavelength of the radiation used to interrogate the sample. Nevertheless, brittle resins may be readily ground, or grinding at sub-ambient temperatures may be employed, or a fine dust may be abraded from a plastic or rubber sample with a diamond powder-impregnated spatula or abrasive disc, or SiC paper. The reader is recommended to see refs 10, 12, 25 for fuller descriptions of most of the above and other approaches. Polytetrafluoroethylene (PTFE) and its copolymers may cold-flow, and can often be cold pressed into self-supporting discs, using an alkali halide disc die and a laboratory hydraulic press.

Procedures based on internal reflection spectroscopy (IRS) (e.g. attenuated total reflectance (ATR) and multiple internal reflectance (MIR)) are also invaluable in the characterisation (particularly of surface layers) and identification of polymers. These techniques rely on the intimate contact of a sample with the surface of a high-refractive-index, IR-transparent prism (see Figure 4.2(a)-4.2(c)). IR radiation entering the prism at an angle greater than the critical is internally reflected at the prism surfaces, but attenuated by absorptions from the sample contact layer [26, 27]. IRS techniques have many applications and are widely used in the study of foams, adhesives, films, rubbers and coatings. For thin coatings on metal substrates, the 'transflectance' [10, 28] (see Figure 4.2(d)), or emission [29] techniques can prove effective in material fingerprinting, or in more specialised cases grazing-incidence reflection–absorption infrared spectroscopy (RAIRS) may be used [10, 12, 30].

The development of modern FTIR spectrometers has been accompanied by a rapid evolution of many novel sampling techniques [31]. These include making routinely practicable some long-established methods such as diffuse reflectance, microscopy and specular reflectance, and the introduction of new procedures such as photoacoustic (PA) measurements. All of these have found extensive uses in polymer studies and characterisations [32]. For instance,

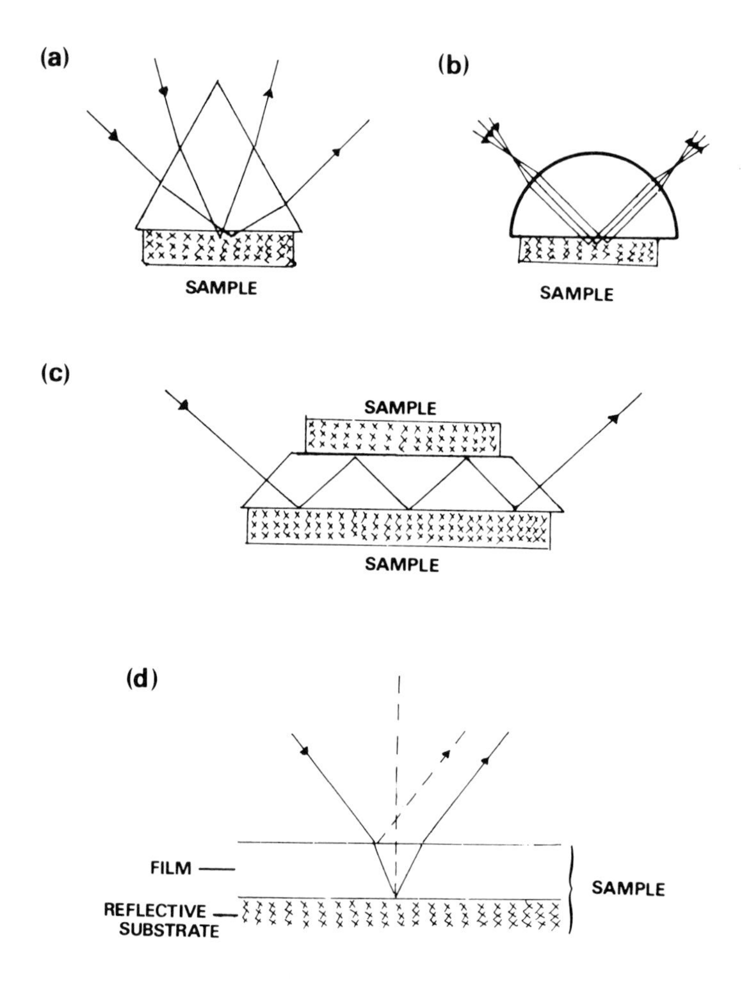

Figure 4.2 Schematic diagrams of some reflection spectroscopy techniques: (a)–(c), internal reflection spectroscopy; (a) and (b), single-reflection ATR; (c), multiple internal reflection (MIR); (d) 'transflectance'. Notes: (i) collimated beam with ATR hemicylinder; (ii) the component of specular reflectance (dotted line) will be superimposed on the double-pass 'transmission' spectrum of the film in the 'transflectance' measurement.

grazing-incidence DRIFT (diffuse reflectance infrared FT) [33–35] and PA-FTIR [36] spectroscopic techniques have proved useful in composite analysis. Examples from a wide range of other investigations include studies on: DRIFT with foams [37, 38], DRIFT with polymer powder [32, 37], PA with polymer powder [39], DRIFT with fibres [40], PA with fibres [39], DRIFT/PA with fibres [41], DRIFT with films [42], PA with films [43] and PA with filled rubber systems [44, 45]. The SiC abrasion technique is a very practical way of obtaining samples for DRIFT identification [46–47].

Raman and IR–microscopy are very powerful tools in polymer science and technology support activities [48]. With spatial resolutions of ~ 1 μm (Raman) and ~ 10 μm (FTIR), they are widely used in bulk and surface contaminant fingerprinting [FTIR: 48,49; Raman: 50,51], laminated structure characterisations [FTIR: 48,52,53; Raman: 51,53], fibre analyses [FTIR: 54,55; Raman: 48,50,55,56], compositional and morphological mapping of processed and fabricated materials [FTIR: 48,57; Raman: 48,51], and the study of intractable solids or difficult samples [FTIR: 48,58; Raman: 48,51].

While DRIFT and PA-FTIR examinations enable the nearly direct study of many intractable, cross-linked, rubber and insoluble polymeric materials, characterising the material from its pyrolysis products is a more traditional method which sometimes proves fruitful [59], particularly when gas chromatography (GC) separation is interposed between the pyrolysis unit and the FTIR spectrometer [10, 12]. Among the many FTIR hybrid techniques developed, increasing use in polymer-cure and other studies is being made of evolved gas analysis (EGA) [60, 61], particularly as part of a combined simultaneous thermogravimmetric (TGA) analysis study [62–65] (see Figure 4.3). The potential value of the synergy of simultaneous spectroscopic and thermodynamic information from a single sample has also been demonstrated when DSC (differential scanning calorimetry) is interfaced to FTIR-microscopy [66, 67].

4.3.3 Data processing

The almost complete inseparability of vibrational spectroscopy spectrometers from computer 'workstations' with dedicated data-manipulation packages, standard protocols for the interchange of data through networks [68], and relatively cheap off-line commercial data processing and analysis routines has over the last decade led to the increasing use and widespread application of sophisticated data treatments of vibrational spectroscopy polymer spectra. Many examples are referenced in this chapter and these extend from derivative and difference spectroscopy, through curve deconvolution and resolution enhancement, to factor analysis and statistical regression methods. In addition, library databases and spectral searching are invaluable assets in many analytical laboratories.

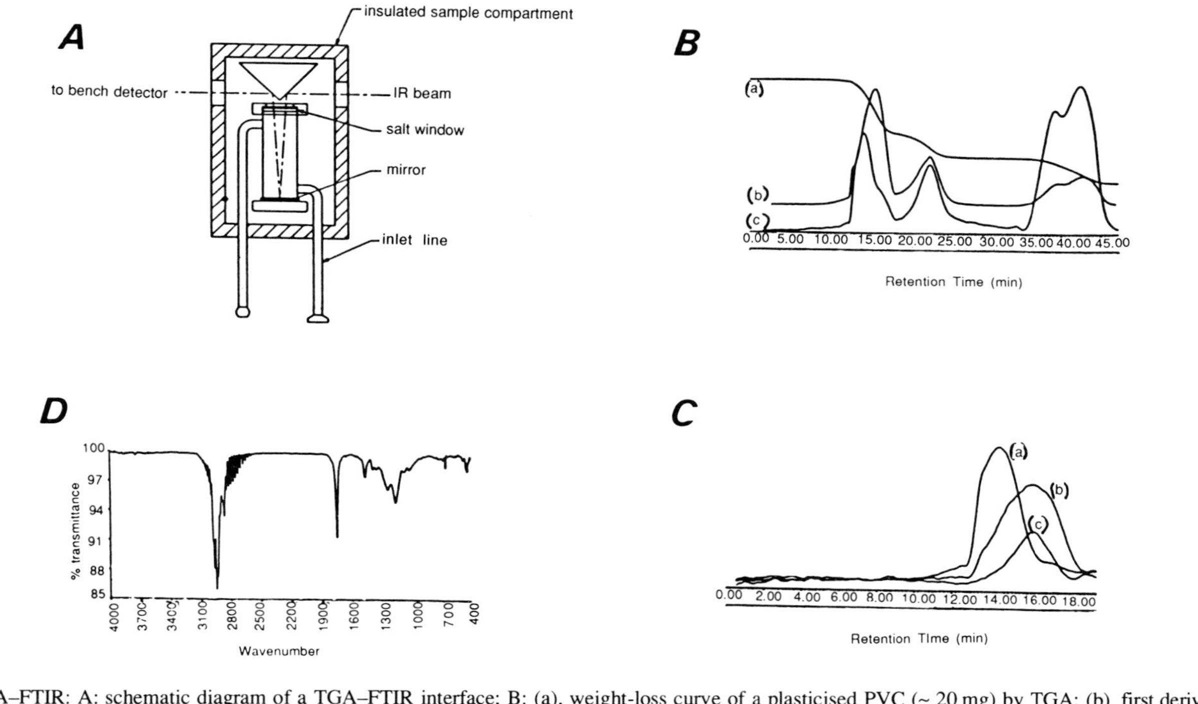

Figure 4.3 TGA–FTIR: A: schematic diagram of a TGA–FTIR interface; B: (a), weight-loss curve of a plasticised PVC (~ 20 mg) by TGA; (b), first derivative of weight-loss curve; (c), Gram–Schmidt reconstruction calculated from spectral data acquired by FTIR analysis; C: Thermograms™, specific reconstruction for three different functional groups, which are a chemoselective record of the evolving gas: (a) carbonyl (1760–1740 cm^{-1}); (b) HCl (2831–2785 cm^{-1}); (c) benzene (684–664 cm^{-1}); D: spectrum obtained from the middle of the first weight-loss region. Reproduced from: R. C. Wieboldt, S. R. Lowry and R. J. Rosenthal (1988) *Mikrochim. Acta I*, (*Recent Aspects of Fourier Transform Spectroscopy*, Vol. 2, *Proc. 6th Int. Conf. Four. Trans. Spec.*), pp. 179–82, by permission of Springer-Verlag, New York.

4.4 Copolymer composition and sequencing

For the analytical spectroscopist working in a quality assurance laboratory or a research technical-support group, the need to provide cost-effective, robust, reproducible and simply operated quantitative methods of compositional and microstructure analysis can be a prime task and important challenge. Vibrational spectroscopy techniques have proved themselves among the most powerful for the routine quantitative analysis of polymers, with the wide variety of sampling procedures usually offering a method (at least semi-quantitative) for even the most intractable of materials.

As was pointed out in section 4.2.4, neither IR nor Raman as generally practised is an absolute method *per se*. Notwithstanding this, spectrometers and analysers, particularly IR, proliferate throughout the polymer industry as compositional determinators or monitors in off-line, near-line, at-line and on-line situations. In the simpler approaches, normalised peak intensities or areas or band ratios are calibrated against independent measurements such as NMR, wet chemistry, mass-balance, radioactive labelled standards, etc. The trend, through computer-coupled instrumentation, is to automate these essentially single- or dual-wavenumber procedures, or to supercede them with more advanced data-processing techniques which utilise or scrutinise much wider ranges of the available data array. With the ready accessability of 'chemometric' data-processing routines such as factor analyses and statistical regression methods, the scope of quantitative applications has undergone a tremendous expansion over the last decade; many complex multicomponent analyses are now performed routinely, reliably and simply. The reader is recommended to refs 69 and 70 for discussions of single-component analysis methods and multicomponent matrix treatments.

Many examples of quantitative vibrational spectroscopy applications will be mentioned in the sections following, but here more specifically copolymer composition and sequencing will be considered. The increasing role and potential which dedicated NIR instrumentation is realising in production and quality-control applications is well covered elsewhere [71–74] and will not be pursued further here, (see Figure 4.4(e)). Many examples of the direct application of Beer's law (see Figure 4.4) to the IR spectra of polymers will be found in refs 75 and 76, which include copolymer composition and additive-level determinations, and procedures for dealing with overlapping absorption bands. Raman and IR are referenced alongside other chemical and physical methods for the analysis of a wide range of plastics in another book [77]. Undoubtedly, new lower-cost, bench-top FT-Raman spectrometers will provide an alternative cost-effective approach to many compositional determinations [22].

The determination of the vinyl acetate (VA) content in ethylene–vinyl acetate (EVA) copolymers serves as a good illustration of the traditional 'key-band' application of Beer's law. Commercial EVA products are available

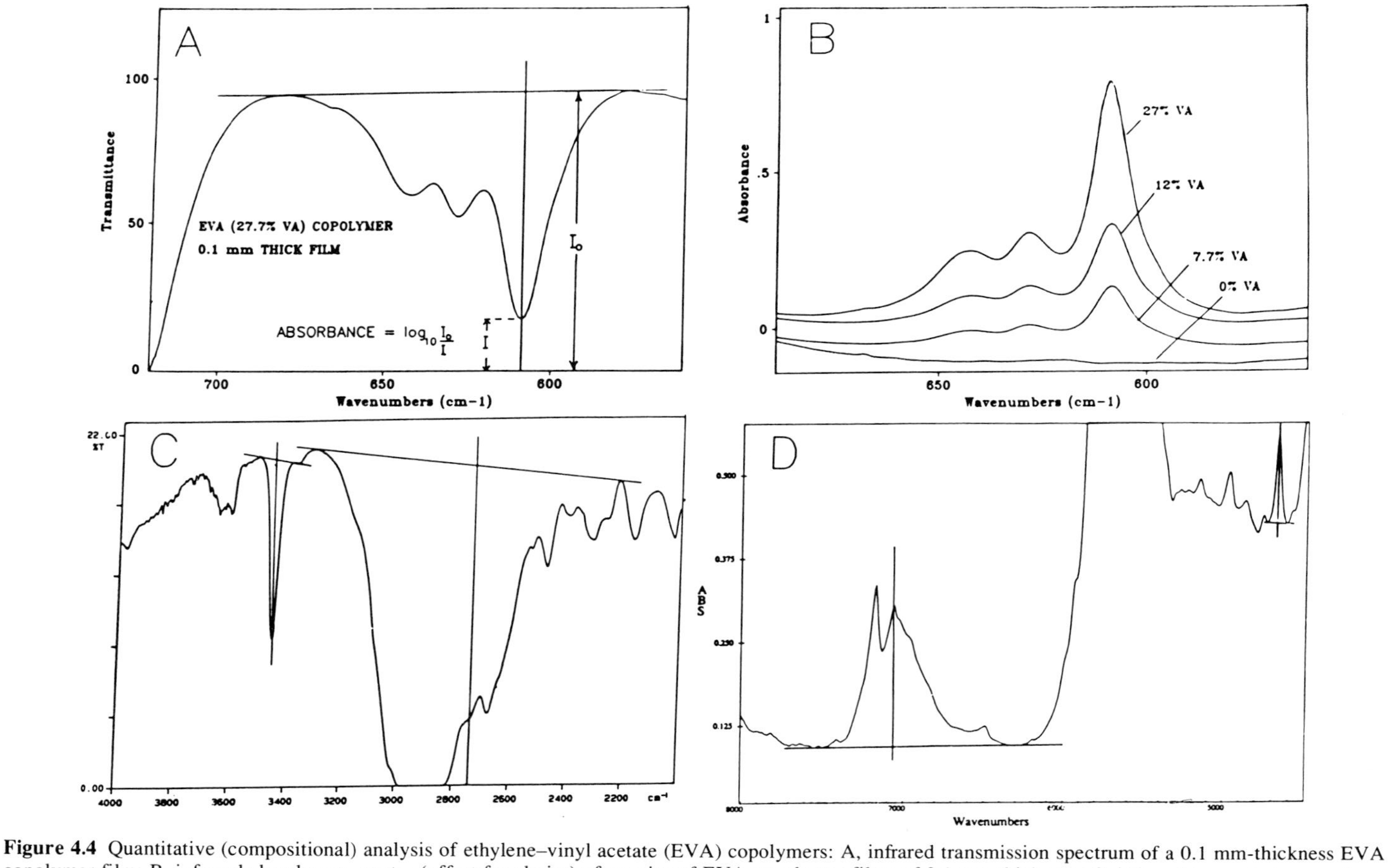

Figure 4.4 Quantitative (compositional) analysis of ethylene–vinyl acetate (EVA) copolymers: A, infrared transmission spectrum of a 0.1 mm-thickness EVA copolymer film; B, infrared absorbance spectra (offset for clarity) of a series of EVA copolymer films of 0.1 mm thickness; C, infrared transmission spectrum of an EVA (32.3%) copolymer sample, 0.25 mm thickness; D, infrared absorbance spectrum of an EVA (17.2%) copolymer sample,~4 mm thickness moulding;

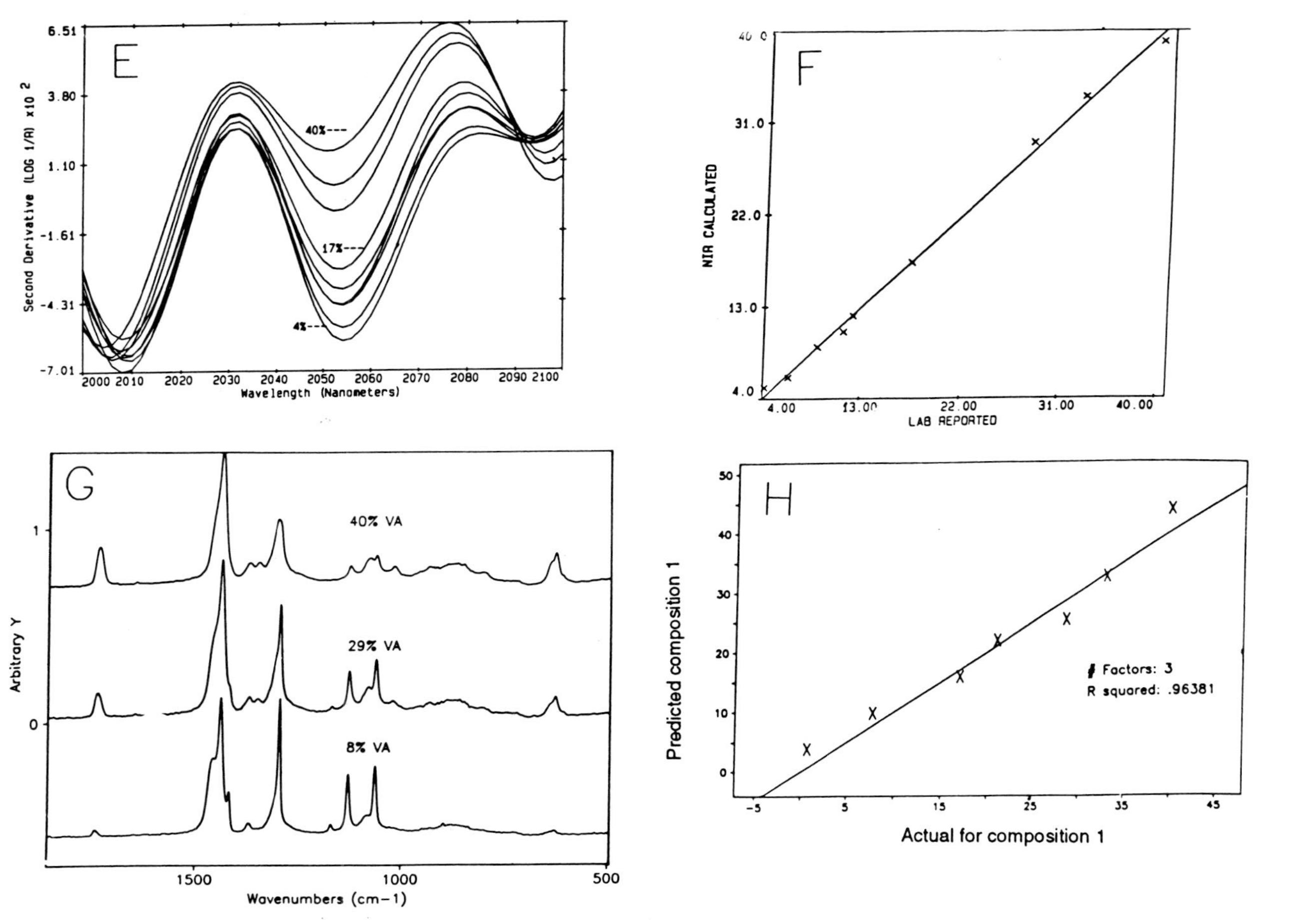

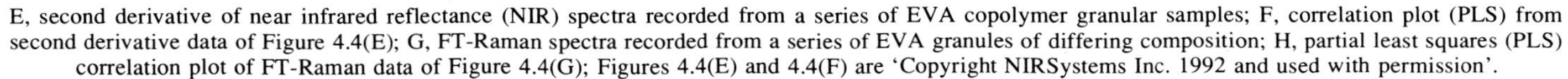

E, second derivative of near infrared reflectance (NIR) spectra recorded from a series of EVA copolymer granular samples; F, correlation plot (PLS) from second derivative data of Figure 4.4(E); G, FT-Raman spectra recorded from a series of EVA granules of differing composition; H, partial least squares (PLS) correlation plot of FT-Raman data of Figure 4.4(G); Figures 4.4(E) and 4.4(F) are 'Copyright NIRSystems Inc. 1992 and used with permission'.

which, according to end-use, cover the range < 1% to > 50% combined VA by weight. For copolymers containing 3–30% VA, the absorbance of the VA band at 16.5 μm (606 cm^{-1}) normalised for specimen thickness has been used [75] (see Figure 4.4(a)). To extend the range of application to 55% VA, other workers [78] investigated measuring the absorbance ratio 3460 cm^{-1}(C=0 1st overtone)/2678 cm^{-1} (Figure 4.4(c)), whereas the present authors have used the ratio 4688 cm^{-1}(C=O combination mode)/7608 cm^{-1} made from spectra of thick plaques (~ 4 mm thickness) [79] (see Figure 4.4(d)). For high VA levels, determinations from solution have also been proposed [75].

Valid, reproducible results can also be derived from ATR studies [80], and more recently both transmission and ATR data have been compared for routine FTIR analysis [81]. The method of choice will be dictated by the relative concentrations of comonomers and consequent strength of absorption bands, and also by local needs and the availability and suitability of sample-preparation procedures and spectrometers [82]. In many circumstances, reproducibility and precision may be of greater priority than absolute accuracy. FT-Raman [83] and NIR now open up the opportunity for simpler, efficient, more direct measurements on granular product [83] (see Figure 4.4(e)–4.4(h)).

The resurgence of quantitative applications generated by the ready availability of advanced data-processing treatments has been accompanied by a growing exploration into the potential of novel techniques for simplifying sampling, particularly those which are considered as essentially non-preparative and non-destructive. Figure 4.5 shows the DRIFT spectra [32, 84] recorded directly from a series of fluoropolymer powders, and compares the band ratios 980 cm^{-1}/935 cm^{-1} determined from the DRIFT and tramsmission measurements. In this instance, DR sampling clearly offers a good prospect for routine quality-control purposes. This and other novel DR and PA-FTIR methods are featured in ref. 84. Relative intensities of bands in both the FTIR and FT-Raman spectra have been shown to be associated with the ratio of components in the copolymers of aryl ether sulphone/aryl ether ether sulphone [85], and the potential demonstrated for each technique directly to characterise a ~30% w/w glass-fibre-reinforced composite material. More recently, improved sensitivity for this copolymer determination has been obtained by utilising a PLS data array treatment [86]. The possibility of rapid non-destructive compositional analysis of acrylic fibres by FTIR-microscopy transmission measurements has been demonstrated [87], while the potential of FT-Raman for characterisation of double-number (1,3 to 10,10) and copolymer analysis of nylons has been raised [88].

The IR determination of the content and sequence distribution of ethylene units in propylene/ethylene (P/E) copolymers has received considerable attention over many years [e.g. 89–97]. Many of these studies have concerned the rocking-mode absorption band for $-(CH_2)_n-$, the position of which depends on the value of n, i.e. the sequence length of methylene units. Absorption bands

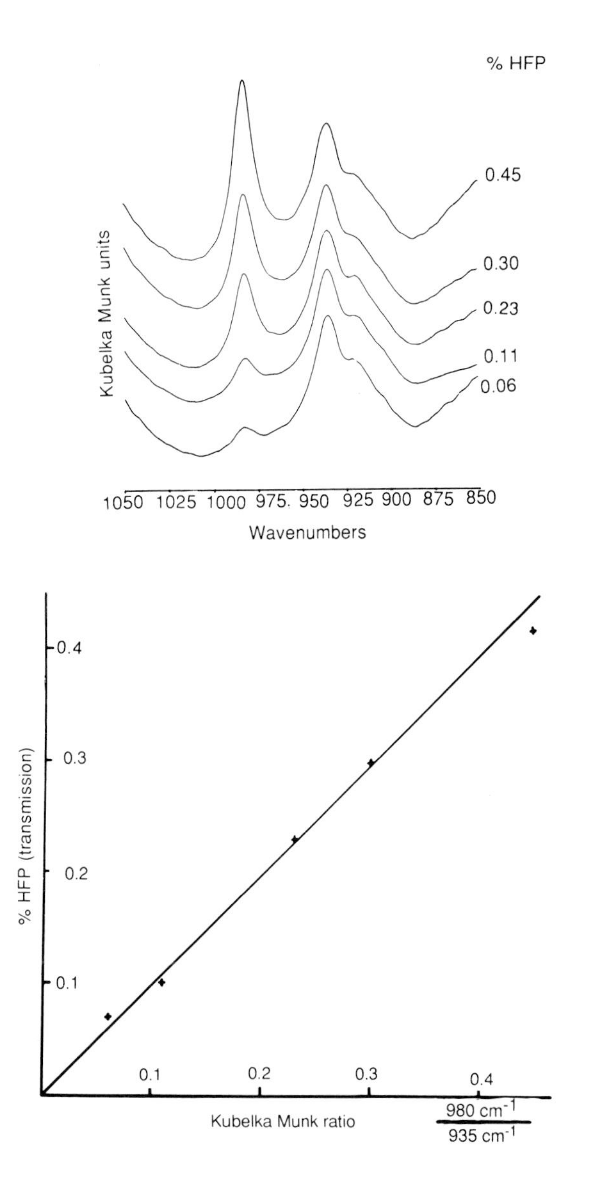

Figure 4.5 Upper: diffuse reflectance FTIR spectra of TFE–HFP copolymer powders. Lower: correlation between 980 cm^{-1}/935 cm^{-1} band ratios measured by transmission and DR spectroscopy. Reproduced from ref. 84, by permission of the American Society for Testing and Materials, Philadelphia, © ASTM.

at 751, 733, 726 and 722 cm^{-1} have been assigned to 2, 3, 4, and ≥ 5 contiguous methylene groups, respectively [89]. The analysis of these overlapping absorption bands is further complicated, since long $-(CH_2)-$ runs in a copolymer can crystallise and the single rocking mode is then replaced by a sharp and highly characteristic doublet at 720 cm^{-1} and 730 cm^{-1}. This added complication can be removed if the spectra are analysed above the melting point of the 'polyethylene' sequences, typically > 140°C [94–97]. Resolution enhancement of the FTIR$-(CH_2)_n-$ rocking-mode band envelope at the elevated temperatures

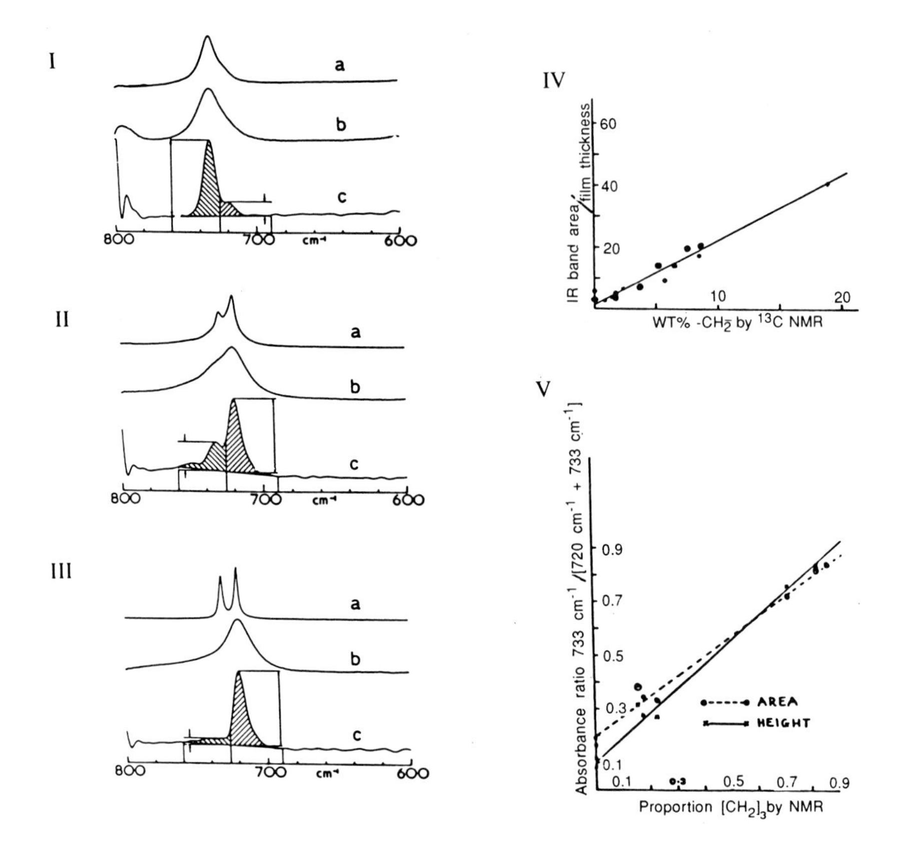

Figure 4.6 Quantitative analysis by FTIR of propylene/ethylene (P/E) copolymers: I: IR absorbance spectrum of P/E with mainly isolated E groups; II: IR absorbance spectrum of P/E copolymer with mainly adjacent E units; III: IR absorbance spectra of blend of polypropylene and linear polyethylene; a represents spectra at room temperature; b represents spectra of 'melt', i.e. at 145°C; c represents resolution enhanced 'melt' spectra; IV: band area per unit thickness measurements of 720 cm^{-1} and 733 cm^{-1} absorption bands versus wt% of $-CH_2-$ as $-(CH_2)_n-$ ($n \geqslant 5$) and $-(CH_2)_3-$ determined from ^{13}C-NMR data. V: quantitative determination of $-(CH_2)_3-$ groups as a proportion of the total combined ethylene content of P/E copolymers—correlation of absorbance ratio 733 cm^{-1}/(720 cm^{-1} + 733 cm^{-1}) with ^{13}C-NMR data. Note: in IV, IR absorbance per methylene appears essentially independent of the length of the E sequence. Reproduced from refs. 96 and 97 by permission of Springer-Verlag, New York, and The Royal Society of Chemistry, London, respectively.

(see Figure 4.6) can provide a sensitive, reliable method of measuring the composition [96] and distribution [97] of E in P/E copolymers, and the possible basis for an on-line melt analysis procedure [98].

Laser Raman spectroscopy has been proposed as a useful technique for probing the microstructure of copolymers. Good correlations were found between the concentrations of some isolated, dyad, triad and tetrad co-monomer sequences in vinyl chloride/vinylidene chloride copolymers and certain scattering intensities [99]. The positions and intensities of particular absorption bands have also been correlated with chain microstructure in an infrared study of ethylene/vinyl chloride copolymers, previously characterised by ^{13}C-NMR analysis [100]. More recently, FTIR spectra have been analysed for monad, dyad and triad monomer sequence-distribution dependencies in random styrene/acrylonitrile copolymers [101]. Changes in peak intensities from normalised spectra were correlated with microstructure probabilities; assignments were given if there existed a linear relationship between peak intensity and the number fraction of a microstructure.

4.5 Polymerisation kinetics and cure studies

IR and Raman spectroscopies are both tools that may be useful for monitoring polymerisation rates and cure mechanisms, particularly sequential time-lapse measurements of functional group concentrations at predefined intervals when utilising multichannel Raman or rapid-scan FTIR or FT-Raman detection.

Spectroscopic probing of the kinetics of the thermal bulk polymerisations of styrene [102,103] and methyl methacrylate [103,104] have been investigated [Raman, 102–104; FTIR, 103]. In addition, an *in situ* Raman study of microemulsion polymerisation of these monomers has been reported [105]. Conversion of the reactants to product may be followed readily by monitoring the loss of the aliphatic –C=C– peak in the region 1600–1650 cm^{-1}, although particular attention must be given to the method of quantifying the observations [102–104], since at equivalent concentrations the intensities of similar vibrational bands in the monomer and polymer will most likely be different. A combined Raman and laser-light scattering apparatus [106,107] has been applied to the on-line monitoring and *in situ* product characterisation of the thermal, melt [106] and solution [107], polymerisation of hexachlorocyclotriphosphazene, a precursor for polyorganophosphazenes, a new class of technologically important polymers.

FTIR spectroscopy techniques have been variously and successfully applied to the study of polyurethane systems. Several studies have used its sensitivity to isocyanate absorption near 2250 cm^{-1} to follow the kinetics of reaction of polyurethane preparations [e.g. 108–110], and monitored changes by ATR techniques in free and hydrogen-bonded urethane and urea groups in cross-linked foams [111, 112]. ATR techniques have also been used to study the effect of isocyanate index on the polyurethane foaming reaction [113], and to measure the evolution of residual –NCO– groups in flexible foams under humid and dry conditions, depletion of bound –NCO– groups relying mainly on reaction with water, with the rate depending on the prevailing relative humidity [114]. The FTIR technique has also proved of value in elucidating the formation of interpenetrating polymer networks of poly(methyl methacrylate) (PMMA) and polyurethane [115, 116].

Resin cure and cross-linking are areas of considerable study and application for vibrational spectroscopic techniques. The potential of laser Raman for distinguishing critical structure differences in cured urea–formaldehyde (UF) resins has been demonstrated from studies made on model compounds [117], while FTIR absorbance spectra of model compounds and semi-solid prepolymer thin films between KBr discs have been used to develop a representation of a space-network UF resin prepolymer [118]. In this latter study a freeze-drying sample-preparation procedure was employed to gain more distinct spectra for prepolymer preparations. However, FTIR approaches proved less informative when used in combination with ^{13}C-NMR in a kinetic study of sequential aliquots withdrawn from a phenolic resin formulation [119]. Model compound studies have also featured strongly in some of the extensive uses made of vibrational spectroscopic techniques to characterise intermediate formation, cure reaction and final product of a range of aromatic imide polymers: FTIR [120–124]; FTIR plus Raman [125]; and FT-Raman [126]. *In situ* mid-IR FTIR probing of a composite curing reaction has been demonstrated by embedding a chalcogenide glass fibre in a graphite fibre reinforced polyimide matrix resin [127].

The techniques, interpretation and data analysis for FTIR and solid-state NMR applications to epoxy resin systems are discussed in ref. 128, while the characterisation by FTIR of thermosetting epoxy glues is reviewed in ref. 129. In addition to monitoring the absorption of the epoxy group at 910 cm^{-1}, the advantages of measuring cross-linking from the epoxy-group NIR absorption at 4532 cm^{-1} are discussed [129]. Other workers have used FTIR techniques to study catalyst mechanisms of epoxy-resin cure systems [130, 131]. NIR absorption-band studies have also been used recently in a demonstration of the feasibility of using remote real-time FTIR sensing of the cure reaction of an epoxy resin used in advanced composite materials [132]. From measurements made by using a stripped silica fibre optic interfaced to a microcapillary cell, it was found that the primary amine band at 5067 cm^{-1} was the most sensitive up to gelation.

An internal-reflection prism cell has been used for reaction-kinetics measurement of styrene–unsaturated polyester resins at elevated curing temperatures and pressures [133]. Other researchers have estimated the cross-linking efficiency in methyl methacrylate–ethylene glycol dimethacrylate copolymer networks by laser Raman spectroscopy [134], while recently

FT-Raman studies of thermally and photochemically induced epoxy curing reactions have been reported [135]. In some instances pulsed sources may be required to overcome thermal emission problems.

4.6 Polymer branching and end-group measurements

The measurement of end groups and side-chain branches on polymer chains has importance, since they can exert a major influence on morphology and properties, and consequently on processibility and performance.

A classic example of this type of determination is measuring the methyl content of polyethylenes by IR spectroscopy [136, 137]. Simply, the method involves comparison (spectral subtraction) between a low-density polyethylene and a polymethylene or a standard polymer of a very high molecular weight and low degree of branching (a high density polyethylene), thereby separating the methyl symmetrical deformation absorption band near 1378 cm^{-1}. This procedure has more recently been developed to identify and quantify the branch type in a series of linear low-density polyethylenes (LLDPEs) [138]. The investigations rely on the wavenumber reproducibility of FTIR measurements. These workers [138] and others [139] suggested that branch type in these ethylene–α-olefin copolymers might also be distinguished, after appropriate bromination of unsaturation groups, by examining the position of the methyl rocking-vibration absorption bands near 900 cm^{-1}. In contrast, Raman measurements are less informative and at best no more than semi-quantitative for methyl and ethyl branches [140]. A recent paper reviews the results and progress obtained in several laboratories, using a variety of IR approaches, in elucidating the fine structure in polyethylene and ethene copolymers [141].

The measurement of end groups may not always be readily accomplished because of their relatively low concentration, and since their absorption bands are frequently overlapped with other structures they do not appear as distinct bands. Here, the method of comparing polymers of differing molecular weights by difference spectroscopy is particularly useful, since end-group concentration falls with increasing molecular weight. Alternatively, absorption bands characteristic of certain end groups may be ascertained by establishing the difference between two polymers of essentially the same molecular weight, but synthesised (e.g. end-capped) in such a manner as to have distinctly different end groups. These procedures have been used to assign an absorption band characteristic of the aryl–F end group in poly(aryl ether ether ketone) (PEEK) [139]. Once established, the normalised intensities of the end-group bands may be correlated with independent or referee methods such as NMR or wet chemistry (e.g. titration).

The technique of heavy-isotopic substitution can also be beneficial in determining the position of –OH (and –NH) stretching frequencies in polymers in which the end-group vibration band cannot be observed readily. The approach is particularly well-suited to the products of polycondensations. Deuteration of –OH end groups can be achieved by immersing a dried polymer film in D_2O, which is then redried before measurement. Figure 4.7 illustrates how subtraction of the 'hydrogenated' spectrum from the 'deuterated' spectrum reveals distinct absorption bands due to –OH and –OD. These methods were used in 1957 to assign and measure the concentration of carboxylic and alcoholic –OH end groups in poly(ethylene terephthalate), PET [142, 143]. The approach has been developed subsequently for direct application to a wide range of PET films [144], and used similarly in studies on poly(butylene terephthalate) (PBT) [145]. Polyesters such as PET and bisphenol-A-Polycarbonate have been shown to be readily nucleated by alkali-metal salts of

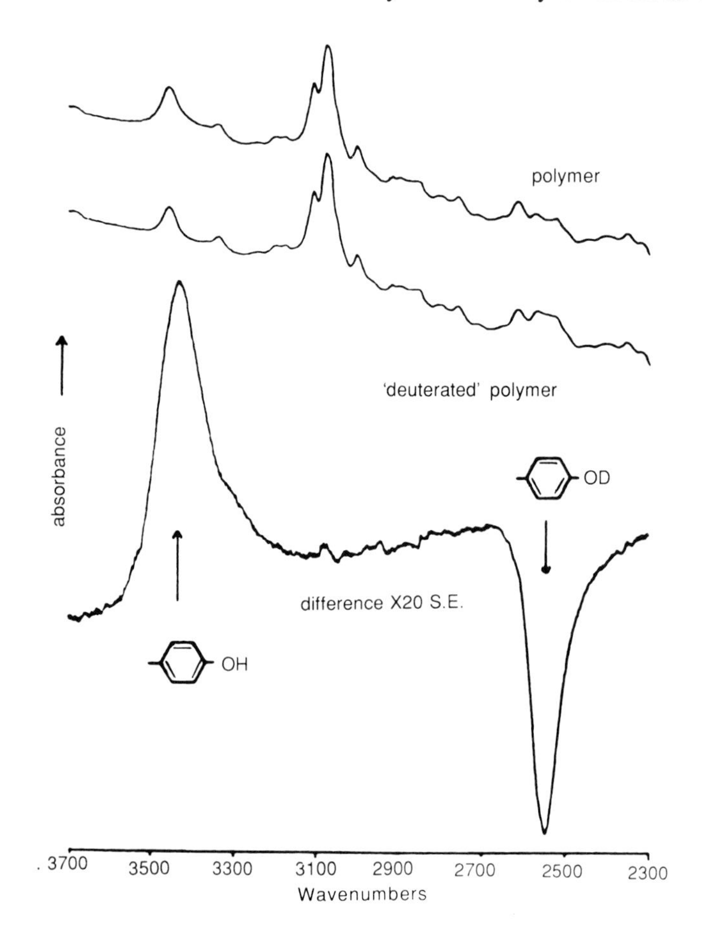

Figure 4.7 FTIR absorbance spectra of aromatic polymer film before and after deuteration, and difference spectrum showing position of phenolic end-group. Reproduced from ref. 139, by permission of the Society of Applied Spectroscopy.

organic acids [146]. The polymers undergo chain scission on reaction with the alkali-metal salts to form polymeric species with ionic end groups. These were confirmed by IR spectroscopy and it was shown, in particular, that for the PET case the ionic chain ends were aggregated in the solid state into the molten polymer.

4.7 Degradation

Understanding and monitoring the mechanisms by which polymers deteriorate is commercially very important, particularly in the search for improved stabilisation and the quest for extended retention of specific properties. Vibrational spectroscopy techniques have a long tradition in elucidating these processes by characterising the molecular species produced as a result of thermal, thermal-oxidative, photo, photo-oxidative and radiation-induced damage. Raman and IR may both be used to observe conformational changes which give rise to alternations in the physical characteristics of polymeric structures (see sections 4.9–4.11) and each has been used extensively to study chemical change. IR has particular sensitivity to oxygenated species, while Raman techniques are uniquely effective for polyene characterisation.

Study of the degradative processes of aliphatic polymers has proven particularly rewarding, with the sensitivities of modern vibrational spectroscopy instrumentation enabling much earlier detection of degradation products or their precursors. In addition, localised deteriorations may be more conveniently mapped by microscopy–spectroscopy techniques (see section 4.3.2). The gross changes in the IR spectrum of polyethylene indicative of both thermal [147] and photo [147, 148] oxidative degradation are well established and readily observed. A primary step in the autocatalytic process is the formation of hydroperoxides, which give rise to an absorption band at 3555 cm^{-1}, prior to the generation of carbonyl and hydroxyl entities [147–151]. These oxidative pathways also effect changes in the concentration of unsaturation groups. Thermal oxidation leads principally to a reduction of vinylidene-type (888 cm^{-1}), while photo-oxidation is characterised primarily by an increase in vinyl (909 and 991 cm^{-1}) end groups [148, 149]. The degradation processes give rise to an envelope of overlapping carbonyl absorption bands in the region 1900–1600 cm^{-1}, and methods for determining [150] or deconvoluting [152] this into constituent parts have been developed. Treatment with SO_2, followed by detailed study of the pendant methylene (vinylidene) concentration using FTIR difference spectroscopy techniques, has been shown to provide a very sensitive analytical method for detecting the early onset of oxidation [151] (see Figure 4.8).

Differences and similarities in the thermal oxidation behaviour of polyethylene and polypropylene have been highlighted through comparison of their IR spectra [153]. FTIR spectra have also provided support for the

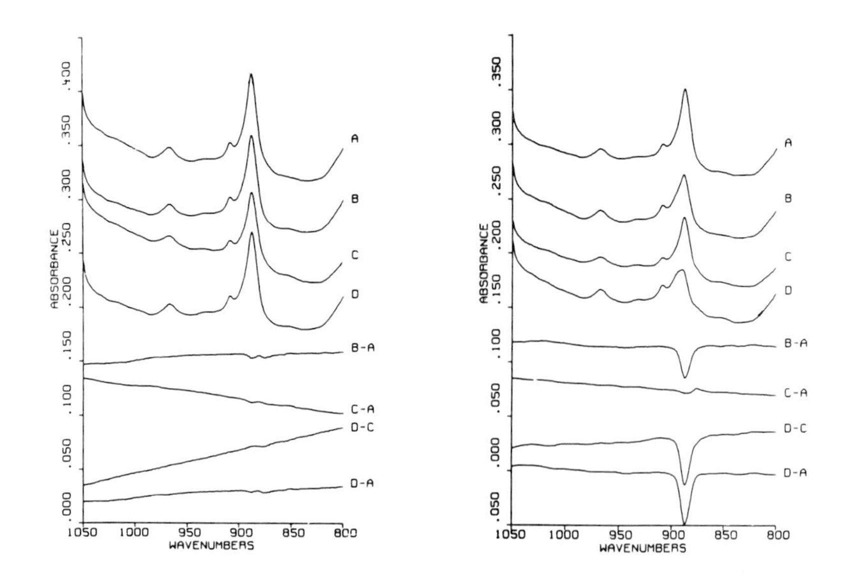

Figure 4.8 Low-density polyethylene (LDPE) suspected to have low levels of thermal degradation (by a deterioration of physical properties) may show negligible differences in the infrared. Initial oxidative reaction takes place adjacent to the vinylidene units [151], and treatment with SO_2 will lead to a reduction in pendant methylene absorption, which can be further intensified by prior heating. This is illustrated in this figure by comparing the treatment effects on a complaint (oxidised, right) and control (left) samples. A, the samples as received; B, after SO_2 treatment; C, heated at 100°C for 18h; D, heated at 100°C for 18h then SO_2-treated. Reproduced from ref. 151, by permission of Elsevier Applied Science Publishers Ltd, Barking.

notion of a free-radical, chain mechanism of oxidative degradation in which hydroperoxide groups accumulate during the induction period for the case of poly(ethylene oxide) [154]. Difference IR spectra have been used to indicate changes caused by crosslinking and chain scission reactions to polyethylene during irradiation [155].

The very high sensitivity of resonance Raman spectroscopy (RRS) to detecting the longer polyene sequences present in degraded poly(vinyl chloride) (PVC) is now well established [156, 157]. When PVC is exposed to solar or ultraviolet (UV) radiation for prolonged periods, an ‘unzipping’ reaction evolving HCl occurs to form conjugated polyenes, which exhibit UV/visible absorptions. By taking advantage of the RRS technique low levels of detection (0.00001%) are attainable. This enables the early stages of natural weathering to be monitored [10, 157], or chemically initiated degradation to be studied [157], such as the accelerated ageing caused by residual amine in polyurethane (PU) foams. Non-flammable PVC is often used to cover PU foams in applications such as floor tiles. FT-Raman studies have extended the range of this application to other fully processed systems [158]. The synergistic stabilising effects on PVC of mixed metal soaps have also been elucidated through FTIR studies [159–161].

The progress of the chemical and physical changes in nylon-6 and -6,6 undergoing photo and photo-oxidative [162] and thermal and thermal-oxidative [163] degradation has been the subject of detailed investigations by FTIR spectroscopy. These have included IR analysis of any evolved gases. FTIR spectroscopy has also proven an excellent tool for studying the degradation of polyacrylonitrile (PAN) and copolymers of PAN in the initial stages where the polymer chain undergoes cyclisation [164–167]. In addition to extensive uses in the textile industry, PAN polymers are employed as precursors to the formation of carbon fibres. These studies complement Raman studies on carbon-fibre graphitisation levels [168, 169]. It is essential for many critical areas of application for modern composite materials that their components have durable mechanical properties. These properties will most likely be adversely affected by degradation processes, and consequently these need to be understood and monitored. PEEK and poly(aryl sulphones) are high-temperature thermoplastics which have been used as matrices for fibre-reinforced composites. Two distinct thermal degradation reactions in PEEK films and PEEK–carbon

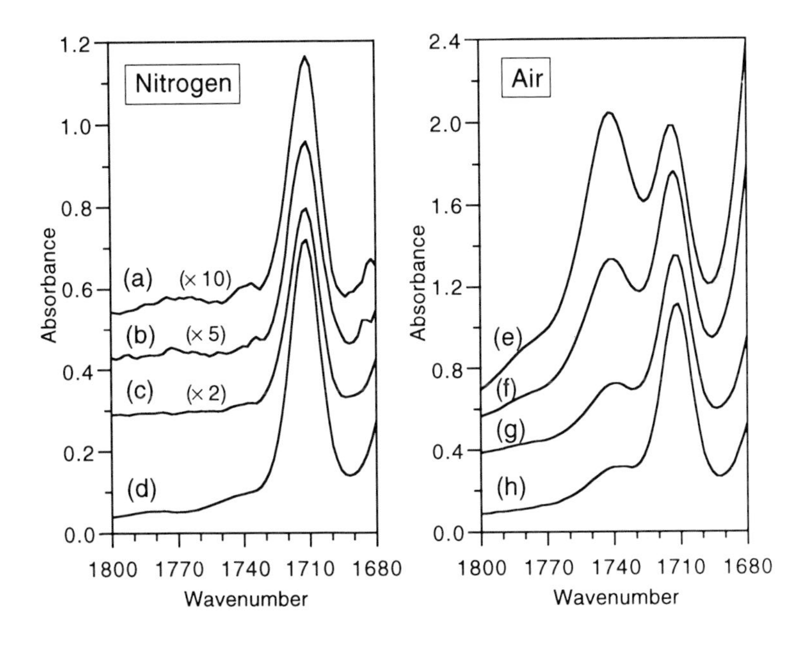

Figure 4.9 Difference FTIR spectra for PEEK film heated in nitrogen atmosphere at (a) 400°C for 480 min, (b) 430°C for 200 min, (c) 460°C for 70 min, (d) 485°C for 40 min, and in air atmosphere at (e) 400°C for 360 min, (f) 430°C for 110 min, (g) 460°C for 45 min, (h) 485°C for 20 min. Spectra are offset on absorbance scale for clarity. Reproduced from ref. 170(a), by permission of the Division of Polymer Chemistry, American Chemical Society.

composites have been identified and followed quantitatively by FTIR spectroscopy [170]. In a nitrogen atmosphere a carbonyl peak arises at 1711 cm^{-1}; in air a similar but faster reaction occurs plus a reaction that results in an absorption at 1740 cm^{-1} (see Figure 4.9). A mechanism for proton-beam irradiation-induced degradation on a commercial polysulphone has been proposed from FTIR studies in combination with ESR and electronic spectroscopy [171]. Aminolysis, rather than hydrolysis, was the preferred chemistry to explain the mode of composite failure of the reinforcement, derived from IR studies of PET fibre–nitrile rubber composites [172].

FTIR techniques in combination with or as complement to other measurement techniques have been used in a wide range of photochemistry studies on polymers. These include: bisphenol-A polycarbonate [173], polycarbonate coatings on mirrors [174], PMMA [175], poly(*n*-butyl acrylate) [176] and polypropylene [177]. DSC and FTIR studies have been used in conjunction to investigate the nature of γ-radiation-induced degradation and its effect on the 19°C and 30°C phase transitions in PTFE [178]. IR studies of the hydrolysis of melamine–formaldehyde crosslinked acrylic copolymer films have shown that copolymer–melamine formaldehyde crosslinks are broken and that crosslinks between melamine molecules are formed [179]. The thermal and photo-degradation mechanisms in an IR study of cured epoxy resins were found to be related to the autoxidative degradation processes for aliphatic hydrocarbons [180].

4.8 Blends

The blending or simple mixing of polymers provides materials that often exhibit advantageous mechanical, thermal or other properties, which enable commercial product ranges to be increased with minimal development costs.

Compatibility or miscibility in polymer blends implies a homogeneous solid solution. In most cases, this requires a specific interaction between components, such as hydrogen bonding or dipolar intermolecular interactions, which can often be detected in their IR spectra. Blends involving completely incompatible polymers (e.g. rubber-modified plastics) may be considered as macroscopically multiphase systems, and their IR spectra are essentially produced by appropriately summing the spectra of the individual constituents. There is an excellent treatise (with many examples) on the subject [181], which details the role and limitations that FTIR spectroscopy has in probing the structure of multicomponent blends. The investigations of interactions rely on the precision of the FTIR technique. Recently, a theoretical equilibrium approach has been adopted to predict phase behaviour in hydrogen-bonding polymers, and related to experimental FTIR data [182,183].

Many published studies consider cases in which one of the components contains an accessible carbonyl group, the stretching vibration band of which broadens and shifts to lower wavenumber in the presence of a second compatible polymer (see an example in Figure 4.10). Recent examples of thermal and spectroscopic analyses of blend systems include: poly(vinyl acetate) (PVA) with PMMA [184]; poly(vinylidene fluoride) (PVDF) with PVA [185, 186]; polybenzimidazole/poly(bisphenol-A-carbonate) [187]; and PMMA–fluorinated polymer blends [188]. All these make use of spectral subtraction routines to observe interactions, but additional data-processing tools have also been used to aid understanding, such as Fourier self-deconvolution [188],

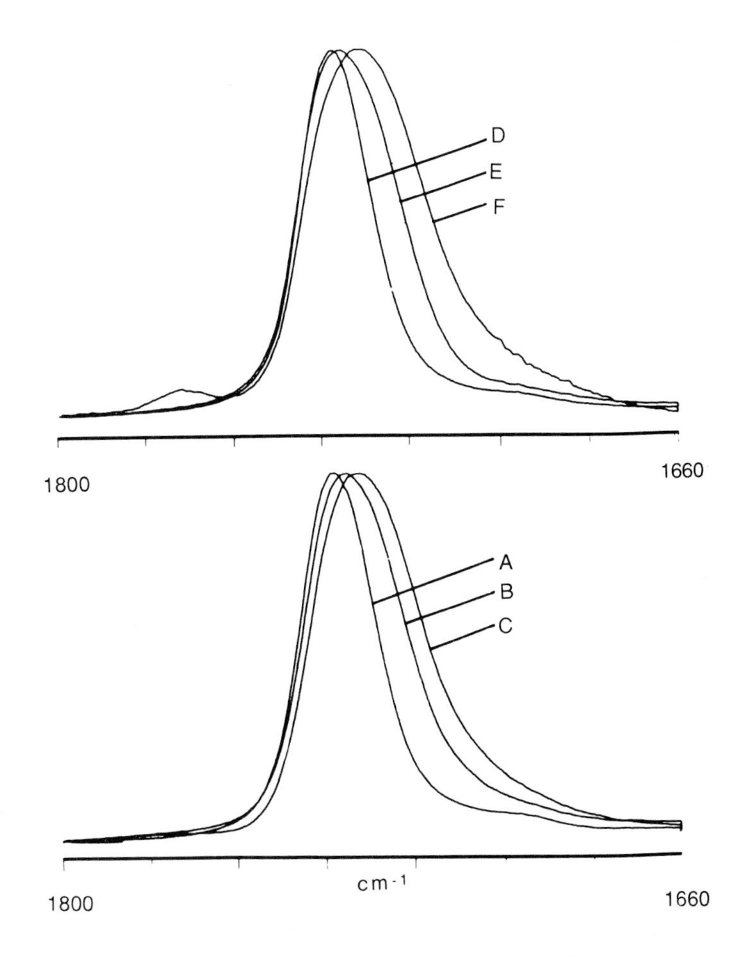

Figure 4.10 FTIR absorbance spectra recorded at room temperature of an ethylene–vinyl acetate (EVA) copolymer blended with chlorinated polyethylene (CPE) and poly(vinyl chloride) (PVC). A, pure EVA; B, 40:60 and C, 80:20 wt% CPE–EVA, respectively; D, pure EVA; E, 40:60 and F, 80:20 wt% PVC–EVA, respectively. Reproduced from ref. 197, by permission of the publishers Butterworth Heinemann Ltd ©.

derivative [187,188], and factor analysis and least-squares curve fitting [186]. These data-analysis software routines have also been important in elucidating interaction mechanisms in urethane binary polymer blend systems, which have been the subject of several investigations. In contrast to the examples above, self-associated carbonyl groups in the pure urethane component are liberated to yield an absorption band characteristic of 'free' carbonyl [189,190] or other hydrogen-bonded structures [191], as the N–H groups partake in intermolecular hydrogen bonding.

Other recent studies which involve and illustrate the power of the FTIR technique include: surface studies of PVC systems with PMMA [192] and poly(ε-caprolactone) (PCL) [193, 194]; PVC with styrene/acrylonitrile copolymers [195]; polyester/nitrocellulose [196]; EVA copolymer with PVC and chlorinated polyethylene (CPE) [197]; and interactions in blends involving *p*-sulphonated polystyrene [198, 199]. FTIR techniques have been used to map the phase diagram of an aromatic polyamide–poly(ethylene oxide) blend [200], while microscopy–FTIR has been used to obtain information on intermolecular interactions and conformational changes in specific domains in functionalised polyolefins with PVC or polystyrene [201]. Segmental motions and microstructure studies from combined DSC and FTIR measurements have been used to interpret solid-state transitions in miscible rubber blends [202].

4.9 Morphology

4.9.1 General

Since vibrational spectra are sensitive to polymer tacticity, configuration, conformation and chain packing, they provide a useful probe for characterising morphology. At the most fundamental level, the symmetry and vibrations of a polymer chain are determined by chain tacticity, configuration and conformation. In addition to the effects of local configuration/conformation, vibrations also arise (specifically as a result of the polymer chain adopting a regular, extended geometry) in which repeat units vibrate 'in phase' along the regular sequence [203]. The number, activity and polarisation properties of these so-called 'regularity bands' can be calculated for a chain of given symmetry using group theory (line-group analysis), and this approach has been used to distinguish between possible polymer structures [13]. Finally, the intermolecular ordering that arises during crystallisation can also modify the vibrational spectrum, but usually to a much smaller extent than changes in chain conformation. The distinction between conformation/configuration, regularity and crystallinity effects was clearly established long ago [203], but confusion and incorrect terminology still occur occasionally in the literature. These points are exemplified below.

4.9.2 Configuration and conformation

Vibrational spectroscopy is of great help in the analysis of both regular and non-regular chains, which can be treated simply as collections of individual units subject to point-group rather than line-group symmetry. For example, stereoisomers can be distinguished and quantified, classic examples being the quantification of *cis* and *trans* isomers in polybutadiene [204] and polyisoprene [205] by Raman spectroscopy, and in polyacetylenes using FTIR [206]. In a similar vein, the influence of tacticity on spectra has been studied for a wide range of polymers, both by analysing the effect on the symmetry of the polymer chain and hence its vibrations [11, 13], and by empirically correlating changes in tacticity with spectral changes [12]. Generally speaking, only isotactic or syndiotactic polymers can crystallise and have sharper and more intense vibrational bands in this form; it is therefore necessary to distinguish between the effects of tacticity and crystallinity on the spectrum. Work on polypropylene [207] and polystyrene [208] provide examples of this extensive area of study.

Although NMR is normally the technique of choice for tacticity analysis, sampling constraints will sometimes favour the application of vibrational spectroscopy. For example, Figure 4.11 shows spectra obtained *in situ* from polypropylene particles on the surface of three different catalysts using a Raman microscope. The difference in tacticity of the polymers is evident. Since no sample work-up was necessary for the analysis, catalyst effectiveness can be assessed easily with minimal amounts of sample.

The analysis of conformation in both crystalline and amorphous polymers has received much attention. For example, quantification of the *trans* and *gauche* conformers of the glycol residue in PET has been demonstrated by using factor analysis of IR spectra [209–211]. This work illustrates the value of spectral subtraction and ratio methods [212] to generate the spectra of 'pure' conformers or phases from materials which are actually mixtures of both, as is usually the case with polymers. Similar studies have been carried out on many systems including PVC, where conformer sequences can be distinguished in the IR [213, 214] and Raman [215] spectra, with the latter work also providing spectral indicators for tacticity and crystallinity. The use of curve fitting and deconvolution is often necessary in such work, owing to overlapping of bands due to different conformer sequences. Again, the use of NCA in aiding interpretion of the effect of polymer conformation is stressed [11].

As an extreme example of conformational study, it has been possible to identify bands arising from folds in polymer chains at crystal boundaries, and to follow how these change during crystal annealing [216–218]. This throws light on how chains re-enter the crystals or lamellae (i.e. tight versus loose folds). There is also potential for measuring the energy of different conformers by measuring conformer distribution as a function of temperature [219].

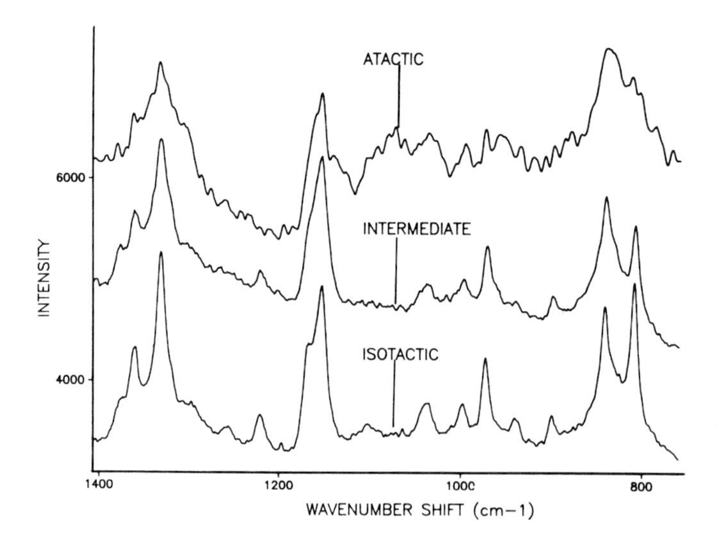

Figure 4.11 Raman spectra of polypropylene particles obtained *in situ* on three different alumina-supported catalysts by using the Raman microscope (488 nm excitation). The microscope allows fluorescence from the alumina support to be avoided. (Interestingly, these samples fluoresced worse when using the FT-Raman system than with the visible laser system, a result found to be quite common with catalyst samples.) The variation in polymer tacticity is evident in the three spectra: the very weak, diffuse upper spectrum is characteristic of atactic material; the strong, sharp lower spectrum indicates highly isotactic polypropylene; and the middle spectrum indicates intermediate tacticity.

4.9.3 Crystallinity

Measurement of 'crystallinity' from vibrational spectra usually entails correlating vibrational features with independent parameters such as density, DSC or X-ray data. Thus, if a conformer specific to the amorphous or crystalline phase gives a distinct vibrational band, its intensity, position or shape can sometimes be correlated with crystallinity changes. Alternatively, a regularity band associated with the extended chain (and hence crystalline phase) can be used for the correlation. However, since these methods are sensitive to intramolecular conformation, they do not directly measure crystallinity, which involves chain packing, and so simply provide an empirical measure. Examples include correlation of carbonyl bandwidth with density in Raman spectra of PET [220] and poly(propylene terephthalate) [221], and correlation of carbonyl band position and IR band absorbance ratios with (X-ray) crystallinity in PEEK [222, 223]. Similar correlations exist for nylon-6,6 [224] and PTFE [225]. More recently, multivariate data-processing techniques have been used to relate spectral changes to crystallinity, including near-IR absorbance and FT-Raman measurements on PET [226, 227], and IR data for PTFE [228].

It has been noted [11] that when unique vibrational bands can be associated with each phase, it is not necessary in principle to calibrate against other standards, although it is often desirable to do this. All that is required is a set of samples of widely varying composition for which the relative band intensities can be determined and extrapolated to the 100% and 0% crystallinity levels. Raman measurement of crystallinity in polyethylene illustrates this approach [229], as does the IR measurement of crystallinity of nylon [224]. Extrapolation of the spectroscopic calibration has sometimes been used to determine the amorphous and crystalline densities of a polymer [224, 225], thereby adding to the information available from the primary calibration standard. However, one must note that vibrational spectroscopy, DSC, X-ray and density actually measure different physical parameters and so a 'crystallinity' determined solely by vibrational spectroscopy may well differ from that obtained by other techniques [230]. Also, with anisotropic materials, molecular orientation can alter band intensities and invalidate calibrations developed using isotropic standards [48, 227, 231].

True 'crystallinity bands' arise from intermolecular packing rather than intramolecular ordering [203]. They can result from vibrations of whole molecules about their equilibrium lattice position, as in the case of several bands in the far-IR spectrum of PTFE [232], and the 65 cm^{-1} band in the far-IR spectrum of PVC [233]. Alternatively, when more than one molecule is present in the unit cell, 'correlation' splitting of bands can occur because of reduction in the symmetry of the chain environment. The classic example is the splitting of the 720 cm^{-1} band in polyethylene, although similar effects have been observed in the spectrum of PET [209] and other polymers. However, for many polymers, splitting is resolved only at low temperatures, if at all, owing to the small interaction between chains [13, 207, 234].

It is stressed that true crystallinity bands have only been proved for a small number of polymers, and so the majority of measurements that actually correlate intramolecular conformation with crystallinity should be interpreted with caution. For example, in the Raman spectrum of highly oriented PET it is possible to observe a high level of *trans* glycol conformers which are nonetheless located in amorphous material [235]. In this case, correlation of conformer content with crystallinity would be misleading, and the carbonyl bandwidth actually gives a better measure. As well as phase quantification, characterisation of phase transitions in response to temperature or mechanical changes has also received considerable attention [236–240].

Aside from a fundamental contribution to understanding polymer morphology, vibrational spectroscopy offers a tool for analysing finished articles or for following changes occurring during processing. For example, vibrational microscopy allows one to map morphological properties with micrometre resolution in inhomogeneous materials or structures. A range of industrially-relevant examples have recently been summarised [48]. Often, several vibrational techniques can be applied to the same problem, with the method of choice being determined by sampling constraints. This is illustrated in Figure 4.12, where the Raman, mid-IR and near-IR spectral changes occurring on crystallisation of a copolyester are shown. In this case both near-IR and Raman data can be obtained from thick (2 mm) polymer plaques, but the mid-IR data must be obtained by transmission through a solution-cast film. Attenuated total-reflection IR measurements are not appropriate in this case since the plaque surface in contact with the ATR crystal showed an excessively high crystallinity compared with the bulk, which was known to be amorphous at the time of measurement (Figure 4.12(d)).

4.9.4 Amorphous regions in polymers

In addition to longitudinal acoustic modes (section 4.10), the low-frequency regions of Raman spectra often contain rather broad, ill-defined bands. While these can sometimes be ascribed to specific intra- or inter-molecular vibrations, work has suggested they may be due to 'fracton' vibrational modes localised in disordered regions or 'blobs' [241]. Analysis of the bandshape has been used to measure the size of the disordered regions; for example, ~50Å 'blobs' were found in amorphous PET and PMMA [242].

4.10 Longitudinal acoustical (accordion) modes (LAMs)

LAMs were first observed in the low-frequency Raman spectra of solid *n*-alkanes as strong, polarised bands with frequency inversely proportional to the chain length [243]. They are due to a symmetrical 'accordon-like' vibration of the extended chain, which can be modelled as an elastic rod of modulus E, length L and density p, with a resultant vibrational frequency $(E/p)^{\frac{1}{2}}/2L$. Therefore, the effect of chain length on frequency is predicted by a simple mechanical model.

Polymers that crystallise as lamellae posess extended chains or stems and so would be expected to exhibit LAMs. For example, in polyethylene LAMs are observed at $< 30\,\mathrm{cm}^{-1}$, with the exact position depending on the stem length (and hence the lamellar thickness) [244]. Measurement of the LAM position is a powerful and widely used tool for following changes in lamellar thickness during processing, where polyethylene has been the most widely studied polymer [245–249]. Most work is aimed simply at determining the average lamellar thickness; for example, a combination of synchrotron X-ray diffraction and Raman measurements [249] has indicated a bimodal distribution in lamellar sizes in polyethylene crystallised for short times. The spread in lamellar thickness in polydisperse polyethylene can also be measured, provided that suitable corrections are applied [250]. However, it should be noted that extraction of subtle morphological information from LAM data

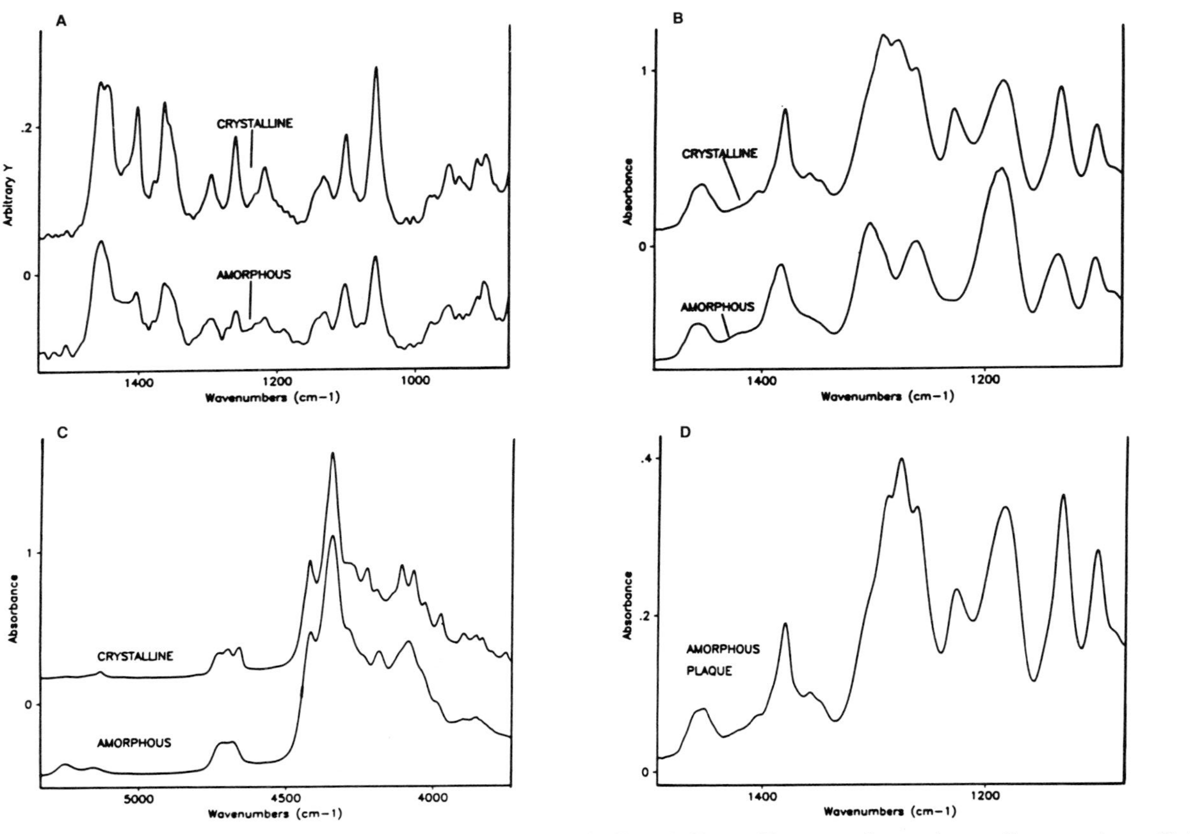

Figure 4.12 Comparison of the influence of crystallinity on the (A) Raman, (B) mid-IR and (C) near-IR spectra of a copolyester. Raman and near-IR data were obtained from ~2 mm-thick plaques, while the mid-IR spectrum was obtained in transmission through a cast film; (D) shows an internal-reflection mid-IR measurement on an amorphous plaque; interestingly, the surface of the plaque appears to be highly crystalline (compare (D) with (B)).

alone is fraught with potential pitfalls. A review covering work up to 1986 has been prepared and provides a useful introduction to this area [251].

Considerable effort has been expended in detailed modelling of LAMs [252–255]. In particular, treatment as amorphous/crystalline composite rods has been quite successful for helical polymers, which give anomalously high frequencies compared to the predictions of the simple elastic-rod model. However, current calculations tend to utilise NCA (section 4.2) to predict LAM frequencies [256], since this also allows prediction of intensities and the effect of interchain interactions and defects [257]. For instance, defects have been calculated to broaden the LAM, thereby making measurement of the lamellar thickness distribution even more difficult [257]. A range of polymers other than polyethylene have been studied, with the emphasis primarily on detection and assignment of LAMs, and semi-empirical studies on morphology changes. Recent efforts have been concentrated on helical polymers such as poly(ethylene oxide) [256], but references to earlier studies of other polymers can be found elsewhere [11, 12, 251].

A different LAM mode, termed the 'disorder' or D-LAM, has also been observed at higher ($\sim$200 cm^{-1}) frequencies in a range of polymers [251,258,259]. This mode appears to be associated specifically with conformationally disordered regions and bears the familiar inverse dependence upon region size.

4.11 Molecular orientation

Characterisation of molecular orientation is important since many physical and mechanical properties of polymers depend upon the extent and uniformity of that orientation. Orientation can be measured by using a variety of techniques [260,261]. Vibrational spectroscopy is particularly attractive since it is widely applicable, it often allows characterisation of amorphous and crystalline phases separately, it simultaneously provides morphological data (section 4.9), and it can be used to map orientation with high spatial resolution.

4.11.1 Orientation functions

Figure 4.13 illustrates the use of polarised absorption spectroscopy to measure orientation for a uniaxially oriented polymer. For a single molecule, the absorbance is proportional to the square cosine of the angle between the vibration transition dipole moment and the IR electric field vector (E). Thus if $\theta = 90°$, the absorbance is zero (note that the measurement is sensitive to the alignment of the vibrational dipole moment rather than the chain axis itself). The dichroic ratio R is calculated by ratioing the absorbance measured with E parallel to X_3 with that measured perpendicular to X_3 ($R = A_{\|}/A_{\perp}$). For an assembly

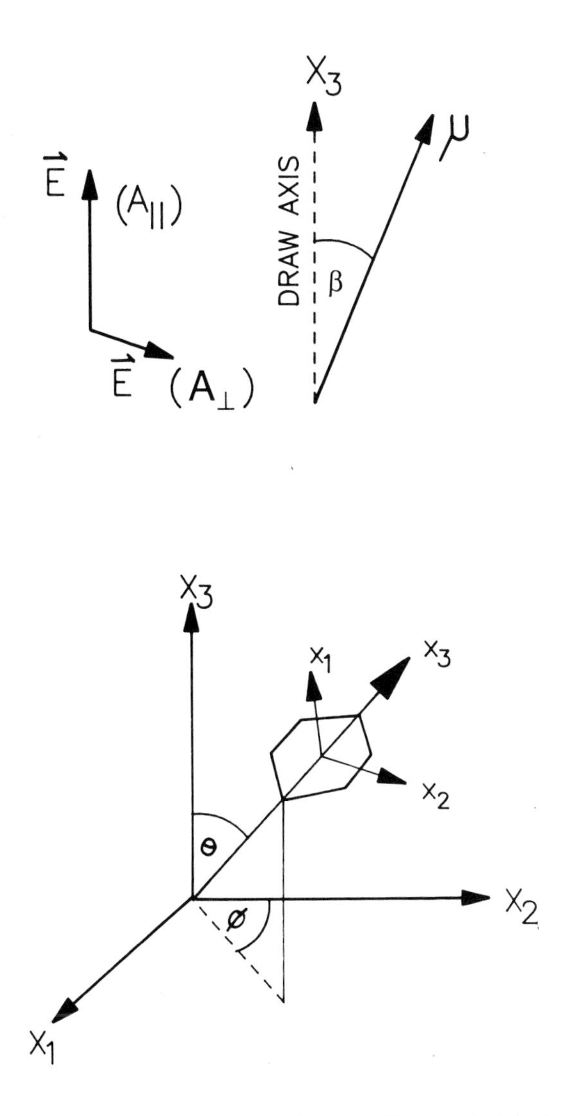

Figure 4.13 Definition of axes used to describe orientation. Upper: uniaxial orientation: only the angle between the draw axis X_3 and the transition moment μ is non-random. Measurements are made with the IR E vector parallel and perpendicular to X_3, yielding absorbances $A_{\parallel}$ and $A\perp$ respectively. Note, in general μ lies at an angle β to the polymer chain, and this angle must be known to calculate chain orientations from the dichroic data. Lower: biaxial orientation: the average angles θ and φ, plus rotation about the chain axis must be considered. Alternatively, direction cosines between molecular axes x_i and sample axes X_j $(\cos^2(x_iX_j))$ are used. In this case, orientation of the *para*-substituted benzene ring of PET is used to illustrate the system, where x_3 lies along the chain axis, x_1 is normal to the plane of the ring and x_2 is orthogonal to x_1 and x_3. The 1018 cm^{-1} IR band is almost parallel to x_3, while the 875 cm^{-1} band is almost parallel to x_1 (see Figure 4.14). Thus both orientation of the chain axis and the rotation about the x_3 axis can be measured.

of molecules aligned uniaxially about X_3, R is related to the orientation function f by eqn (4.1) [262]:

$$f = 0.5 \langle 3\cos^2\theta - 1 \rangle = \left[\frac{R-1}{R+2}\right]\left[\frac{2\cot^2\alpha + 2}{2\cot^2\alpha - 1}\right] \tag{4.1}$$

where $\langle\,\rangle$ denotes an average over all θ, and α is the angle between the chain axis and the transition moment. The function f is actually the second moment of the orientation distribution function (ODF) [12, 260, 261, 263], and varies from -0.5 for perfect alignment perpendicular to X_3, through 0 for random orientation to 1 for perfect alignment along X_3; α can be determined by comparing IR results with those of independent techniques on the same sample, by measuring perfectly aligned materials (e.g. crystals), or by calculation. For convenience, α is often assumed to be 0° (parallel band) or 90° (perpendicular band) for a given vibration in the absence of other data. Polarised Raman intensities can also be used to measure ODFs [264]. Both second and fourth moments of the ODF can be determined from Raman measurements, with the higher-order components being useful in relating mechanical or deformation properties to orientation. Unfortunately, processing of the Raman data is considerably more complex than of the IR data. For example, individual scattering tensor components must be known (these are not usually readily available), and it is unclear exactly how Raman intensities should be corrected to allow for factors such as internal field effects in highly birefringent materials [265]. The reader is referred elsewhere for further details [264], but it suffices to say that neither measurement nor analysis of Raman orientation data should be undertaken lightly! As a qualitative example, Figure 4.14 shows dichroic IR and Raman spectra of an oriented PET film. The 1615 cm^{-1} Raman and 1017 cm^{-1} IR bands have parallel character, while the 875 cm^{-1} IR band has perpendicular character.

4.11.2 Experimental

Measurement of IR dichroic ratios necessitates measurement of band absorbances with the IR E vector parallel and perpendicular to the sample draw axes. Detailed experimental arrangements are discussed elsewhere [12]. Weakly absorbing materials can be examined by transmission through a film, which is tilted to measure out-of-plane orientation in biaxial material. The tilted film method and appropriate data corrections have been discussed elsewhere [266]. Strongly absorbing polymers can be examined by microtoming a thin section and then measuring transmission spectra with an IR microscope [267]. This also allows orientation mapping on a small scale. Polarised DRIFTS [268], photoacoustic [269] and ATR spectroscopy [270, 271] are also useful.

Raman sampling is simpler, since sample thickness is not an issue. Measurements of the Raman scatter parallel and perpendicular to the laser E vector are made, with the laser vector parallel to the sample draw axes [264, 266].

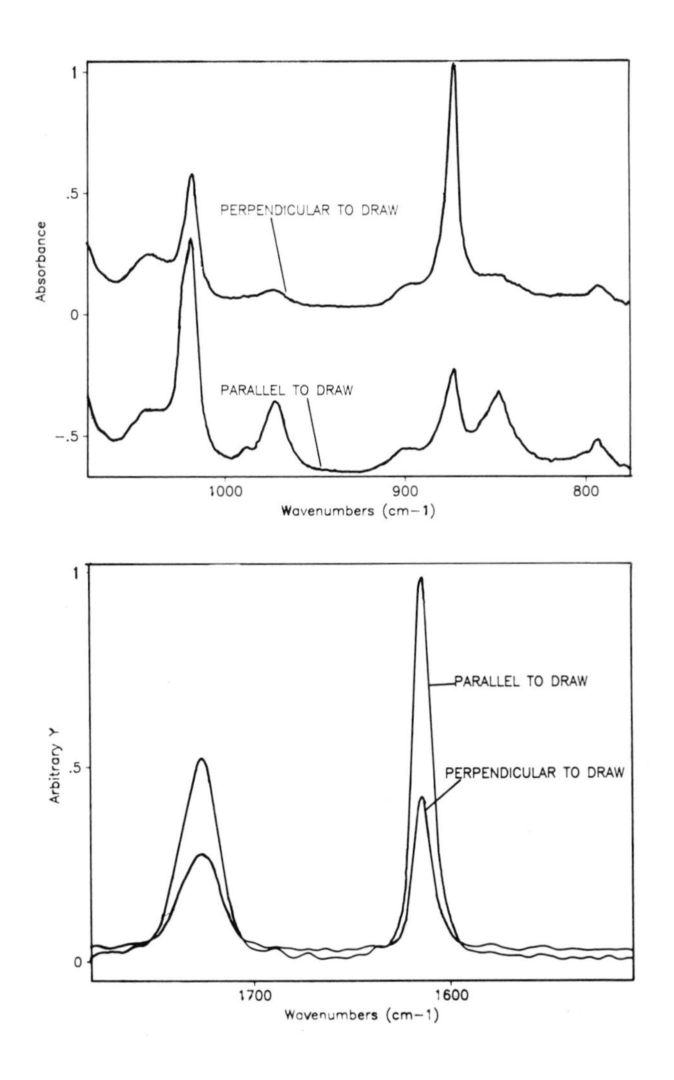

Figure 4.14 Comparison of dichroic IR and Raman spectra of drawn PET. Upper: IR spectra measured with E parallel and perpendicular to the draw axis. Lower: Raman spectra with laser polarised parallel and perpendicular to the draw axis. Note how the 1615 cm^{-1} and 1018 cm^{-1} bands have parallel character, while the 875 cm^{-1} band has perpendicular character with respect to the draw direction (and hence the polymer chain axis).

However, samples must be of good optical quality to avoid polarisation scrambling, and both the laser and the Raman scatter must be polarised parallel to sample optical axes if complex corrections are to be avoided [264, 266]. Practical details have been given for measuring accurate polarised Raman intensities [272].

4.11.3 Applications

The aim of most investigations is to characterise orientation as a function of process conditions. Raw IR or Raman dichroic ratios can be used for qualitative purposes, for example in Raman work on injection-moulded polycarbonate discs [273] or IR measurements on uniaxial PET [274]. More often, ODFs are calculated. Typical studies include IR [275] and Raman [276] measurements of crystalline and amorphous orientation in polyethylene, utilising the principle that different conformers or phases have distinct bands, allowing the orientation of each phase to be distinguished. Orientation of individual phases has been measured for many polymers including PTFE [277], polypropylene [278, 279], PVC [280], polyisoprene [281], polybutadiene [282] and PET [283]. It is often found that the amorphous phase orients less than the crystalline phase and that changes in conformer or phase distribution occur during drawing. The influence of plasticisers has been studied for PVC and poly(vinylidene chloride) (PVDC) [280, 284], and alignment of the plasticiser molecule itself has been observed [280]. ATR measurements on nylon-6,6 fibres [285] and PET [270, 271], PTFE [286] and polypropylene [270, 279] films have shown that bulk and surface orientations can differ. Measurement of orientation relaxation allows models for polymer dynamics to be tested [287, 288], while second and fourth moments of the ODF have been measured for PET and used to test deformation models [289]. Individual components in blends [290, 291] and copolyurethanes [292] have been shown to orient differently as a function of draw.

Orientation measurement is clearly a powerful tool for investigating changes in microstructure during deformation. A simple example is shown in Figure 4.15, where for uniaxial PET film f is plotted as a function of draw ratio. Results obtained from magic-angle solids NMR [293], polarised fluorescence [294], birefringence [260, 261] and IR spectroscopy are shown to be in good agreement. It is interesting to note that a small level of orientation was detected by IR micro-spectroscopy for the non-drawn film; this has been attributed to microtome-induced orientation [295].

4.11.4 Biaxial orientation

Characterisation of biaxial orientation is more complicated, both experimentally and theoretically, and is only briefly considered here. Figure 4.13(b) illustrates the problem, using as an example an oriented PET chain. Random alignment about any axis cannot be assumed and so the distribution of the polar and azimuthal angles θ and ϕ must be specified, as well as the rotation of the chain about its own axis [266]. Alternatively, the averages of the square direction cosines $\cos^2 x_3 X_3$, $\cos^2 x_3 X_2$ and $\cos^2 x_3 X_1$ together define the orientation of the chain axis x_3 in space. In general, large values of $\langle \cos^2 x_i X_j \rangle$ imply high orientation of molecule axis i along sample axis j. Equations have been given that relate the moments of the ODF and the $\langle \cos^2 x_i X_j \rangle$ to IR and Raman dichroic ratios [263, 264, 266], and have been used in an in-depth study of

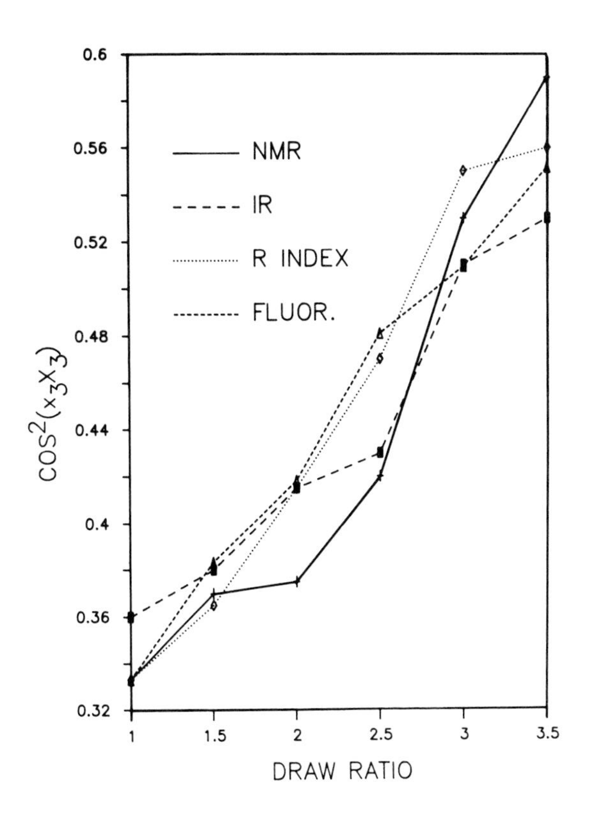

Figure 4.15 Comparison of orientation in uniaxially drawn PET as a function of draw ratio, as measured by four independent techniques (see text for details). The residual orientation detected in the undrawn sample using IR dichroism is believed to arise as a result of microtome-induced orientation.

biaxial orientation in PET films [266]. The present authors have used these methods, in conjunction with IR microscopy, to map orientation in blow-moulded PET bottles and biaxially drawn films in an attempt to understand process–property relationships for real products [48].

4.12 Rheo-optical studies

Studies of deformation-induced changes in polymers by vibrational spectroscopic techniques can lead to a better understanding of structure–property relationships. They may provide insights into the microstructural mechanisms of orientation, relaxation and conformational rearrangement.

A rheo-optical measurement is considered to be a combination of a mechanical test with a simultaneous optical measurement. In rheo-optical FTIR

(ROFTIR), the relations between stress, strain and spectral variations, sometimes under elevated temperature conditions, are established. The interested reader is recommended to refs 296–298, which include descriptions of electromechanical apparatus for simultaneous measurement of FTIR spectra and stress–strain diagrams, and many examples of its application. During the mechanical treatment, spectra in small intervals are acquired with unpolarised radiation or radiation polarised alternatively parallel and perpendicular to the stretching direction (see an example in Figure 4.16). Dynamic IR measurements during stressing typically require much less than 1 min per spectrum [296–308], while, for cyclic events, using special time-resolved techniques, spectroscopic capture of the deformation processes can be of the order of 100 ms or less [309–312].

The potential of ROFTIR has been exploited in a wide variety of polymer systems. These include: the orientation processes during elongation and relaxation of polyethylene [299–301, 312]; uniaxial deformation phenomena in polypropylene [302, 303, 309–311], PET [304, 305] and polystyrene [306]; hard and soft segmental orientations during stretching of polyurethanes [307, 308]; and the stress-induced reversible α–β phase transition in crystalline PBT [312–314] and uniaxial deformation of amorphous PBT [315].

4.13 Optical properties and constants

Knowledge of the optical properties of polymers is of interest in many applications. Measurement of the optical properties (transmittance, reflectance and absorbance) with respect to thermal radiation and other sources may be important for determining their behaviour in end-uses such as glazing, horticulture, textiles, energy-conversion systems and radiative cooling, or in predicting their response to processing variables (for example, heaters). Relatively simple practical methods of deriving these properties from transmission and reflection spectra have been developed. For example, the solar radiative properties of a range of fibres have been determined from measurements made in the near-IR region using an integrating sphere [316]. The use of conventional IR transmission and reflection spectroscopy has enabled quantitative estimates to be made of the optical properties with respect to thermal radiation for a number of polymer films [317]. Since these materials had minimal scatter, a compromise but readily applicable option of using a microspecular reflectance accessory was taken.

An in-sample compartment reflection accessory was also used in determining the optical constants (n, real refractive index, and k, absorption coefficient) for polymer films: PET [318], polycarbonate [319], and PAN [320]. Many polymers are used as lens or window materials, because of their transparency, in the far-IR (10–400 cm^{-1}) as well as in the visible. The optical constants of polymers such as TPX (poly-4-methylpent-1-ene) [321] and PET [322] have

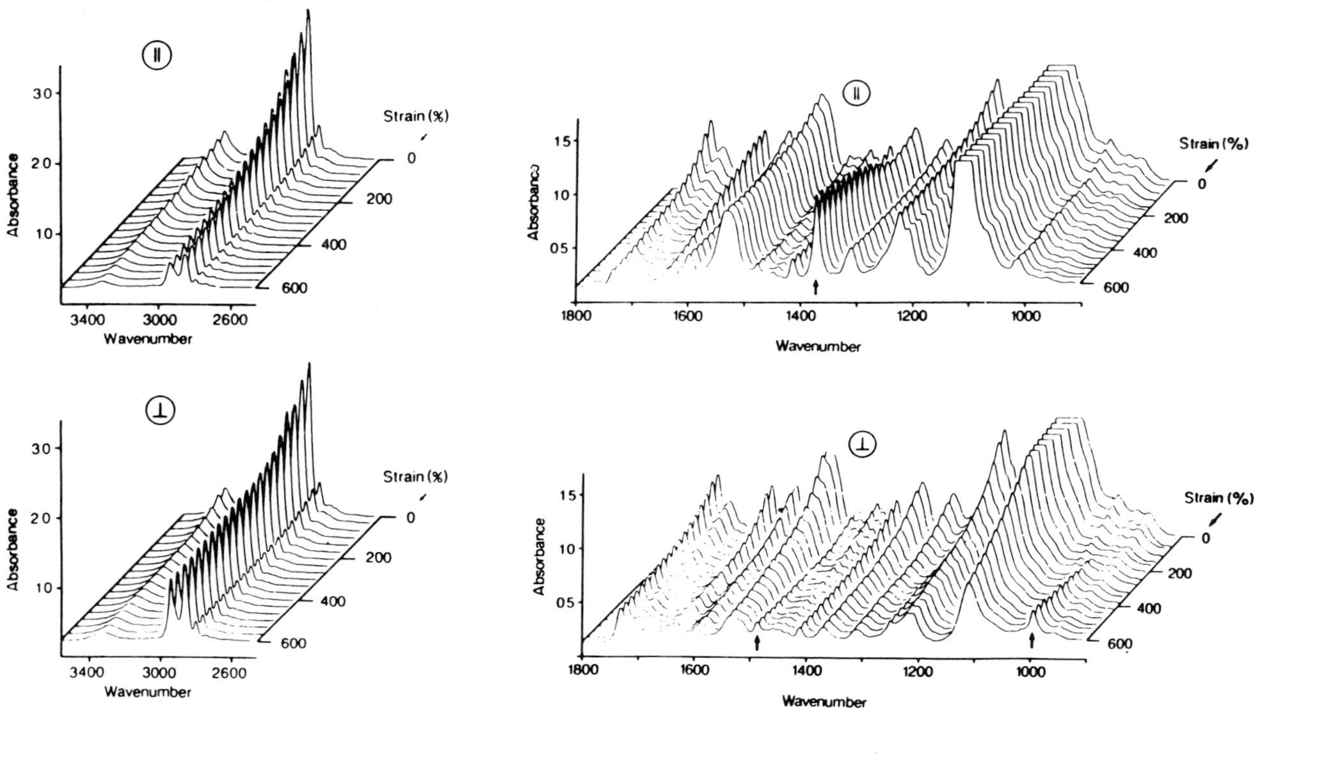

Figure 4.16 Polarisation FTIR absorbance spectra of a polyether–polyurethane recorded during elongation. Alternate polarised spectra were taken as 4 cm^{-1}, 12-scan spectra in 9-s intervals. Reproduced from ref. 298, by permission of John Wiley & Sons Ltd, Chichester.

been made by the specialised technique of dispersive Fourier transform spectroscopy [323]. Other 'low loss' material evaluations for millimetre and submillimetre wavelength uses have included IR studies on polypropylene [324] and polythylene, TPX and polypropylene [325].

4.14 Conclusion

Vibrational spectroscopy techniques will continue to be key tools in polymer characterisation and structure elucidation. The general applicability of FTIR techniques to all morphological and processed forms has been enhanced by the arrival of the simply operated FT-Raman systems, which supplement a wide range of established and complementary uses of Raman spectroscopy.

Unfortunately, space constraints have prevented detailed discussions of an array of techniques that further enhance the utility of vibrational spectroscopy in polymer studies. These include surface enhanced Raman spectroscopy (SERS) and waveguide Raman spectroscopy for the study of surfaces, and for quality-control and process studies. The latter technique, particularly in conjunction with multivariate analysis software routines, is rapidly becoming important in monitoring and controlling modern production facilities. However, discussion of these and other areas can be found in many of the references cited in this chapter.

Acknowledgements

We are indebted to ICI plc for granting permission to publish this chapter. We also gratefully acknowledge the help and inspiration given to us by many past and present ICI colleagues and collaborators, which enabled us to prepare this chapter. In particular, we wish to extend our special thanks to Jeff Eaves, Wendy Gaskin, Janet Lumsdon, Gary Manton and Ian Priestnall. Nigel Clayden and Simon Allen are thanked for providing NMR and refractive-index data for the orientation analysis of uniaxial PET.

References

1. G. Herzberg (1945). *Infrared and Raman Spectra of Polyatomic Molecules*, Van Nostrand, New York.
2. E. B. Wilson, J. C. Decius and P. C. Cross (1955). *Molecular Vibrations*, McGraw-Hill, New York.
3. N. B. Colthup, L. H. Daly and S. E. Wiberly (1990). *Introduction to Infrared and Raman Spectroscopy*, Academic Press, Boston.
4. L. Woodward (1976). *Introduction to the Theory of Molecular Vibrations and Vibrational Spectroscopy*, Oxford University Press, Oxford.
5. F. A. Cotton (1971). *Chemical Applications of Group Theory*, 2nd edn, Wiley-Interscience, New York.
6. L. J. Bellamy (1958). *The IR and Raman Spectra of Complex Molecules*, Methuen, London.
7. D. Lin-Vien, N. B. Colthup, W. G. Fateley and J. G. Grasselli (1991). *The Handbook of Infrared and Raman Characteristic Frequencies of Organic Molecules*, Academic Press, Boston.

8. L. Piseri, B. M. Powell and G. Dolling (1973). *J. Chem. Phys.*, **58**(1), 158–71.
9. J. G. Grasselli and B. J. Bulkin (eds) (1991). *Analytical Raman Spectroscopy*, John Wiley and Sons, New York.
10. H. A. Willis, J. H. van der Maas and R. G. J. Miller (eds) (1987). *Laboratory Methods in Vibrational Spectroscopy*, 3rd edn, John Wiley and Sons Ltd, Chichester.
11. M. C. Painter, M. M. Coleman and J. Koenig (1982). *The Theory of Vibrational Spectroscopy and its Applications to Polymeric Materials*, Wiley-Interscience, New York.
12. H. W. Siesler and K. Holland-Moritz (1980). *Infrared and Raman Spectroscopy of Polymers*, Practical Spectroscopy Series 4, Marcel Dekker Inc., New York.
13. J. L. Koenig (1971). *Appl. Spectrosc. Rev.*, **4**(2), 233–306.
14. P. J. Hendra, W. F. Maddams, I. A. M. Royaud, H. A. Willis and V. Zichy (1990). *Spectrochim. Acta*, **46A**(5), 747–56.
15. J. Stokr, B. Schneider, D. Dostocilova, J. Lovy and P. Sedalacek (1982). *Polymer*, **23**, 714–21.
16. P. R. Griffiths and J. A. de Haseth (1986). *Fourier Transform Infrared Spectrometry*, Wiley-Interscience, New York.
17. P. R. Griffiths (1975). *Chemical Infrared Fourier Transform Spectroscopy*, Wiley-Interscience, New York.
18. R. J. Bell (1972). *Introductory Fourier Transform Spectroscopy*, Academic Press, New York.
19. P. R. Griffiths and M. P. Fuller (1982). In *Advances in Infrared and Raman Spectroscopy*, Vol. 9, R. J. H. Clark and R. E. Hester (eds), Heyden, London, pp. 63–129.
20. D. J. Gardiner and P. R. Graves (eds) (1989). *Practical Raman Spectroscopy*, Springer-Verlag, Berlin and Heidelberg.
21. D. B. Chase (1991). In Ref. 9, Chapter 2, pp. 21–43.
22. P. J. Hendra, C. Jones and G. Warnes (1991). *Fourier Transform Raman Spectroscopy: Instrumentation and Chemical Applications*, Ellis Horwood, Chichester.
23. P. J. Hendra (ed.) (1990). Applications of Fourier transform Raman spectroscopy, *Spectrochim. Acta*, **46A**(2), special issue.
24. P. J. Hendra (ed.) (1991). Applications of Fourier transform Raman spectroscopy II, *Spectrochim. Acta*, **47A**(9/10), special issue.
25. R. G. J. Miller and B. C. Stace (eds) (1979). *Laboratory Methods in Infrared Spectroscopy*, 2nd edn, Heyden & Son Ltd., London.
26. N. J. Harrick (1967). *Internal Reflection Spectroscopy*, John Wiley & Sons Inc., New York.
27. F. M. Mirabella and N. J. Harrick (1985). *Internal Reflection Spectroscopy: Review and Supplement*, Harrick Scientific Corp., New York.
28. T. J. Allen (1992). *Vibrational Spectroscopy*, **3**, 217–37.
29. F. J. DeBlase and S. J. Compton (1991). *Appl. Spectrosc.*, **45**(4), 611–18.
30. H. X. Nguyen and H. Ishida (1986). *Makromol. Chem., Macromol. Symp.* **5**, 135–49.
31. J. M. Chalmers (1990). *Spectroscopy World*, **2**(5), 18–26.
32. J. M. Chalmers and M. W. Mackenzie (1988). Chapter 4 in *Advances in Applied Fourier Transform Infrared Spectroscopy*, M. W. Mackenzie (ed.), John Wiley & Sons Ltd, Chichester, pp. 105–88.
33. K. C. Cole, D. Noel, J.-J. Hechler and D. Wilson (1990). *J. Appl. Polym. Sci.*, **39**, 1887–902.
34. P. R. Young and A. C. Chang (1986). *SAMPE J.*, March–April, 70–74.
35. P. R. Young and A. C. Chang (1984). *Natl. SAMPE Tech. Conf.*, **16**, 136–47.
36. J. M. Chalmers and J. Wilson (1988). *Mikrochim. Acta II*, 109–11 (*Recent Aspects of Fourier Transform Spectroscopy*, Vol. 3, Proc. 6th Int. Conf. Four. Trans. Spec.).
37. J. M. Chalmers and M. W. Mackenzie (1985). *Appl. Spectrosc.*, **39**(4), 634–41.
38. E. Suzuki and R. Gresham (1987). *J. Forensic Sci.*, **32**, 377–95.
39. C. Q. Yang, R. R. Bresee and W. G. Fateley (1990). *Appl. Spectrosc.*, **44**(6), 1035–39.
40. E. G. Chatzi, S. L. Tidrick and J. L. Koenig (1988). *J. Polym. Sci. B, Polym. Phys.*, **26**, 1585–93.
41. M. W. Urban, E. G. Chatzi, B. C. Perry and J. L. Koenig (1986). *Appl. Spectrosc.*, **40**(8), 1103–07.
42. S. R. Culler, M. T. McKenzie, L. J. Fina, H. Ishida and J. L. Koenig (1984). *Appl. Spectrosc.*, **38**(6), 791–95.
43. R. M. Dittmar, J. L. Chao and R. A. Palmer (1991). *Appl. Spectrosc.*, **45**(7), 1104–10.
44. R. O. Carter, M. C. Paputa Peck, M. A. Samus and P. C. Killgoar (1989). *Appl. Spectrosc.*, **43**(8), 1350–54.

45. T. T. Nguyen, (1989). *J. Appl. Polym. Sci.*, **38**, 765–78.
46. R. A. Spragg (1984). *Appl. Spectrosc.*, **38**(4), 604–05.
47. W. D. Mazzella and C. J. Lennard, (1991). *J. Forensic Sci.*, **36**, 556–64.
48. J. M. Chalmers, L. Croot, J. G. Eaves, N. Everall, W. F. Gaskin, J. Lumsdon and N. Moore (1990). *Spectrosc. Int. J.*, **8**, 1–6, 13–41.
49. R. G. Messerschmidt and M. A. Harthcock (eds) (1988). *Infrared Microspectroscopy. Theory and Applications*, Marcel Dekker Inc., New York.
50. G. D. Ogilvie and L. Addyman (1980). *L'Actualite Chimique*, Avril, 51–54.
51. J. D. Louden (1989). In Ref. 20, Chapter 6, pp. 119–51.
52. J. A. Reffner, (1989). *Microbeam Anal.*, **24**, 167–70.
53. T. Jawhari and J. M. Pastor (1992). *J. Mol. Struct.*, **266**, 205–210.
54. M. W. Tungol, E. G. Bartick and A. Montaser (1991). *J. Forensic Sci.*, **36**(4), 1027–43, and *Appl. Spectrosc.*, **44**(4), 543–49 (1990).
55. J. M. Pastor (1991). *Makromol. Chem. Macromol. Symp.*, **52**, 57–73.
56. N. Everall (1991). *Spectroscopy World*, **3**(4), 28–29.
57. S. M. Stevens (1991). *Analytical Applications of Spectroscopy II*, A. M. C. Davies and C. S. Creaser (eds), The Royal Society of Chemistry, Cambridge, pp. 79–84.
58. M. Claybourn, P. Colombel and J. M. Chalmers (1991). *Appl. Spectrosc.*, **45**(2), 279–86.
59. J. W. Washall and T. P. Wampler (1991). *Spectroscopy Int.*, **3**(4), 36–40.
60. L. A. Sanchez (1987). *Appl. Spectrosc.*, **41**(6), 1019–23.
61. C.-Y. Kuo and T. Provder (1990). *Adv. Chem. Ser.*, **227**, 345–55, and *Polym. Mater. Sci. Eng.*, **59**, 474–79 (1988).
62. R. C. Wieboldt, G. E. Adams, S. R. Lowry and R. J. Rosenthal (1988). *Am. Lab.*, **20**, 70–76.
63. M. B. Maurin, L. W. Dittert and A. W. Hussain (1991). *Thermochimica Acta*, **186**, 97–102.
64. D. A. C. Compton, D. J. Johnson and M. L. Mittleman (1989). *Res. Dev.*, **31**, 68–73.
65. J. Khorami, A. Lemieux, H. Menard and D. Nadeau (1988). *ASTM Spec. Tech. Publ.*, STP **997**, 147–159.
66. F. M. Mirabella and M. J. Shankernarayanan (1988). *Microbeam Anal.*, **23**, 233–235.
67. D. J. Johnson, D. A. C. Compton and P. L. Canale (1992). *Thermochimica Acta*, **195**, 5–20.
68. R. S. McDonald and P. A. Wilks (1988). *Appl. Spectrosc.*, **42**(1), 151–62; see also A. N. Davies (1991). *Spectroscopy Int.*, **3**(2), 16–18.
69. G. L. McClure (1987). In Ref. 10, Chapter 7, pp. 145–201.
70. P. R. Griffiths and J. A. de Haseth (1986). In ref. 16, Chapter 10, pp. 338–68.
71. L. G. Weyer (1985). *Appl. Spectrosc. Rev.*, **21**(1&2), 1–43.
72. L. G. Weyer, (1986). *J. Appl. Polym. Sci.*, **31**, 1–14.
73. C. E. Miller, P. G. Edelman and B. D. Ratner, (1990). *Appl. Spectrosc.*, **44**(4), 576–80, C. E. Miller, P. G. Edelman, B. D. Ratner and B. E. Eichinger (1990). *Appl. Spectrosc.*, **44**(4), 581–86.
74. C. E. Miller, B. E. Eichinger, T. W. Gurley and J. G. Hermiller (1990). *Anal. Chem.*, **62**(17), 1778–85.
75. J. Haslam, H. A. Willis and D. C. M. Squirrel (1980). *Identification and Analysis of Plastics*, Heyden & Son Ltd, London.
76. V. J. I. Zichy (1979). In Ref. 25, Chapter 5, pp. 48–70.
77. T. R. Compton (1984). *The Analysis of Plastics*, Pergammon Press, Oxford.
78. R. J. Koopmans, R. Dommisse, R. van der Linden, E. F. Vansant and E. F. Alderweireld, (1983). *J. Adhesion*, **15**, 117–24.
79. J. M. Chalmers and W. F. Gaskin (1992). Unpublished work.
80. F. M. Mirabella (1982). *J. Polym. Sci. Polym. Phys. Ed.*, **20**, 2309–15.
81. S. C. Pattacini, T. J. Porro and J. Pavlik, (1991). *Am. Lab.*, June, 38–41.
82. R. J. Koopmans, R. van der Linden and E. F. Vansant (1982). *Polym. Eng. Sci.*, **22**(14), 878–82.
83. J. M. Chalmers and N. J. Everall (1992). Unpublished work.
84. H. A. Willis, J. M. Chalmers, M. W. Mackenzie and D. J. Barnes (1987). In *Computerised Quantitative Infrared Analysis*, G. L. McClure (ed.), ASTM STP 934, American Society for Testing and Materials, Philadelphia, pp. 58–77.
85. G. Ellis, A. Sanchez, P. J. Hendra, H. A. Willis, J. M. Chalmers, J. G. Eaves, W. F. Gaskin and K.-N. Kruger (1991). *J. Mol. Struct.*, **247**, 385–95.
86. H. J. Luinge, J. A. de Koeijer, J. H. van der Maas, J. M. Chalmers and P. J. Tayler, (1992). *Vibrational Spectroscopy*, in the press.

87. G. C. Pandey, (1989). *Analyst*, **114**, 231–32.
88. W. F. Maddams and I. A. M. Royaud (1991). *Spectrochim. Acta*, **47A**(9/10), 1327–33.
89. G. Bucci and T. Simonazzi (1964). *J. Polym. Sci. C*, **7**, 203–12.
90. H. V. Drushel, J. S. Ellerbe, R. C. Cox and L. H. Lane (1968). *Anal. Chem.*, **40**(2), 370–79.
91. J. M. Chalmers (1976). *Proc. 3rd Int. IFAC Conf. Instrum. Autom. Pap. Rubber Plast. Ind.*, pp. 51–60.
92. I. J. Gardner, C. Cozewith and G. Ver Strate (1971). *Rubber Chem. & Tech.*, **44**, 1015–24.
93. R. M. Paroli, J. Lara, J.-J. Hechler, K. C. Cole and I. S. Butler (1987). *Appl. Spectrosc.*, **41**(2), 319–20.
94. P. C. Ng, P.-L. Yeh, M. Gilbert and A. W. Birley (1984). *Polym. Commun.*, **25**, 250–51.
95. P. Simak (1988). *Kunststoffe*, **78**(3), 234–36.
96. J. M. Chalmers, A. Bunn, H. A. Willis, C. Thorne and R. Spragg (1988). *Mikrochim. Acta I*, 287–90 (*Recent Aspects of Fourier Transform Spectroscopy*, Vol. 2, Proc. 6th Int. Conf. Four. Trans. Spec.).
97. H. A. Willis, J. M. Chalmers, A. Bunn, C. Thorne and R. Spragg, (1988). In *Analytical Applications of Spectroscopy*, C. S. Creaser and A. M. C. Davies (eds), Royal Society of Chemistry, London, pp. 188–200.
98. J. M. Chalmers (1992). Unpublished work.
99. M. Meeks and J. L. Koenig, (1971). *J. Polym. Sci. A-2*, **9**, 717–29.
100. A. E. Tonelli and T. N. Bowmer (1987). *Polym. Prep.*, 28(1), 23–26.
101. M. Sargent, J. L. Koenig and N. L. Maecker (1991). *Appl. Spectrosc.*, **45**(10), 1726–32.
102. B. Chu, G. Fytas and G. Zalczer (1981). *Macromolecules*, **14**, 395–97.
103. E. Gulari, K. McKeigue and K. Y. S. Ng (1984). *Macromolecules*, **17**, 1822–25.
104. B. Chu and D.-C. Lee (1984). *Macromolecules*, **17**, 926–37.
105. L. Feng and K. Y. S. Ng, (1990). *Macromolecules*, **23**, 1048–53.
106. D.-C. Lee, J. R. Ford, G. Fytas, B. Chu and G. L. Hagnauer (1986). *Macromolecules*, **19**, 1586–92.
107. B. Chu and D.-C. Lee (1986). *Macromolecules*, **19**, 1592–603.
108. G. M. Carlson, C. M. Neag, C. Kuo and T. Provder (1984). *Adv. Urethane Sci. Technol.*, **9**, 47–64.
109. E. Margalit, H. Dodiuk, E. M. Kosower and A. Katzir, (1990). *Surface and Interface Analysis*, **15**, 473–78.
110. Z. H. Xu, Z. X. Huang and H. J. Xu (1989). *Proc. 7th Int. Conf. Four. Trans. Spect.*, SPIE **1145**, 346–47.
111. R. D. Priester, J. V. McClusky, R. E. O'Neill, R. B. Turner, M. A. Harthcock and B. L. Davis (1990). *J. Cellular Plast.*, **26**(4), 346–67.
112. B. L. Davis, M. A. Harthcock, C. P. Christenson and R. B. Turner (1989). *Proc. 7th Int. Conf. Four. Trans. Spect.*, SPIE **1145**, 542–43.
113. K. C. Cole, P. van Gheluwe, C. L. Dueck, P. Martineau and J. Leroux (1988). *Proc. SPI 31st Ann. Tech./Mark. Conf. (Polyurethanes 88)*, Philadelphia, Oct. 18–21, pp. 2–10.
114. K. Cole, P. van Gheluwe, M. J. Hebrard and J. Leroux (1987). *J. Appl. Polym. Sci.*, **34**, 395–407.
115. G.-C. Wang, J.-X. Wang and Z.-P. Zhang (1991). *Polymer Int.*, **25**, 237–43.
116. S. R. Jin and G. C. Meyer (1986). *Polymer*, **27**, 592–96.
117. C. G. Hill, A. M. Hedren, G. E. Meyers and J. A. Koutsky (1984). *J. Appl. Polym. Sci.*, **29**, 2749–62.
118. S. S. Jada (1988). *J. Appl. Polym. Sci.*, **35**, 1573–92.
119. S. A. Sojka, R. A. Wolfe and G. D. Guenther (1981). *Macromolecules*, **14**, 1539–43.
120. R. W. Snyder, B. Thomson, B. Bartges, D. Czerniawski and P. C. Painter (1988). *Macromolecules*, **22**, 4166–72.
121. C. Di Giulio, M. Gautier and B. Jasse (1984). *J. Appl. Polym. Sci.*, **29**, 1771–79.
122. P. R. Young and A. C. Chang (1986). SAMPE J., March/April, 70–74.
123. D. Garcia and T. T. Serafini (1987). *J. Polym. Sci. B, Polym. Phys.*, 25, 2275–82.
124. R. W. Snyder and P. C. Painter (1988). *Polym. Mater. Sci.*, **59**, 57–60.
125. S. F. Parker, J. A. Lander, D. L. Gerrard, H. J. Bowley and J. N. Hay (1989). *High Performance Polymers* **1**(4), 311–21.
126. S. F. Parker, S. M. Mason and K. P. J. Williams (1990). *Spectrochim. Acta*, **46A**(2), 315–21.

127. D. A. C. Compton, S. L. Hill, N. A. Wright, M. A. Druy, J. Piche, W. A. Stevenson and D. W. Vidrine (1988). *Appl. Spectrosc.*, **42**(6), 972–79.
128. E. Mertzel and J. L. Koenig (1986). *Adv. Polym. Sci.*, **75**, 73–112.
129. R. O. Allen and P. Sanderson (1988). *Appl. Spectrosc. Reviews*, **24**(3&4), 175–87.
130. R. E. Smith, F. N. Larsen and C. L. Long (1984). *J. Appl. Polym. Sci.*, **29**, 3713–26.
131. M. K. Antoon and J. L. Koenig (1981). *J. Polym. Sci., Polym. Chem.*, **19**, 549–70.
132. G. A. George, P. Cole-Clarke, N. St. John and G. Friend (1991). *J. Appl. Polym. Sci.*, **42**, 643–57.
133. Y. S. Yang, L. J. Lee, S. K. T. Lo and P. J. Menardi (1989). *J. Appl. Polym. Sci.*, **37**, 2313–30.
134. M. B. Moran and G. C. Martin (1983). *Polym. Prep.*, **24**(2), 141–42.
135. J. R. Walton and K. P. J. Williams (1991). *Vib. Spectrosc.*, **1**, 339–45.
136. A. H. Willbourn (1959). *J. Polym. Sci.*, **34**, 569–97.
137. ASTM (1991). Standard test methods for absorbance of polyethylene due to methyl groups at 1378 cm^{-1}. Designation D2238–68. *ASTM 1991 Annual Book of Standards*, Vol 08.02, Plastics II, pp. 207–13, The American Society for Testing and Materials, Pheladelphia.
138. T. Usami and S. Takayama (1984). *Polymer J.*, **16**(10), 731–38.
139. J. M. Chalmers, M. W. Mackenzie and H. A. Willis (1984). *Appl. Spectrosc.*, **38**(6), 763–73.
140. D. L. Gerrard, W. F. Maddams and K. P. J. Williams (1984). *Polym. Commun.*, **25**, 182–84.
141. A. Solti, D. O. Hummel and P. Simak, (1986). *Makromol. Chem., Macromol. Symp.*, **5**, 105–33.
142. D. Patterson and I. M. Ward (1957). *Trans. Faraday Soc.*, **53**(3), 291–94.
143. I. M. Ward (1957). *Trans. Faraday Soc.*, **53**(11), 1406–12.
144. R. L. Addleman and V. J. I. Zichy (1972). *Polymer*, **13**, 391–98.
145. P. G. Kosky, R. S. McDonald and E. A. Guggenheim (1985). *Polym. Eng. Sci.* **25**(7), 389–94.
146. R. Legras, C. Bailly, M. Daumerie, J. M. Dekoninck, J. P. Mercier, V. J. I. Zichy and E. Nield (1984). *Polymer*, **25**, 835–44.
147. J. P. Luongo (1960). *J. Polym. Sci.*, **42**, 139–50.
148. D. L. Wood and J. P. Luongo (1961). *Modern Plastics*, March, 132–144, 198–209.
149. R. Arnaud, J.-Y. Moisan and J. Lemaire (1984). *Macromolecules*, **17**, 332–36.
150. Z. S. Fodor, M. Iring, F. Tudos and T. Kelen (1984). *J. Polym. Sci., Polym. Chem. Ed.*, **22**, 2539–50.
151. T. J. Henman (1985). Chapter 3 in *Developments in Polymer Degradation-6*, N. Grassie (ed.), Elsevier Applied Science, Barking, pp. 105–145.
152. W.-D. Domke and H. Steinke (1986). *J. Polym. Sci. A, Polym. Chem. Ed.*, **24**, 2701–05.
153. M. Iring, S. Laszlo-Hedvig, K. Barabas, T. Kelen and F. Tudos (1978). *Eur. Polym. J.*, **14**, 439–42.
154. J. Scheirs, S. W. Bigger and O. Delatycki (1991). *Polymer*, **32**(11), 2014–19.
155. D. L. Tabb, J. J. Sevcik and J. L. Koenig (1975). *J. Polym. Sci., Polym. Phys. Ed.*, **13**, 815–24.
156. A. Baruya, D. L. Gerrard and W. F. Maddams (1983). *Macromolecules*, **16**, 578–80.
157. D. L. Gerrard and W. F. Maddams (1975). *Macromolecules*, **8**(1), 54–58, G. Martinez, C. Mijangos, J.-L. Millan, D. L. Gerrard and W. F. Maddams (1979). *Makromol. Chem.*, **180**, 2937–45, and H. J. Bowley, D. L. Gerrard and I. S. Biggin (1988). *Polym. Degrad. Stab.*, **20**, 257–69.
158. K. P. J. Williams and D. L. Gerrard (1990). *Eur. Polym. J.*, **26**(12), 1355–58.
159. Z. Vymazal, E. Czako, K. Volka and J. Stepak, Chapter 3 in *Developments in Polymer Degradation-4*, N. Grassie (ed.), Elsevier Applied Science, Barking, pp. 71–100.
160. Z. Vymazal, K. Volka, M. W. Sabaa and Z. Vymazalova (1983). *Eur. Polym. J.*, **19**, 63–69.
161. M. W. Mackenzie, H. A. Willis, R. C. Owen and A. Michel, (1983). *Eur. Polym. J.*, **19**(6), 511–17.
162. C. H. Do, E. M. Pearce, B. J. Bulkin and H. K. Reimschuessel (1987). *J. Polym. Sci.* A, *Polym. Chem.*, **25**, 2301–21.
163. C. H. Do, E. M. Pearce, B. J. Bulkin and H. K. Reimschuessel (1987). *J. Polym. Sci.* A, *Polym. Chem.*, **25**, 2409–24.
164. M. M. Coleman and G. T. Sivy (1981). *Carbon*, **19**, 123–26.
165. G. T. Sivy and M. M. Coleman (1981). *Carbon*, **19**, 127–31.
166. M. M. Coleman and G. T. Sivy (1981). *Carbon*, **19**, 133–35.
167. G. T. Sivy and M. M. Coleman (1981). *Carbon*, **19**, 137–39.
168. F. Tunistra and J. L. Koenig (1970). *J. Chem. Phys.*, **53**, 1126–30.

169. G. Katagiri, H. Ishida and A. Ishitani (1988). *Carbon*, **26**, 565–71.
170. (a) K. C. Cole and I. G. Casella (1991). *Polym. Preprints* **32**(3), 685–86, and (b) K. C. Cole and I. G. Casella (1992). *Polym. Preprints*, **33**(1), 455–56.
171. D. R. Coulter, M. V. Smith, F.-D. Tsay, A. Gupta and R. E. Fornes (1985). *J. Appl. Polym. Sci.*, **30**, 1753–65.
172. K. R. Carduner, M. C. Paputa Peck, R. O. Carter and P. C. Killgoar (1989). *Polym. Degrad. Stab.*, **26**, 1–10.
173. A. Rivaton, D. Sallet and J. Lemaire (1986). *Polym. Degrad. Stab.*, **14**(1), 1–22.
174. J. D. Webb, P. Schissel, A. W. Czandema, D. Smith and A. R. Chughtai (1982). *Polym. Preprints*, **23**(1), 213–14.
175. A. Gupta, R. Liang, F. D. Tsay and J. Moacanin (1980). *Macromolecules*, **13**, 1696–700.
176. H. R. Dickinson, C. E. Rogers and R. Simha (1982). *Polym. Preprints*, **23**(1), 217–18.
177. A. Torikai, K. Suzuki and K. Fueki (1983). *Polym. Photochem.*, **3**, 379–90.
178. H. Vanni and J. F. Rabolt (1980). *J. Polym. Sci., Polym. Phys. Ed.*, **18**, 587–96.
179. D. R. Bauer (1982). *J. Appl. Polym. Sci.*, **27**, 3651–62.
180. S. C. Lin, B. J. Bulkin and E. M. Pearce (1979). *J. Appl. Polym. Sci., Polym. Chem. Ed.*, **17**, 3121–48.
181. M. M. Coleman and P. C. Painter (1984). *Appl. Spectrosc. Rev.*, **20**(3&4), 255–346.
182. P. C. Painter, Y. Park and M. M. Coleman (1988). *Macromolecules*, **21**, 66–72.
183. Y. Xu, J. Graf, P. C. Painter and M. M. Coleman (1991). *Polymer*, **32**(17), 3103–18.
184. M. Song and F. Long (1991). *Eur. Polym. J.*, **27**(9), 983–86.
185. R. E. Belke and I. Cabasso (1988). *Polymer*, **29**, 1831–42.
186. M. Sargent and J. L. Koenig (1991). *Vib. Spectrosc.*, **2**, 21–28.
187. P. Musto, L. Wu, F. E. Karasz and W. J. Macknight (1991). *Polymer*, **32**(1), 3–11.
188. C. Leonard, J. L. Halary and L. Monnerie (1985). *Polymer*, **26**, 1507–13.
189. S. Sack and G. Haudel (1990). *Angew. Makromol. Chemie*, **180**, 131–43.
190. M. M. Coleman, D. J. Skrovanek, J. Hu and P. C. Painter (1988). *Macromolecules*, **21**, 59–65.
191. M. A. Harthcock (1989). *Polymer*, **30**, 1234–42.
192. J. J. Schmidt, J. A. Gardella and L. Salvati (1989). *Macromolecules*, **22**, 4489–95.
193. M. B. Clark, C. A. Burkhardt and J. A. Gardella (1989). *Macromolecules*, **22**, 4495–501.
194. M. B. Clark, C. A. Burkhardt and J. A. Gardella (1991). *Macromolecules*, **24**, 799–805.
195. J. H. Kim, J. W. Barlow and D. R. Paul (1989). *J. Polym. Sci. B, Polym. Phys.*, **27**, 2211–27.
196. J.-J. Jutier, E. Lemieux and R. E. Prud'homme (1988). *J. Polym. Sci. B, Polym. Phys.*, **26**, 1313–29.
197. M. M. Coleman, E. J. Moskala, P. C. Painter, D. J. Walsh and S. Rostami (1983). *Polymer*, **24**, 1410–14.
198. R. Tannenbaum, M. Rutkowska and A. Eisenberg (1987). *J. Polym. Sci. B, Polym. Phys.*, **25**, 663–71.
199. A. Garton, S. Wang and R. A. Weiss (1988). *J. Polym. Sci. B, Polym. Phys.*, **26**, 1545–48.
200. D. E. Bhagwagar, P. C. Painter, M. M. Coleman and T. D. Krizan (1991). *J. Polym. Sci. B, Polym. Phys.*, **29**, 1547–58.
201. E. Benedetti, F. Galleschi, A. D'Alessio, G. Ruggeri, M. Aglietto, M. Pracella and F. Ciardelli (1989). *Makromol. Chem., Macromol. Symp.*, **23**, 265–75.
202. K. Yamada and Y. Funayama (1990). *Rubber Chem. Technol.*, **63**, 669–82.
203. G. Zerbi, F. Ciampelli and V. Zamboni (1964). *J. Polym. Sci. C*, 141–51.
204. S. W. Cornell and J. L. Koenig (1969). *Macromolecules*, **2**, 540–45.
205. S. W. Cornell and J. L. Koenig (1969). *Macromolecules*, **2**, 546–49.
206. Z. Chen, M. Liu, M. Shi and Z. Shen (1987). *Makromol. Chem.*, **188**, 2687–95.
207. G. V. Fraser, P. J. Hendra, D. S. Watson, M. J. Gall, H. A. Willis and M. E. A. Cudby (1973). *Spectrochimica Acta*, **29A**, 1525–33.
208. G. Guerra, P. Musto, F. E. Karasz and W. J. Macknight (1990). *Makromol. Chem.*, **191**, 2111–19.
209. L. D'Esposito and J. L. Koenig (1976). *J. Polym. Sci., Polym. Phys.*, **14**, 1731–41.
210. S. B. Lin and J. L. Koenig (1982). *J. Polym. Sci., Polym. Phys.*, **20**, 2277–95.
211. S. B. Lin and J. L. Koenig (1983). *J. Polym. Sci., Polym. Phys.*, **21**, 2365–78.
212. T. Hirschfeld (1976). *Anal. Chem.*, **48**, 721–23.
213. S. Krimm, A. R. Berens, V. L. Folt and J. J. Shipman (1963). *J. Polym. Sci. A*, **1**, 2621–50.
214. M. Theodorou and B. Jasse (1983). *J. Polym. Sci., Polym. Phys.*, **21**, 2263–74.

215. M. E. R. Robinson, D. I. Bower and W. F. Maddams (1978). *Polymer*, **19**, 773–84.
216. J. L. Koenig and M. J. Hannon (1967). *J. Macromol. Sci.* (Phys.), **B1**(1), 119–45.
217. J. L. Koenig and M. C. Agboatwalla (1968). *J. Macromol. Sci. (Phys.)*, **B2** 391–420.
218. P. C. Painter, J. Havens, W. W. Hart and J. L. Koenig (1977). *J. Polym. Sci., Polym. Phys.*, **15**, 1223–35.
219. J. M. O'Reilly and R. A. Mosher (1981). *Macromolecules*, **14**, 602–08.
220. A. J. Melveger (1972). *J. Polym. Sci. A-2*, **10**, 317–22.
221. J. S. Kim, M. Lewin and B. J. Bulkin (1986). *J. Polym. Sci., Polym. Phys.*, **24**, 1783–89.
222. J. D. Louden (1986). *Polymer Comm.*, **27**, 82–84.
223. J. M. Chalmers, W. F. Gaskin and M. W. Mackenzie (1984). *Polym. Bull.*, **11**, 433–35.
224. H. W. Starkweather and R. E. Moynihan (1956). *J. Polym. Sci.*, **22**, 363–68.
225. R. E. Moynihan (1959). *J. Amer. Chem. Soc.*, **81**, 1045–50.
226. C. E. Miller and B. E. Eichinger (1990). *Appl. Spectrosc.*, **44**, 496–504.
227. R. Ferwerda, J. H. van der Maas, N. Everall, P. Tayler and J. M. Chalmers (1992). Paper in preparation.
228. H. W. Starkweather, R. C. Ferguson, D. B. Chase and J. M. Minor (1985). *Macromolecules*, **18**, 1684–86.
229. G. R. Strobl and W. Hagerdorn (1978). *J. Polym. Sci., Polym. Phys.*, **16**, 1181–93.
230. E. Foldes, G. Keresztury, M. Iring and F. Tudos (1991). *Angew. Makromol. Chemie*, **187**, 87–99.
231. N. J. Everall, J. Lumsdon, J. M. Chalmers and N. Mason (1991). *Spectrochim. Acta*, **47A**, 1305–11.
232. J. F. Rabolt (1983). *J. Polym. Sci., Polym. Phys.*, **21**, 1797–805.
233. M. Goldstein, D. Stephenson and W. F. Maddams (1983). *Polymer*, **24**, 823–26.
234. D. R. Beckett, J. M. Chalmers, M. W. Mackenzie, H. A. Willis, H. G. M. Edwards, J. S. Lees and D. A. Long (1985). *Eur. Polym. J.*, **21**, 849–52.
235. F. Adar and H. Noether (1985). *Polymer*, **26**, 1935–43.
236. L. A. Hanna, P. J. Hendra, W. Maddams, H. A. Willis, V. Zichy and M. E. A. Cudby (1988). *Polymer*, **29**, 1843–47.
237. J. Green and J. F. Rabolt (1987). *Macromolecules*, **20**, 456–57.
238. J. Friedrich and J. F. Rabolt (1987). *Macromolecules*, **20**, 1975–79.
239. P. C. Gillette, J. B. Lando and J. L. Koenig (1985). *Polymer*, **26**, 235–40.
240. P. J. Hendra, J. Vile, H. A. Willis, V. Zichy and M. E. A. Cudby (1984). *Polymer*, **25**, 785–90.
241. A. Boukenter, E. Duvel and H. M. Rosenberg (1988). *J. Phys. C, Solid State Phys.*, **21**, 541–47.
242. T. Achibat, A. Boukenter, E. Duval, G. Lorentz and S. Etienne (1991). *J. Chem. Phys.*, **95**, 2949–54.
243. S. Mizushima and T. Shimanouchi (1949). *J. Amer. Chem. Soc.*, **71**, 1320–24.
244. W. L. Peticolas, G. W. Hibler and J. L. Lippert (1971). *Appl. Phys. Lett.*, **18**, 87–89.
245. H. G. Olf, A. Peterlin and W. L. Peticolas (1974). *J. Polym. Sci., Polym. Phys.*, **12**, 359–84.
246. I. G. Voigt-Martin, A. J. Peacock and L. Mandelkern (1989). *J. Polym. Sci., Polym. Phys.*, **27**, 957–65.
247. L. H. Wang, S. Ottani and R. S. Porter (1991). *J. Polym. Sci., Polym. Phys.*, **29**, 1189–92.
248. L. H. Wang, R. S. Porter, H. D. Stidham and S. L. Hau (1991). *Macromolecules*, **24**, 5535–38.
249. J. Martinez-Salazar, P. J. Barham and A. Keller (1985). *J. Mat. Sci.*, **20**, 1616–24.
250. R. G. Snyder and J. R. Scherer (1980). *J. Polym. Sci., Polym. Phys.*, **18**, 421–28.
251. D. L. Gerrard and W. F. Maddams (1986). *Appl. Spectrosc. Rev.*, **22**, 251–334.
252. S. L. Hsu and S. Krimm (1976). *J. Appl. Phys.*, **47**, 4265–71.
253. S. L. Hsu, G. W. Ford and S. Krimm (1977). *J. Polym. Sci., Polym. Phys.*, **15**, 1769–78.
254. S. L. Hsu and S. Krimm (1977). *J. Appl. Phys.*, **48**(10), 4013–18.
255. A. Peterlin (1982). *J. Polym. Sci., Polym. Phys.*, **20**, 2329–56.
256. K. Song and S. Krimm (1990). *J. Polym. Sci., Polym. Phys.*, **28**, 35–50.
257. C. Chang and S. Krimm (1983). *J. Appl. Phys.*, **54**, 5526–40.
258. L. Mandelkern, R. Alamo, W. L. Mattice and R. G. Snyder (1986). *Macromolecules*, **19**, 2404–08.
259. H. Nishida, M. Oshyanagi, O. Okada and E. Tsuchida (1986). *Macromolecules*, **19**, 496–98.
260. R. J. Samuels (1974). *Structured Polymer Properties*, Wiley-Interscience, New York.
261. G. Wilkes (1971). *Adv. Polym. Sci.*, **8**, 91–136.

262. R. D. B. Fraser (1956). *J. Chem. Phys.*, **24**, 89–95.
263. I. M. Ward (1985). *Adv. Polym. Sci.*, **66**, 81–118.
264. D. I. Bower (1972). *J. Polym. Sci., Polym. Phys. Ed.*, **10**, 2135–53.
265. E. L. V. Lewis and D. I. Bower (1987). *J. Raman Spectrosc.*, **18**, 61–70.
266. D. A. Jarvis, I. J. Hutchinson, D. I. Bower and I. M. Ward (1980). *Polymer*, **21**, 41–54.
267. H. Tadokoro, S. Seki, I. Nitta and R. Yamadera (1958). *J. Polym. Sci.*, **28**, 244–47.
268. J. A. J. Jansen, J. H. van der Maas and A. Posthuma de Boer (1991). *Macromolecules*, **24**, 4278–80.
269. K. Krishnan, S. Hill, J. P. Hobbs and C. S. P. Sung (1982). *Appl. Spectrosc.*, **36**, 257–59.
270. J. P. Hobbs, C. S. P. Sung, K. Krishnan and S. Hill (1983). *Macromolecules*, **16**, 193–99.
271. P. Yuan and C. S. P. Sung (1991). *Macromolecules*, **24**, 6095–103.
272. J. R. Scherer (1991). In ref. 9, Chapter 3, pp. 47–57.
273. M. Takeshimo and N. Funakoshi (1986). *J. Appl. Polym. Sci.*, **32**, 3457–68.
274. M. Ito, J. R. C. Periera, S. L. Hsu and R. S. Porter (1983). *J. Polym. Sci., Polym. Phys. Ed.*, **21**, 389–400.
275. A. R. Wedgewood and J. C. Seferis (1983). *Pure Appl. Chem.*, **55**, 873–92.
276. M. Pigeon, R. E. Prud'homme and M. Pezolet (1991). *Macromolecules*, **24**, 5687–94.
277. C. K. Yeung and B. Jasse (1982). *J. Appl. Polym. Sci.*, **27**, 4587–97.
278. Y. V. Kissin (1983). *J. Polym. Sci., Polym. Phys. Ed.*, **21**, 2085–96.
279. F. M. Mirabella, M. Shankernarayanan and P. L. Fernando (1989). *J. Appl. Polym. Sci.*, **37**, 851–60.
280. M. Theodorou and B. Jasse (1986). *J. Polym. Sci. B*, **24**, 2643–54.
281. B. Amram, L. Bokobza, J. P. Quesiel and L. Monnerie (1986). *Polymer*, **27**, 877–82.
282. B. Amram, L. Bokobza, L. Monnerie and J. P. Quesiel (1988). *Polymer*, **29**, 1155–60.
283. F. Rietsch and B. Jasse (1984). *Polymer Bull.*, **11**, 287–92.
284. G. Hinrichsen, H. J. Jahr and H. Springer (1988). *Polym. Comm.*, **29**, 199–200.
285. S. R. Samanta, W. W. Lanier, R. W. Miller and M. E. Gibson (1990). *Appl. Spectrosc.*, **44**, 286–89.
286. G. V. Saidov, Y. V. Bernshtein and L. Y. Barbanel (1982). *Polym. Sci. USSR*, **24**, 1611–16.
287. B. Lefebvre, B. Jasse and L. Monnerie (1983). *Polymer*, **24**, 1240–44.
288. C. W. Lantman, J. F. Tassin, P. Sergot and L. Monnerie (1989). *Macromolecules*, **22**, 483–85.
289. J. Purvis, D. I. Bower and I. M. Ward (1973). *Polymer*, **14**, 398–400.
290. H. Saito and T. Inoue (1987). *J. Polym. Sci. B*, **25**, 1629–36.
291. C. Bouton, V. Arrondel, V. Rey, P. Sergot, J. L. Manguin, B. Jasse and L. Monnerie, (1989). *Polymer*, **30**, 1414–18.
292. N. Reynolds (1991). *Adv. Mater.*, **3**, 614–16.
293. N. J. Clayden, private communication; see also ref. 263 for a general treatment.
294. M. Hennecke, A. Kud, K. Kurz and J. Fuhrmann (1987). *Colloid and Polymer Sci.*, **265**, 674–80.
295. J. G. Eaves, private communication.
296. H. W. Siesler (1979). In *Proc. 5th European Symposium on Polymer Spectroscopy*, D. O. Hummel (ed.), Verlag Chemie, Weinheim, pp. 137–80.
297. H. W. Siesler (1984). *Adv. Polym. Sci.*, **65**, 1–77.
298. H. W. Siesler (1988). Chapter 5 in *Advances in Applied Fourier Transform Infrared Spectroscopy*, M. W. Mackenzie (ed.), John Wiley & Sons, Chichester, pp. 189–246.
299. K. Holland-Moritz, I. Holland-Moritz and K. van Werden (1981). *Colloid & Polym. Sci.*, **259**(2), 156–62.
300. K. Holland-Moritz and K. van Werden (1981). *Makromol. Chem.*, **182**, 651–55.
301. H. W. Siesler (1984). *Infrared Phys.*, **24**(2/3), 239–44.
302. G. Bayer, W. Hoffman and H. W. Siesler (1980). *Polymer*, **21**, 235–38.
303. Y.-L. Lee, R. S. Bretzlaff and R. P. Wool (1984). *J. Polym. Sci., Polym. Phys. Ed.*, **22**, 681–98.
304. V. B. Gupta, C. Ramesh and H. W. Siesler (1985). *J. Polym. Sci., Polym. Phys. Ed.*, **23**, 405–11.
305. S. S. Sheiko, I. S. Vainilovitch and S. N. Magonov (1991). *Polym. Bull.*, **25**, 499–506.
306. S. N. Magonov, I. S. Vainilovitch and S. S. Sheiko (1991). *Polym. Bull.*, **25**, 491–98.
307. H. W. Siesler (1983). *Polym. Bull.*, **9**(8/9), 382–89.
308. H. W. Siesler (1983). *Polym. Bull.*, **9**(10/11), 471–78.

309. J. A. Graham, W. M. Grim and W. G. Fateley (1984). *J. Mol. Struct.*, **113**, 311–12.
310. W. M. Grim, J. A. Graham, R. M. Hammaker and W. G. Fateley (1984). *Am. Lab.*, **16**, 22–32.
311. W. G. Fateley and J. L. Koenig (1982). *J. Polym. Sci., Polym. Lett. Ed.*, **20**, 445–52.
312. J. E. Lasch, D. J. Burchell, T. Masoaka and L. S. Hsu (1984). *Appl. Spectrosc.*, **38**(3), 351–58.
313. H. W. Siesler (1979). *J. Polym. Sci., Polym. Lett. Ed.*, **17**, 433–58.
314. H. W. Siesler (1979). *Makromol. Chem.*, **180**(9), 2261–63.
315. K. Holland-Moritz and H. W. Siesler (1981). *Polym. Bull.*, **4**, 165–70.
316. M. Papini (1989). *Appl. Spectrosc.*, **43**(8), 1475–81.
317. B. J. Stay, J. M. Chalmers, M. W. Mackenzie and D. R. Moseley (1985). *Appl. Spectrosc.*, **39**(3), 412–17.
318. C. A. Sergides, A. R. Chughtai and D. M. Smith (1987). *Appl. Spectrosc.*, **41**(1), 154–57.
319. J. D. Webb, G. Jorgensen, P. Schissel, A. W. Czanderna, A. R. Chughtai and D. M. Smith (1983). *Polymers for Solar Energy Utilization*, G. C. Gebelein, D. J. Williams and R. Deanin (eds), American Chemical Society, Washington, DC, pp. 143–67.
320. C. A. Sergides, A. R. Chughtai and D. M. Smith (1985). *J. Polym. Sci., Polym. Phys. Ed.*, **23**, 1573–84.
321. J. R. Birch and E. A. Nicol (1984). *Infrared Phys.*, **24**(6), 573–75.
322. J. R. Birch, G. W. Chantry, D. B. Shenton, R. E. Hills, M. E. A. Cudby and H. A. Willis (1977). *Infrared Phys.*, **17**, 425–26.
323. See Chapter 11 in ref. 16, and references therein.
324. G. W. Chantry, J. W. Fleming, G. W. F. Pardoe, W. Reddish and H. A. Willis, (1971). *Infrared Phys.*, **11**, 109–18.
325. G. W. Chantry, J. W. Fleming, P. M. Smith, M. Cudby and H. A. Willis, (1971). *Chem. Phys. Letters*, **10**(4), 473–477.

5 Molecular weight determination

K. KAMIDE, M. SAITO and Y. MIYAZAKI

5.1 Introduction

For characterisation of macromolecules the first step is to determine the degree of polymerisation (DP) or the molecular weight (M) of the polymeric compounds. The methods usually employed for this purpose are, without exception, based on the physical chemistry of polymer solutions (Table 5.1) and have been popularised quickly and widely owing to the recent remarkable development of automatic apparatus.

Table 5.1 Various methods for determination of average molecular weights

Method	Detecting physical quantity	Average MW[a]	Measurable upper limit of MW range (approx.)
1. End-group assay	Number of end groups in polymeric chains	M_n^b	2×10^4
2. Vapour-pressure osmometry	Temperature difference between solution and solvent drops in saturated solvent vapour	M_n	4×10^5
3. Membrane osmometry	Osmotic pressure of solvent	M_n	5×10^5
4. Ebulliometry	Boiling point of solution	M_n	2×10^4
5. Cryoscopy	Freezing point of solution	M_n	2×10^4
6. Light-scattering including SAXS[d] and SANS[e]	Intensity of scattered visible light, X-rays and neutrons	M_w^c	1×10^8
7. Viscometry	Time for passing through capillary or pressure difference	M_v^f	1×10^8
8. Ultracentrifugation	Concentration of solute under centrifugation gravity	ES[g]; $M_n\, M_w\, M_z^h$ ...	1×10^7

[a] Molecular weight, [b] number-average molecular weight, [c] weight-average molecular weight, [d] small-angle X-ray scattering, [e] small-angle nuetron scattering, [f] viscosity-average molecular weight, [g] equilibrium sedimentation method, [h] z-average molecular weight.

In Table 5.1, membrane osmometry, cryoscopy, ebulliometry, end-group analysis, light-scattering and ultracentrifugation belong to the category of so-called absolute methods. Vapour pressure osmometry, viscometry and GPC can be classified as relative methods, which require samples of known molecular weight for calibration. GPC is discussed in chapter 6, and will not be dealt with here.

5.2 Average molecular weights

When we deal with a pure polymer sample which is composed of molecules differing only in degree of polymerisation (DP), the various average molecular weights, such as the number-average molecular weight M_n, the weight-average molecular weight M_w, and the z-average molecular weight M_z are defined mathematically by:

$$M_n = \Sigma N_i M_i / \Sigma N_i = \Sigma n_i M_i \tag{5.1}$$

$$M_w = \Sigma N_i M_i^2 / \Sigma N_i M_i = \Sigma w_i M_i \tag{5.2}$$

$$M_z = \Sigma N_i M_i^3 / \Sigma N_i M_i^2 \tag{5.3}$$

where: N_i is the number of molecules with DP $= i$ (i is a positive integer); M_i is the molecular weight of the molecules with DP $= i$; n_i and w_i are the molar and weight fractions of the molecules with DP $= i$ respectively. M_n can be determined by end-group assay and by using the colligative properties of the polymer solutions. M_w is usually evaluated by the light-scattering method (including small-angle X-ray and neutron-scattering methods) and the sedimentation equilibrium method. M_z can be determined by the sedimentation equilibrium method and by pulse-induced critical scattering (PICS) developed first by Gordon *et al.* [1].

By generalisation of eqns (5.1)–(5.3), the kth moment of M, $M(k)$ is defined as:

$$M(k) = \Sigma w_i M_i^k \tag{5.4}$$

Then, $M_n = M(0)/M(-1)$, $M_w = M(1)/M(0)$, and $M_z = M(2)/M(1)$

Between the limiting viscosity number $[\eta]$ (see eqns (5.39) and (5.40)) of linear polymer solutions and the molecular weight the following semi-empirical relation, the Mark[2]–Houwink[3]–Sakurada[4] (MHS) equation, holds over a very wide range of molecular weights:

$$[\eta] = K_m M^a \tag{5.5}$$

where K_m and a are the characteristic parameters predominantly determined by the combination of polymer and solvent at constant temperature. Through eqn (5.5), the viscosity-average molecular weight M_v of a polydisperse polymer is obtained. This is defined by the relation:

$$M_v = \{\Sigma N_i M_i^{a+1} / \Sigma N_i M_i\}^{1/a} = \{\Sigma w_i M_i^a\}^{1/a} = \{M(a)\}^{1/a} \tag{5.6}$$

In the case of $a = 1$, M_v precisely reduces to M_w. In deriving eqn (5.6), it is assumed that eqn (5.5) is established for a series of absolutely monomolecular samples. If not, some correction is necessary to eqn (5.6).

When the samples are polydisperse, the following relation holds:

$$M_n < M_v \leqslant M_w < M_z \tag{5.7}$$

5.3 Colligative methods

5.3.1 Theoretical background

The chemical potential μ (i.e. the partial molar free energy) is the most fundamental physical quantity to describe the thermodynamic properties of polymer solutions. If μ is represented as functions of absolute temperature (T), pressure (P) and the composition (for example, the polymer concentration for a binary mixture), one can determine not only the molecular weight, M, of the solute (the polymer), but also many other thermodynamic quantities, such as the partial molar entropy, the partial molar enthalpy and the partial molar volume of each component.

The difference between the chemical potential of the solvent in the solution μ_0 and that of pure solvent μ_0^0, chosen as the standard state, is usually denoted by $\Delta\mu_0$, which is simply referred to as the chemical potential of the solvent, relating directly to the activity a_0 through the equation:

$$\Delta\mu_0(\equiv \mu_0 - \mu_0^0) = RT \ln a_0 \tag{5.8}$$

where R is the gas constant and T the Kelvin temperature. When a_0 is equal to the molar fraction of the solvent n_0, the solution is defined as an ideal solution. It can be shown that, for the two-component ideal solution, $\ln n_0$ is inversely proportional to M_n at sufficiently low solute concentration:

$$\ln n_0 \simeq -V_0 c/M_n \tag{5.9}$$

where V_0 is the molar volume of solvent in the solution and c is the solute concentration, expressed in g/cm^{-3}. In an ideal solution, no volumetric change occurs by mixing solute and solvent and V_0 is equal to the molar volume of pure solvent V_0^0.

It has been widely confirmed with numerous experiments that polymer solutions reveal remarkable deviation from the behaviour expected for an ideal solution [5]. According to the Flory–Huggins lattice theory of polymer solutions [6,7], $\Delta\mu_0$ is given by:

$$\Delta\mu_0/RT = V_0^0\left\{(c/M_n) + \left[\left(\tfrac{1}{2}\right) - \chi\right](N_A V_0^0 c^2/m^2 + \ldots)\right\} \tag{5.10}$$

where χ is the polymer–solvent interaction parameter, N_A the Avogadro number and m the molecular weight of the repeating unit of the polymer.

When the polymer solution is contacted with pure solvent through a semipermeable membrane, which allows free passage of solvent only (Figure 5.1(a)), and these two phases are in equilibrium at temperature T, the pressure difference between the two phases is defined as the osmotic pressure Π, which relates to a_0 and M_n through eqn (5.11):

$$\Pi = -\Delta\mu_0/V_0^0 = -(RT/V_0^0) \ln a_0 = RT\,\{(c/M_n) + A_{2,0}\,c^2 + A_{3,0}\,c^3 + \ldots\} \tag{5.11}$$

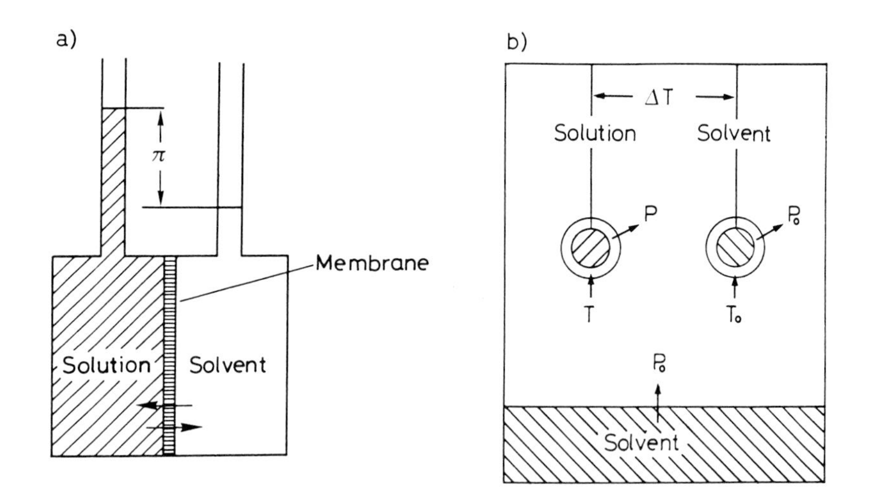

Figure 5.1 Schematic diagrams of apparatus for membrane osmometry (a) and for vapour-pressure osmometry (b). Reprinted from ref. 5 by permission of Pergamon Press plc.

where $A_{2,0}$ and $A_{3,0}$ are the osmotic second and third virial coefficients.

When the polymer solution is placed under an atmosphere of saturated vapour phase and both phases are in equilibrium, the vapour pressure of the solvent component in the solution P_0 is smaller than the vapour pressure of the pure solvent P_0^0. Then their ratio P_0/P_0^0 is equal to a_0 and eqn (5.12) holds:

$$\ln (P_0/P_0^0) = \ln a_0 = V_0^0\{(c/M_n) + A_{2,0}c^2 + A_{3,0}c^3 + \ldots\} \qquad (5.12)$$

Accurate determination of P_0 and P_0^0 for solutions with various values of c allows one to evaluate M_n by extrapolating the plot of $\ln(P_0/P_0^0)$ versus c to $c = 0$. Unfortunately, the difference between P_0 and P_0^0 is extremely small in dilute solution, and the maximum molecular weight (M_n) determinable by the vapour-pressure depression is never beyond 1000. As will become clear, this limit is overcome by an indirect measurement technique.

Suppose that at a certain time, t, a drop of pure solvent at temperature T_0 and a drop of solution at temperature T are placed in a saturated vapour phase whose temperature and vapour pressure are T_0 and $P_0(T_0)$, respectively (vapour-pressure osmometry (VPO); see Figure 5.1(b)).

The temperature difference in the steady state, $(T - T_0)_s$, can be expressed as a power series of the concentration [8,9]:

$$(T - T_0)_s = K_s\{(c/M_n) + A_{2,v}c^2 + A_{3,v}c^3 + \ldots\} \qquad (5.13)$$

where K_s is given by:

$$K_s = K_e / \{1 + RT_0^2(k_1 A_1' + k_2 A_2')/(k_3 A_1' P_0(T_0)\Delta H_2)\} \tag{5.14}$$

and

$$K_e = RT_0^2 V_0^0 / \Delta H \tag{5.15}$$

where: k_1 and k_2 are the coefficients of the surface heat transfer at the interface between atmosphere and liquid and between liquid and thermistor, respectively, and have the dimensions of J cm^{-2} s^{-1} K^{-1}; A_1' is the surface area of the solution drop; A_2' is the area of contact between the solution drop and the thermistor or thermocouple, including lead wire; k_3 (mol $g^{-1}s^{-1}$) is the mass transfer coefficient, which varies depending on the detailed mechanism of diffusion; ΔH is molar heat of condensation (J mol^{-1}); K_s is a calibration parameter (cm^3 K mol^{-1}), depending on the combination of solute and solvent and on the dimensions of the apparatus (in the hypothetical case of $k_1 = k_2 = 0$, K_s reduces to K_e); $A_{2,v}$ and $A_{3,v}$ are the second and third virial coefficients given by [9]:

$$A_{2,v} = A_{2,0} + (V_0/M_1)\left\{\left[(RT_0/\Delta H) - \left(\frac{1}{2}\right)\right](K_s/K_e)^2 + (K_s/K_e) - \left(\frac{1}{2}\right)\right\} \tag{5.16}$$

$$\begin{aligned} A_{3,v} = {} & A_{3,0} + 2A_{2,v}(V_0/M_n)\left\{\left[(RT_0/\Delta H) - \left(\frac{1}{2}\right)\right](K_s/K_e)^2 + (K_s/K_e) - \frac{1}{2}\right\} \\ & + (V_0^2/M_n^3)\left\{(RT_0/\Delta H - \tfrac{1}{2})\left[2\left(RT_0/\Delta H - \tfrac{1}{2}\right)(K_s/K_e)^2 - (K_s/K_e) - \tfrac{1}{2}\right](K_s/K_e)^2 \right. \\ & \left. - 2(K_s/K_e)^3 + (K_s/K_e)^2 - (K_s/K_e) + \left(\frac{1}{6}\right)\right\} \end{aligned} \tag{5.17}$$

Note that $A_{2,v}$ does not coincide with $A_{2,0}$ except in the case $K_s/K_e = 1/\left\{1 + (2RT_0/\Delta H)^{\frac{1}{2}}\right\}$.

The temperature difference at equilibrium, $(T - T_0)_e$, which is never realised because of unavoidable heat loss due to thermal conductivity from the solution drop to thermistor, including lead wire, and thermal radiation from the solution drop to the vapour phase, is given by:

$$(T - T_0)_e = -K_e \ln a_0 \tag{5.18}$$

Kamide *et al.* have given the theoretical background of VPO in the cases of an unsaturated vapour phase [10] and of a ternary mixture of solute, solvent 1 and solvent 2 [11].

The other methods for determining M_n by using the colligative nature of the polymer solutions are ebulliometry and cryoscopy, by both of which the maximum molecular weight obtainable is about 2×10^4. The detailed principles and practical techniques for determination of M_n by these methods are described in refs 12 and 13. Note that these two methods are available only at specific (boiling or melting) temperatures for given solutions and are limited in applications.

5.3.2 Apparatus for determination of molecular weight by colligative methods

5.3.2.1 Vapour pressure osmometry (VPO). At present commercially available apparatus for VPO is: the Corona molecular-weight apparatus models 114 and 117 manufactured by the Corona Co., which have been produced in the past by Hitachi Co. Ltd; Knauer Vapour Pressure Osmometer No. 731.110000 (Dr. H. Knauer Wissenschaftliche Geräte, AG); Jupiter Model 233 Vapour Pressure Osmometer (Jupiter Instrument Co. Inc.); and Gonotec model 070 (Gonotec Gesellschaft für MeB- und Regeltechnik m6H). A typical VPO apparatus consists of: (1) two thermistors covered with glass, to which solution and solvent drops are attached; (2) a cell, the interior of which is saturated with solvent vapour; (3) a solvent vessel, placed at the lower part of the cell; and (4) an electronic circuit. The instrument is usually thermostatted.

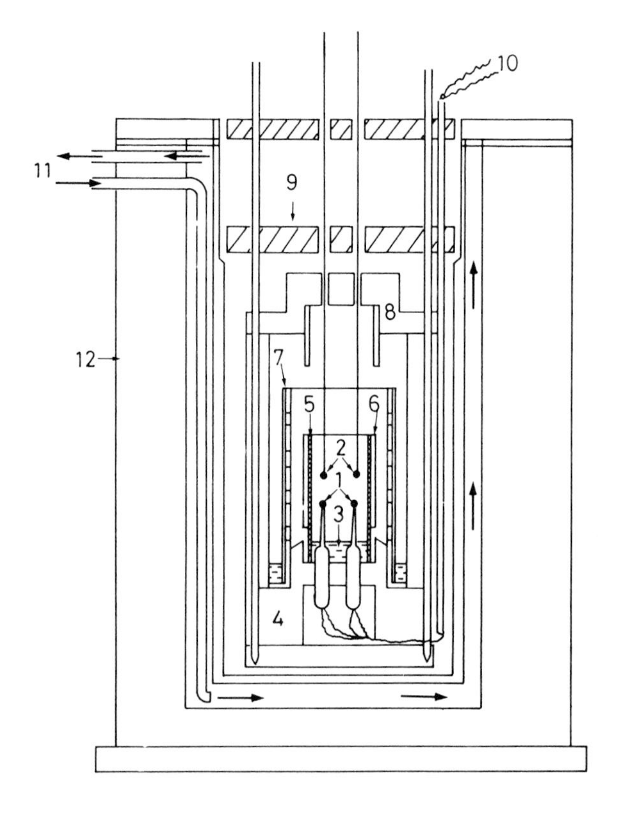

Figure 5.2 Sectional view of vapour-pressure osmometer Corona model 117: (1) thermistor; (2) pipe for dropping in solution and solvent; (3) reservoir; (4) gasket ring; (5) filter paper; (6) inside cap; (7) cap of filter paper; (8) outside cap; (9) pipe holder; (10) lead wire; (11) pipe for circulating water; (12) insulator. Reprinted from ref. 5 by permission of Pergamon Press plc.

Figure 5.2 shows a schematic sectional view of the chamber of the Corona model 117 [15].

Except for the Corona model 117, two carefully matched thermistors, on which the solution and solvent drops are placed, are installed in a chamber saturated with solvent vapour. The temperature difference $(T - T_0)$ is detected as a change in the resistance of the thermistors. A large factor governing the accuracy of the temperature measurement of the drops by the thermistors in VPO is the unmatched resistance of the thermistor bead-temperature relations between two thermistors. Even if an aged and well-matched pair of thermistors are chosen at a specific temperature, these do not always match at a different temperature. That is, even if the temperature of the surroundings fluctuates only to a small extent, the conventional Wheatstone bridge is kept in a balanced state only when [14]

$$B_2 = B_1 \tag{5.19}$$

where B_2 and B_1 are the so-called thermistor constants for the pair of thermistors 1 and 2 in the relation:

$$\mathrm{R}' = \mathrm{R}'_\infty \exp(B_i/T) \qquad (i = 1, 2) \tag{5.20}$$

where R' is the resistance at T and R'_∞ is the limiting value of R' at $T = \infty$. However, from the experimental viewpoint, eqn (5.19) sets an unrealisable condition. Therefore, a detectable limit of $(T - T_0)_s$ by the commercially available VPO apparatus (except for the Corona model 117) is actually about 1×10^{-4} °C. Accordingly, the measurable upper limit of molecular weight is of the order of 10^4.

The Corona model 117 is based on the principle established by Kamide *et al.* [14], who constructed a modified version of the solvent-vapour chamber and thermistor assembly of Dohner *et al.* [15], and installed them in a Hitachi model 115 to obtain a new prototype VPO apparatus with increased sensitivity. With the conventional Wheatstone bridge circuit modified by introducing a matching circuit that has two precision variable resistors, the measurements can be made at any desired temperature without changing the thermistors. This enables a temperature difference of about 6×10^{-6} °C to be detected.

Eqn (5.14) indicates that the ratio K_s/K_e, which is a measure of the efficiency of the measurement, will become larger under the following conditions: (1) large drop size; (2) solvent with large ΔH; (3) solvent with high vapour pressure; (4) low temperature; and (5) reduced total pressure.

Using the prototype of the Corona model 117 VPO, the calibration parameter K_s of solutions of unbranched alkanes, such as octadecane, and tristearin in various solvents, such as *n*-hexane and *n*-octane, was determined from the intercept of the graph of $(T - T_0)_s/c$ against c through the relation [14]:

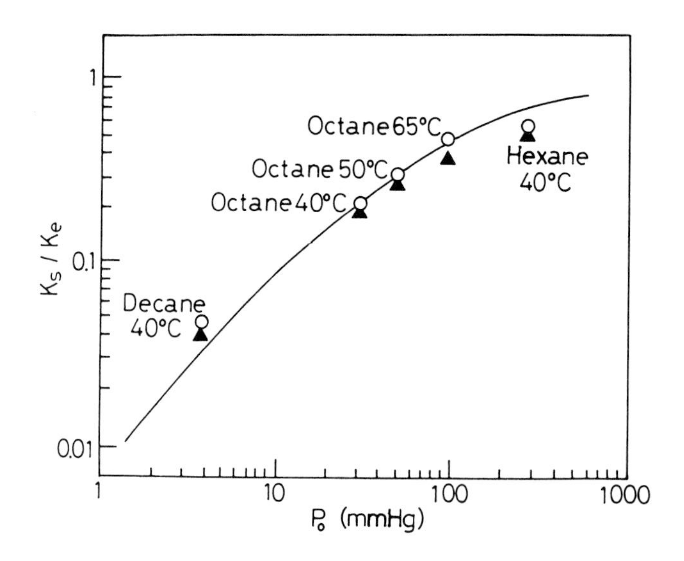

Figure 5.3 The dependence of the ratio K_s/K_e on the solvent vapour pressure P_0 ○, Experimental data for unbranched alkanes and tristearin in various solvents; ▲, calculated points from eqn (5.14) by putting $(k_1A_1 + k_2A_2)/(k_3A_1) = 6 \times 10^7$ cal dyn^{-1} mol^{-1} cm^{-2} K^{-1} together with the experimental values of ΔH and T; full line, theoretical curve obtained from eqn (5.14) by putting $(k_1A_1 + k_2A_2)/(k_3A_1) = 6 \times 10^7$ cal dyn^{-1} mol^{-1} cm^{-2} K^{-1}, $\Delta H = 10^4$ cal mol^{-1} and $T = 313\,K$. Reprinted from ref. 14.

$$K_s = \left\{ \lim_{c \to 0} (T - T_0)_s / c \right\} \Big/ M \tag{5.21}$$

where M is the solute molecular weight. Figure 5.3 shows the plot of the ratio K_s/K_e against the solvent vapour pressure $P_0(T)$ [14]. K_s/K_e monotonically increases with increasing $P_0(T)$. The full line is the theoretical curve calculated from eqn (5.14) by putting $(k_1A_1 + k_2A_2)/(k_3A_1) = 6 \times 10^7$ cal dyn^{-1} mol^{-1} cm^{-2} K^{-1}, $\Delta H = 10^4$ cal mol^{-1} and $T = 313$K. The closed triangle marks are calculated by putting the same value of $(k_1A_1 + k_2A_2)/(k_3A_1)$ and the experimental values of ΔH and T into eqn (5.14). The experimental values fit well with the theoretical curve, indicating that the ratio $(k_1A_1 + k_2A_2)/(k_3A_1)$ is approximately constant regardless of the solute molecular weight, the nature of the solvent (in this case, ΔH), and the temperature, when the same VPO apparatus is employed. Eqn (5.21) can be generalised for the case of a multicomponent polymer solution to yield:

$$K_s = \left\{ \lim_{c \to 0} (T - T_0)_s / c \right\} \Big/ M_n \tag{5.22}$$

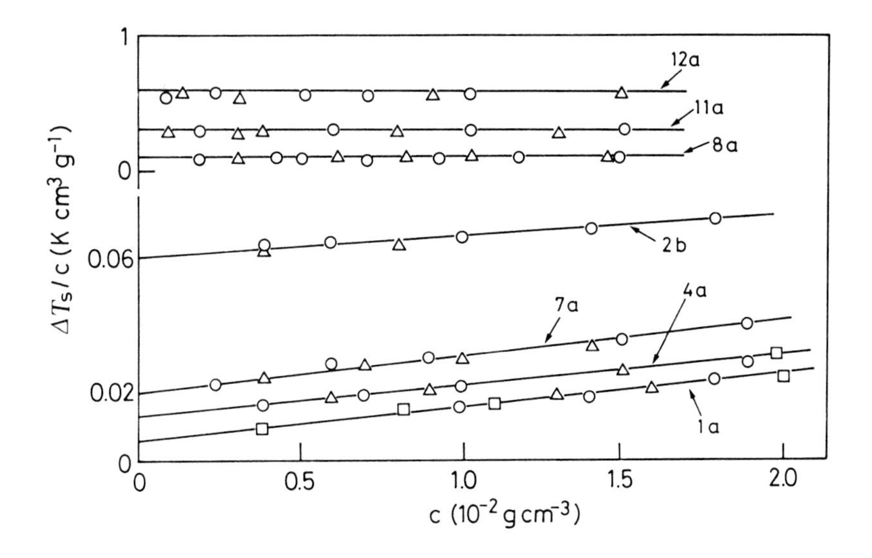

Figure 5.4 Plot of the ratio of steady-state temperature difference $(T-T_0)_s$ and concentration, c, against c for solutions of atactic polystyrene standards in benzene. The plot indicates data obtained in duplicate runs by the use of the prototype vapour-pressure osmometer of Kamide *et al.* at 40°C. Polymer codes are shown on the lines (see Table 5.2). Redrawn and adapted from ref. 14.

Using eqn (5.22), one can determine K_s for polymers with M_n evaluated by the other methods in advance. Figure 5.4 shows plots of $(T-T_0)_s/c$ versus c based on data obtained for solutions of some atactic polystyrenes in benzene at 40°C [14]. Circles and triangles denote duplicate sets of measurements, which gave $\lim_{c\to 0}(T-T_0)_s/c$ reproducible to ± 1% or better. These data, obtained by using the apparatus of Kamide *et al.* [14], can be accurately represented by straight lines with positive slopes, drawn by the least-squares method. No deviations from the straight lines were observed even at low concentrations. In Table 5.2 the data for K_s and $\lim_{c\to 0}(T-T_0)_s/c$ are shown for solutions of atactic polystyrene in benzene at 40°C [14]. The approximate constancy of K_s over a wide range of M_n is verified, but in the strict sense K_s has a tendency to increase gradually as M_n of the solute increases. However, to a good approximation K_s is equal to $0.5K_e$.

M_n for atactic polystyrene in benzene can be estimated from $\lim_{c\to 0}(T-T_0)_s/c$ by utilising either K_s for low molecular weight compounds, such as benzil, or $K_s = 0.5K_e$. Values of M_n obtained in these ways are listed in Table 5.2. Agreement between values of M_n given by the supplier and those evaluated by VPO are fairly reasonable (to an accuracy of ± 10% or less except for the two samples 11a and 1a), particularly in the case where $K_s = 0.5K_e$ is employed. From the foregoing discussion it is clear that M_n of unknown materials can be accurately determined by VPO over a wide M_n range.

Kucharikova [16] noticed that the agreement between the values of M_n by VPO and those given by the manufacturers in Table 5.2 was better than that which had been first indicated by Kamide *et al.*, because the molecular weights used by Kamide *et al.* [14] were not M_n but M_w. The M_n value in the parentheses for sample 1a in Table 5.2 was determined by Kucharikova [16] by using a Knauer membrane osmometer in toluene at 37°C, and that for sample 11a was obtained by Adams *et al.* [17] using GPC. These values are in good agreement with the M_n determined by VPO reported by Kamide *et al.* [14].

Table 5.2 Limiting value for the ratio of steady-state temperature difference $(T-T_0)_s$ and concentration c, the calibration parameter K_s, and the number-average molecular weight M_n, determined by vapour-pressure osmometry (VPO) for a solution of atactic polystyrene (PS) in benzene at 40°C [5]

PS sample code no.	$\lim_{c\to 0}(T-T_0)_s/c$ (K cm^3 g^{-1})	$K_s\times 10^3$ (K cm^3 mol^{-1})	$M_w\times 10^{-4\,a}$	$M_n\times 10^{-4}$		
				As informed^b	VPO^c	VPO^d
12a	0.576	1.169	0.203	0.205	0.188	0.194
11a	0.310	1.448	0.480	0.440	0.348	0.361(0.370)^e
8a	0.110	1.165	1.050	1.030	0.982	1.018
2b	0.059	1.200	2.04	1.930	1.83	1.90
7a	0.0192	0.979	5.10	—	5.63	5.83
4a	0.0124	1.21	9.72	9.17	8.71	9.03
1a	0.0052	0.83	16.00	15.10	20.77	21.50(19.60)^f
3a	0.0030	1.23	41.10	38.80	36.00	37.30

^{a} By supplier, ^{b} M_n as informed by the manufacturer, ^{c} M_n determined by putting $K_s=1.08\times 10^3$, which was obtained for benzil, ^{d} M_n determined by putting $K_s=0.5\,K_e$ and $K_e=2.24\times 10^3$, ^{e} ref. 17, ^{f} ref. 16.

In order to estimate the experimental error in the VPO method, Kamide *et al.* repeatedly determined M_n values of atactic polystyrene samples and cellulose diacetate samples with their apparatus [18]. They obtained an M_n value for a polystyrene sample of 7.22×10^4 within an accuracy of $\pm 0.54\times 10^4$ at the 95% significance level for polystyrene, and that for a cellulose diacetate sample, with total degree of substitution 2.46, of $3.7\times 10^4 \pm 0.36\times 10^4$ at the same significance level as polystyrene. The VPO method is reproducible to ±7–9% or better, which is of the same order of magnitude as that observed in membrane osmometry (MO) or GPC. Of course, the relative experimental error will differ greatly depending on the nature of the solvent (especially the vapour pressure and the heat of condensation), and the temperature.

5.3.2.2 Membrane osmometry (MO). In the 1940s four types of manual osmometers were reported in the literature: the static osmometer (Schulz design [19]), the dynamic osmometer (Fuoss–Mead design [20]), the static–dynamic osmometer (Zimm–Meyerson design [21]) and the osmotic balance (Jullander design [22]). All of them were commonly used until the mid-1960s, but they are now less common due to the very high degree of skill needed for operation, the time involved, and low reproducibility of data.

Self-balancing electronic membrane osmometers came on to the market during the 1960s: Hewlett-Packard's Mechrolab (originally marketed by Mechrolab Co., but thereafter by Hewlett-Packard Co.) and Schell osmometers [23] (Dohrman Instruments Co., Hallikainen Instruments, and J. V. Stavin Co., under license from Schell Development Co.). The former in particular shortened measurement time considerably (to about one hour) and was widely used until the 1980s. Regrettably, this type of osmometer is currently out of production. At present three types of osmometer, manufactured by Jupiter Instruments Co., Dr. H. Knauer, and Gonotec, are commercially available.

Figure 5.5 shows a schematic view of the Jupiter Recording Membrane Osmometer model 230 and 231 (initially commercialised by Wescan Instrument Co.) [5]. Here, the upper solution cell is separated from the lower solution cell by the semi-permeable membrane. Transfer of solvent from the solvent cell to the solution cell generates a negative pressure in the former, which can be detected and measured by a stainless-steel diaphragm attached to a strain gauge transducer. The precision of measurement is said to be 0.5% and the stability is better than 0.02 cm H_2O over a temperature range of 5–130°C.

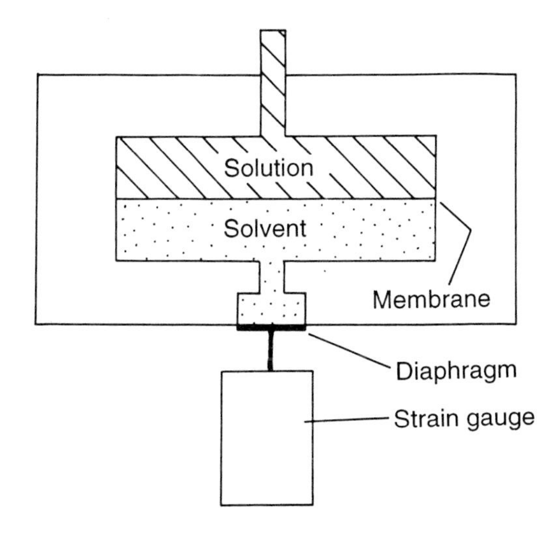

Figure 5.5 Schematic representation of a Jupiter model 230 or 231 membrane osmometer. From ref.5, reprinted by permission of Pergamon Press plc.

Semi-permeable membranes for the osmotic pressure measurement of polymer solutions are manufactured by Sartorius GmbH and Schleicher & Schuell Co. Generally, for measurements using organic solvents, the membranes are made of regenerated cellulose (Sartorius Co. SM11539; Schleicher

& Schuell, RC51–54). For water, cellulose acetate (SM11739; AC61–64) and cellulose nitrate (SM12134; E1, 41, 42) are used. The cut-off molecular weight of these membranes is about 1×10^4 and the range of M_n, which can be accurately determined, lies between 1.5×10^4 and 5×10^5. The membranes are supplied in a dry or wet state, in the latter case usually in a water–alcohol mixture. In the case of wet membranes, a slow and careful conditioning before use is required, while dry membranes can be used directly in the osmotic solvent. If the solvent to be used is not water, the wet membrane is conditioned from water–alcohol to alcohol, followed by a successive substitution of solvent from alcohol to the osmotic solvent. At each stage the membrane is immersed for about one day.

In actual measurements the following should be carefully confirmed: no leakage of solution and/or solvent at the cell or bulbs; no contamination by air bubbles in either cell; and no deformation or swelling of the membrane when installed in the apparatus. The temperature must be controlled to at least ± 0.01°C.

Figure 5.6 demonstrates the plot of Π/c versus c for the atactic polystyrene–toluene system at 25°C measured by the Jupiter Recording Membrane Osmometer Model 231 [18]. Excellent linearity is obtained at polymer concentrations less than about 1% for all samples investigated. This enables the evaluation of M_n with good accuracy by extrapolation of Π/c data to $c = 0$. The second virial coefficient $A_{2,0}$ is given as the slope of the linear plot. Sometimes the plot shows upward curvature when the c^3 and higher terms are not negligible (i.e. at higher c values in a good solvent). In this case the initial slope should be carefully determined from the plot.

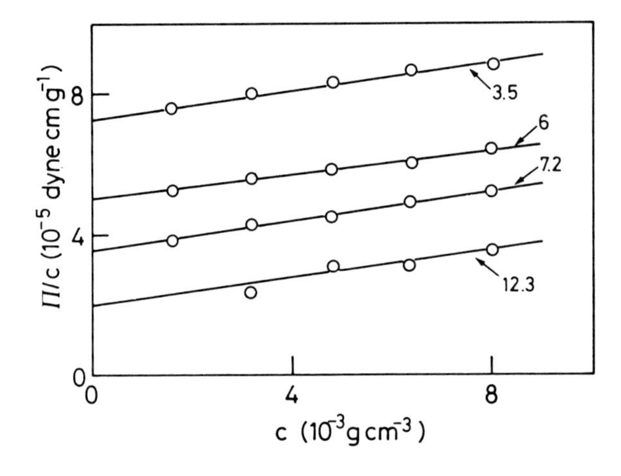

Figure 5.6 Plot of the ratio of osmotic pressure π and concentration, c, against c for atactic polystyrene in toluene at 25°C. The numbers attached to the lines denote the number-average molecular weights ($M_n \times 10^{-4}$) of the samples. From ref. 18, reprinted by permission of Elsevier Applied Science Publishers Ltd, Barking.

5.4 End-group analysis

Chemical methods can also be applied to determine M_n in cases where the chemical structure of the polymer, especially its end group, is known in advance. The requirements are that a definite end group exists in each molecule, the polymer is linear and the molecular weight of the polymer is relatively low (up to about 10^4). Polyamides (mainly nylon-6 and nylon-6, 6) and poly (ethylene terephthalate) are good examples.

Condensation polymerisation of nylon salt or an addition polymerisation of ε-caprolactam yields polymers having a carboxyl group and an amino group at either end of the chain. If we express the mole number of the carboxyl and amino groups per one gram of the sample by [COOH] and $[NH_2]$, M_n is given by:

$$M_n = 1/[COOH] = 1/[NH_2] \tag{5.23}$$

Eqn (5.23) is derived for the case when $[COOH] = [NH_2]$. In the case when the number of moles of carboxyl groups formed is not equivalent to the number of amino groups formed ($[CCOH] \neq [NH_2]$), M_n is given by:

$$M_n = 2/([COOH] + [NH_2]) \tag{5.24}$$

5.4.1 Determination of the carboxyl and the amino groups of a polyamide

Carboxyl group concentration is determined by titration of polyamide solutions in benzyl alcohol or a phenol–methanol mixture with 0.1 M aqueous potassium hydroxide solution, using phenolphthalein as indicator. The amino group concentration is evaluated by titration of the polyamide solution with 0.1 M aqueous hydrochloric acid. This method is not applicable to the analysis of polyamides with M_n higher than about 10^5, which cannot be dissolved in benzyl alcohol or phenol–methanol mixture with 0.1 M aqueous potassium hydroxide solution. In order to overcome this serious drawback, a new method has been proposed [24]. In this, a mixture of $CaCl_2$ and methanol with potassium hydroxide as a solvent and HCl solution alone as titrating reagent are used to analyse the carbonyl and the amino groups concurrently.

Typical titration curves of the a nylon-6, 6 solution (broken line) and the reference solution (solid line) which includes no sample are schematically indicated in Figure 5.7. In Figure 5.7, a and c show the amounts (in units of milligram-equivalent) of HCl consumed to attain the first and second neutralisation points for the reference solution, respectively, while b and d are those for the nylon-6,6 solution. Expressing a–d in milligram equivalents of HCl, the numbers of milligram-equivalents of the end groups of $-NH_2$ and $-COOH$ (x and y) are estimated through the relations:

$$x = (d - c)/W \tag{5.25}$$

$$y = \{(a - b) + (d - c)\}/W \tag{5.26}$$

where W is the sample weight (kg).

Application of the end-group method to other polymers, such as polyesters, polyurethanes and cellulose, is reviewed in ref. 25.

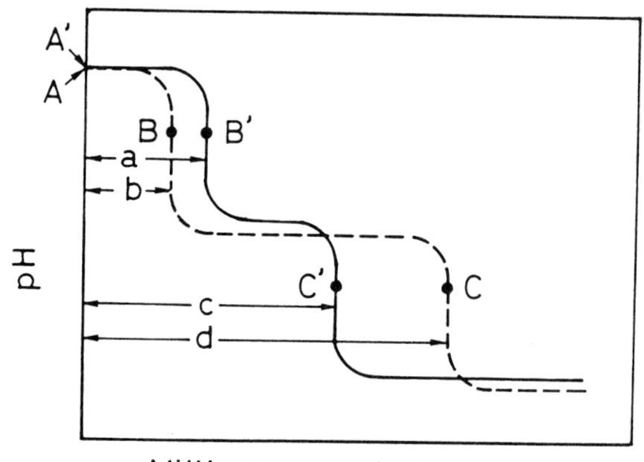

Figure 5.7 Schematic representation of the titration curve of nylon-6, 6–$CaCl_2$–methanol system. Solid line, reference solution without nylon-6,6; broken line, sample solution.

5.5 Light-scattering methods

5.5.1 Theoretical background

The light-scattering method has been extensively applied to determine the weight-average molecular weight, M_w, as well as the molecular dimensions of polymer molecules and the degree of polymer–solvent interaction in polymer solutions. The theoretical basis of light scattering from polymer solutions was first established by Zimm *et al.* [26] in the 1940s. They derived a relation between the intensity of scattered light from the polymer solution and the M_w, size and shape of the solute molecule in the solution.

In light-scattering investigations, one measures the reduced scattering intensity at a scattering angle θ, R_θ, sometimes simply referred to as Rayleigh's ratio, which is defined as [27]:

$$R_\theta = i_\theta r^2/(I_0 V) = I_\theta/(I_0 V) \tag{5.27}$$

$$I_\theta = i_\theta r^2 \tag{5.28}$$

where: i_θ is the intensity of scattered light per unit area (erg s^{-1} cm^{-2}) at a distance (r) of the observer from the sample solution with volume V; I_0 is the intensity of the incident beam and I_θ the radiant intensity of the scattered light (erg s^{-1} per unit solid angle).

For a dilute solution consisting of a monodisperse polymer with molecular weight M and a solvent, R_θ is expressed by [27]:

$$Kc/R_\theta = 1/\{MP(\theta)\} + 2A_{2,L}c + 3A_{3,L}c^2 + \dots \tag{5.29}$$

where: K (see eqns (5.30) and (5.31)) is the optical constant; $P(\theta)$, defined as a ratio $I_\theta/I_{\theta=0}$, is the particle scattering factor expressing an interference of the scattered light from a polymeric chain as solute; and $A_{2,L}$ and $A_{3,L}$ are the second and third virial coefficients determined by the scattering method, respectively.

In the case when unpolarised and vertically polarised light are used as the incident beam, K (in eqn (5.29)) becomes K_u and K_v, respectively, and these are given by:

$$K_u = (2\pi^2/\lambda_0^4 N_A) n_0^2 (\partial n/\partial c)^2 (1 + \cos^2\theta) \tag{5.30}$$

and

$$K_v = (4\pi^2/\lambda_0^4 N_A) n_0^2 (\partial n/\partial c)^2 \tag{5.31}$$

where λ_0 is the wavelength in vacuo of the incident beam, and n_0 and n are the refractive indices of the solvent and the solution, respectively.

For a Gaussian coil $P(\theta)$ is related to the mean radius of gyration of the polymeric chain $\langle S \rangle$ by:

$$P(\theta) = 1 - (1/3\,!)\langle S^2 \rangle \mu^2 + (1/5\,!)\langle S^4 \rangle \mu^5 + \dots \tag{5.32}$$

where:

$$\mu = (4\pi/\lambda)\sin(\theta/2) \tag{5.33}$$

and λ $(= \lambda_0/n)$ is the wavelength of the incident beam in the solution.

For a solution of a Gaussian polymer with polydispersity, eqns (5.29) and (5.32) are modified as:

$$Kc/R_\theta = 1/\{M_w P(\theta)\} + 2A_{2,L}c + 3A_{3,L}c^2 + \dots \tag{5.34}$$

and

$$P(\theta) = 1 - (1/3\,!)\langle S^2 \rangle_z \mu^2 + (1/5\,!)\langle S^4 \rangle_z \mu^5 + \dots \tag{5.35}$$

where $\langle S^2 \rangle_z$ and $\langle S^4 \rangle_z$ are the z-average mean-square and mean-fourth radius of gyration.

When the term $\langle S^2 \rangle_z \mu^2$ is far smaller than unity, combination of eqns (5.34) and (5.35) leads to the equation:

$$Kc/R_\theta = (1/M_w)\,\{1 + (1/3\,!)\langle S^2 \rangle_z \mu^2 + \dots\} + 2A_{2,L}c + \dots \tag{5.36}$$

In order to estimate M_w, $\langle S^2 \rangle_z$ and $A_{2,L}$ from experimental data of R_θ versus c for solutions at various scattering angles θ at constant temperature, it is convenient to use a graphical method, proposed by Zimm [28]. In this method, Kc/R_θ is plotted against $(\sin^2(\theta/2) + kc)$, where k is an arbitrary constant chosen to produce a reasonable spread of data points. In Zimm's method, $\langle S^2 \rangle_z$ and $A_{2,L}$ are estimated from the relations of $\lim_{c \to 0} Kc/R_\theta$ versus $\sin^2(\theta/2)$ and $\lim_{\theta \to 0} Kc/R_\theta$ versus c, respectively. M_w is obtained as the inverse of the value of the intercept of a line of $\lim_{c \to 0} Kc/R_\theta$ at $\theta = 0°$ or that of a line of $\lim_{\theta \to 0} Kc/R_\theta$ at $c = 0$. Note that the intercepts of the two lines coincide with each other. Zimm's procedure is a double-extrapolation method.

5.5.2 Light-scattering instruments

From the late 1950s to the early 1960s light-scattering instruments came on to the market. The most popular ones were the Brice-Phonenix type 2000, the Sofica model PGD42000 and the Shimadzu model PG-21, and these instruments are briefly described in ref. 29. But now all of these are out of production, because the high-pressure mercury arc lamps employed as the light source are no longer available.

Light-scattering instruments available commercially now include the Dynamic (and Static) Light Scattering Spectrophotometer DLS700 (SLS-600R) supplied by Otsuka Electronics Co., the Dawn-B and -F photometers from Wyatt Technology Co., the Light Scattering Mono- (and Duo-) photometer models 6000 and 6200 from C. N. Wood Mfg. Co., and the BI-200SM Goniometer from Brookhaven Instruments Co. All of these instruments use a laser as light source. This has many advantages, such as a precisely monochromated wavelength, high coherency, high stability of light intensity and low power consumption.

The DLS700, which was based on the prototype photometer model SLS 600, enables static and dynamic light-scattering measurements to be made. A side view of the DLS700 is shown in Figure 5.8. A 5 mW helium–neon laser (1) is employed as the light source and the intensity of the incident beam is monitored by a photodiode (2). The cylindrical cell in section (4) is dipped in a non-volatile immersion liquid, such as di-*n*-butyl phthalate, and thermostatted at an accuracy of $\pm 0.2°C$. The light scattered from the sample enters the photomultiplier (9) through a polariser (7) and an attenuator (8). Two kinds of cells with diameters of 21 mm (standard) and 12 mm (optional) are available for the instrument. The goniometer is designed to scan automatically by a stepping motor drive. The manufacturer claims that this instrument is capable of determining M_w values from 3×10^2 to 2×10^7 at a low scattering angle of 5°. The apparatus uses a microcomputer to analyse the data by Zimm [28], Berry [30] and Debye [31] methods.

The Dawn-B photometer, shown in Figure 5.9, is a multiangle light-scattering photometer for the analysis of relatively large samples ($10\ cm^3$) in batch mode,

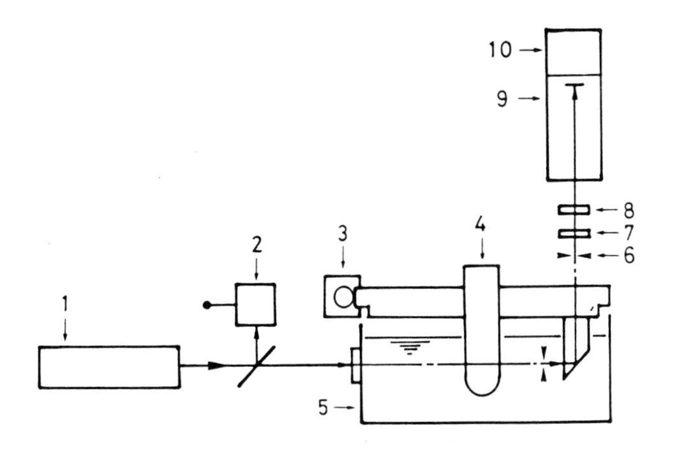

Figure 5.8 A side view of the dynamic light-scattering spectrometer DLS-700 (Otsuka Electronics Co.), from the brochure of Otsuka Electronics Co. (1) He–Ne laser; (2) monitor; (3) motor; (4) cell; (5) bath; (6) pinhole; (7) polariser; (8) ND filter; (9) photomultiplier; (10) preamplifier discriminator.

while the Dawn-F is used for GPC purposes by the use of a sample flow-through system. In the former type, the sample is placed in a cuvette at the centre of the stage and is illuminated by a fine laser beam. The cuvette stage is surrounded by an array of fifteen detectors with high-gain photodiodes. The detector angles are set equidistantly in $\sin(\theta/2)$ from 0.2 to 0.9. Employing the multiangle detectors, the measurement time for static light scattering is

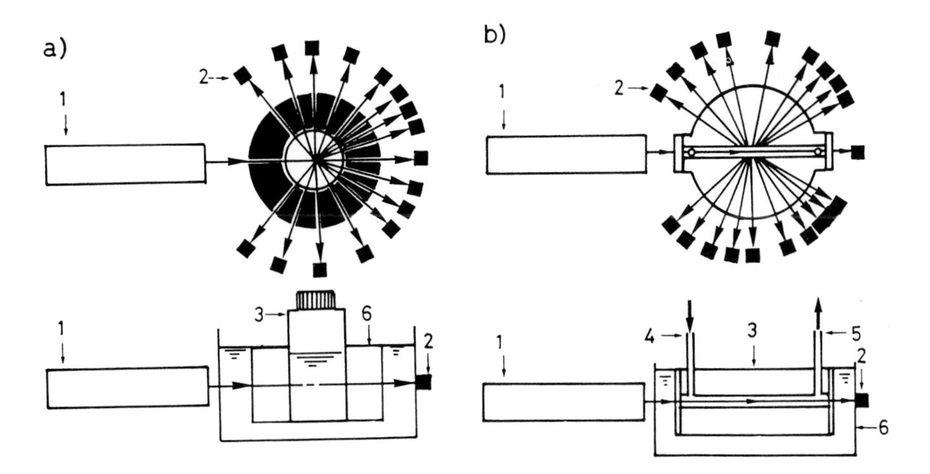

Figure 5.9 Top and side views of the Dawn-B and -F type photometer (Wyatt Technology Co.) from the brochure of Wyatt Technology Co. (1) He–Ne laser; (2) detector; (3) cell; (4) sample inlet; (5) sample outlet; (6) read head.

reduced by between 30 and 90% compared with measurements using a scanning-type apparatus. The analysing software AURORA® allows one automatically to estimate M_w, $\langle S^2\rangle_z$ and $A_{2,L}$ by Zimm plots.

The specifications of the commercially available instruments are briefly summarised in Table 5.3. Note that the instrument of C. N. Wood can use both laser and mercury lamps and so it is useful for light-scattering measurement at different wavelengths.

Table 5.3 Specifications of commercially available light-scattering instruments

Type	Manufacturer	Light source	Wave-length (nm)	Detector	Available cell[a] (mmϕ)	Angle range (degrees)	Software for analysis
Model 6000	C. N. Wood Mfg.	Hg-vapour lamp & laser	436,546, 632.8	Photomult.[b]	26, 36	0–150	—[c]
DLS-700	Otsuka Elec.	He–Ne laser	632.8	Photomult.	12, 21	5–150	Zimm, Berry, Debye plots
Dawn-B & -F	Wyatt Tech.	He–Ne laser	632.8	Photodiodes	B, 10 ml; F, 67 μl	B, 23–129; F, 9–171	Zimm and Debye plots
BI-200SM	Brookhaven Instrument	He–Ne laser	632.8	Photomult.	12, 26	8–162	Zimm plot

[a] Cylindrical cell except for Dawn-B and -F types, [b] photomultiplier tube, [c] not informed.

5.5.3 *Differential refractometry*

In order to determine molecular weight by light scattering, the refractive index n_0 of the solvent and the refractive index increment $(\partial n/\partial c)$ for the solution at a constant temperature must be known. The former is readily measured with a simple laboratory refractometer such as the Abbe type. Determination of $(\partial n/\partial c)$ of the solution with high accuracy can be achieved by using two types of refractometers: divided-cell differential refractometers and interferometers.

The original differential refractometers were designed by Brice and Halwer [32] (referred to as the Brice type) and by Debye [33]. An optical diagram of the Brice type is schematically drawn in Figure 5.10. A beam generated from a mercury arc source (1) goes through a collimating lens (2) and a filter (3), which monochromatises the beam, and impinges vertically on the centre of a cubic glass cell (6), which is partitioned into two compartments. The cell is immersed in a water bath (7) to maintain a constant temperature. The refracted beam converges into the focus of a microscope (8) equipped with an eyepiece. The horizontal displacement, Δd, of the beam in the focal plane (10) is measured with a micrometer (9).

The difference between refractive indices of the solution (n) and the solvent (n_0), Δn ($\equiv n - n_0$), is calculated by the relation:

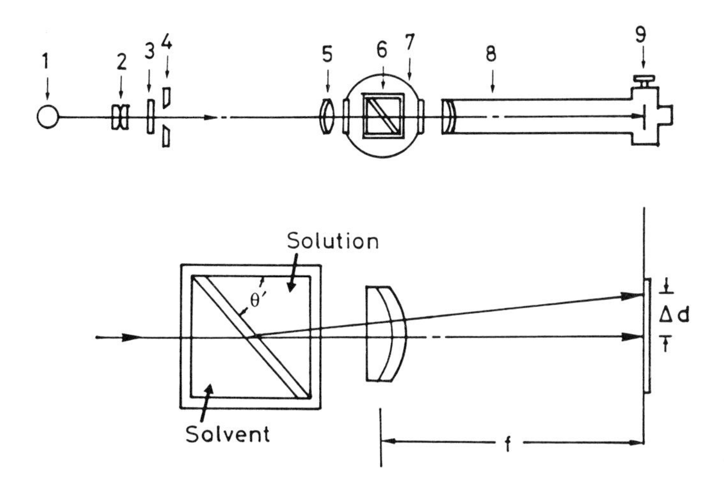

Figure 5.10 Optical diagram of a differential refractometer of the Brice model. (1) Light source; (2) collimating lens; (3) filter; (4) slit; (5) planocylindrical lens; (6) cell; (7) water bath; (8) telescope; (9) micrometer.

$$\Delta n = \Delta d/(f \tan\theta') \tag{5.37}$$

Where f is the focal length of the microscope and θ' is the angle between the thin diagonal glass partition and the lateral face of the cell. The refractive-index increment of the solution is obtained as the initial slope of the plot of Δn versus c for solutions with various polymer concentrations. Brice-type differential refractometers are marketed by Otsuka Electronics Co. (type DRM-1020) and C. N. Wood Co. (type RF-600).

The Wyatt/Optilab 903 interferometer (Wyatt Technology Co.) is also commercially available. A side view of this model is shown in Figure 5.11. A collimated light source (1) is plane-polarised at 45° to the horizontal. The first Wollaston prism (4) splits the beam into two components. The beam passing through the sample in the cell (6) is vertically polarised, while that through

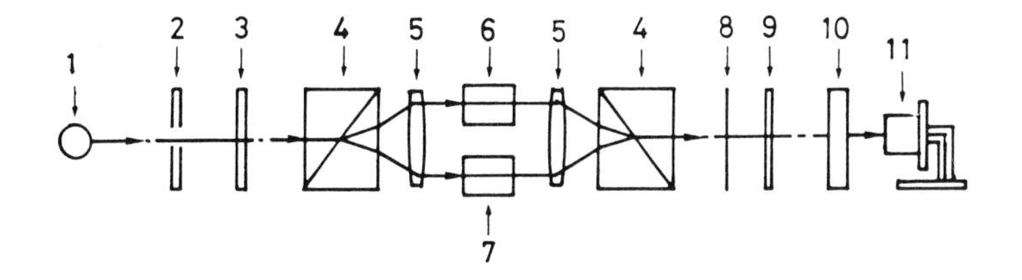

Figure 5.11 Side view of the Wyatt/Optilab interferometer (Wyatt Technology Co.) from the brochure of Wyatt Technology Co. (1) Light source; (2) mask; (3) polariser; (4) Wollaston prism; (5) lens; (6) sample cell; (7) reference cell; (8) quarter-wave plate; (9) analyser; (10) interference filter; (11) detector.

the reference cell (7) is horizontally polarised. The beams are recombined in the second Wollaston prism, quarterwave plate and analyser to yield a plane-polarised beam rotated at an angle $\psi/2$ with respect to the original 45°. The phase difference ψ is directly proportional to Δn.

5.5.4 Experimental aspects of light scattering and application to polymer solutions

Optically clean solutions, especially aqueous solutions (including pure water used as solvent), must be carefully prepared for light-scattering measurements, to be free of dust particles and other extraneous scatterers. Filtration through a membrane filter or centrifugation is widely used to clarify solutions. This concern for cleanliness also extends to glassware, especially scattering cells.

Instrument calibration can be accomplished by using pure liquids, solutions and colloidal dispersions. Among them, pure liquids with known reduced scattering intensity R_θ, such as benzene, are the most commonly employed standards. Detailed procedures for filtration and calibration methods are described in ref. 27. Data on R_θ of several pure liquids, such as benzene, were collected by Kratochvil *et al.* [34] for incident light wavelengths λ_0 of 436 and 549 nm and by Pike *et al.* [35] for an incident light wavelength λ_0 of 633 nm.

As an example of the application of light scattering to polymer solutions, some results for an atactic polyacrylonitrile fraction in dimethylformamide are shown in Figure 5.12 [36]. Here the scattering intensity at scattering angles

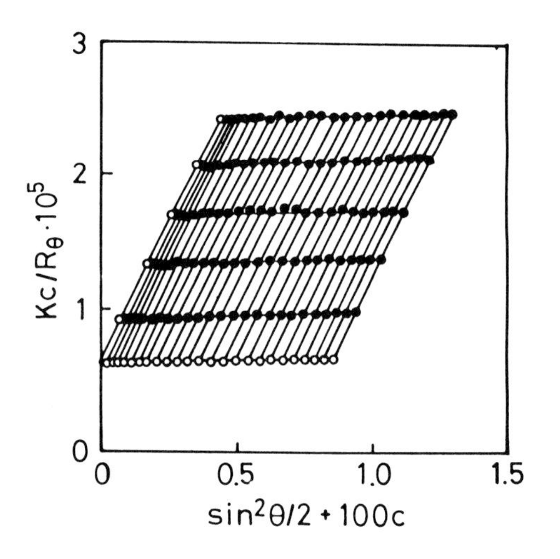

Figure 5.12 A typical Zimm plot of atactic polyacrylonitrile in dimethylformamide at 25°C. From ref. 36.

θ ranging from 20 to 135° was measured on the Otsuka Electronics model DLS-700 at 25°C. The Zimm plot of this system does not display any significant distortion. The z-average radius of gyration $\langle S^2 \rangle_z$ and the second virial coefficient $A_{2,L}$ were estimated from the initial straight lines representing the relations $\lim_{c \to 0} Kc/R_\theta$ versus $\sin^2(\theta/2)$, and $\lim_{\theta \to 0} Kc/R_\theta$ versus c, and were found to be 23.6 nm and 2.55×10^{-3} mol cm^3 g^{-2}, respectively. M_w calculated from the intercept at $\theta = 0$ and $c = 0$ was 1.58×10^5.

5.6 Viscometric methods

5.6.1 Limiting viscosity number

Molecular-weight determination by viscometry is principally based on the fact that the viscosity of a polymer solution, η, is generally larger than that of the solvent, η_0, and depends on the molecular weight of the polymer (assuming that concentration, temperature, and shear rate are kept constant). This method, first applied by Staudinger *et al.* [37] to cellulose and cellulose derivatives, played an important role in establishing the concept of macromolecules and even now is one of the most familiar methods for measuring molecular weight. It is widely used in fundamental research in polymer science as well as in industrial fields.

Viscosity, η, can be expressed, in the region of low solute concentration c, in the form:

$$\eta = \eta_0 \{1 + [\eta]c + \mathrm{K}_\eta c^2 + \text{higher term of } c^3\} \tag{5.38}$$

where K_η is a concentration-independent parameter and $[\eta]$ is the limiting viscosity number (or intrinsic viscosity) defined by eqn (5.39) or (5.40):

$$[\eta] = \lim_{c \to 0} \eta_{sp}/c \tag{5.39}$$

or

$$[\eta] = \lim_{c \to 0} (\ln \eta_r)/c \tag{5.40}$$

where

$$\eta_{sp} = (\eta - \eta_0)/\eta_0 = \eta_r - 1 \tag{5.41}$$

Where η_{sp} is the specific viscosity and η_r $(= \eta/\eta_0)$ is the viscosity ratio (or relative viscosity).

The Mark–Houwink–Sakurada (MHS) equation (eqn (5.5)) offers a convenient means of determining the molecular weight of a polymer which is soluble in a solvent. It has been experimentally confirmed that the parameters K_m and a in the MHS equation are constant over a wide range of molecular weights under the constraints of zero shear rate at given temperature for a given polymer–solvent system; Figure 5.13 demonstrates the plot of [η] against M_w for atactic polystyrene with a relatively narrow molecular-weight distribution in cyclohexane at 34.5°C and in benzene at 25°C and 30°C [38]. For the atactic polystyrene–solvent system, the constancy of the parameters K_m and a is confirmed experimentally over four orders of magnitude of M_w. Note that this cannot be interpreted by any existing theories. Cyclohexane at 34.5°C is known for atactic polystyrene as a Flory Θ-solvent, in which the excluded-volume effect has apparently disappeared (i.e. the second virial coefficient obtained by both light scattering and membrane osmometry is zero). The exponent a in this solvent is almost 0.5. In benzene at 25°C and 30°C the exponent a in the region of $M_w \geqslant 10^4$ is 0.75; in a good solvent, the exponent a for polystyrene reported in the literature [39] lies in the range 0.5–0.8. For extremely rigid polymers, such as fully aromatic polyamides, the exponent a determined experimentally is larger than unity. In Figure 5.13, a slight upward discrepancy from the straight line is observed at the region of $M_w \leqslant 10^4$, suggesting that the MHS equation, established using low molecular weight samples, is not always applicable to the high molecular weight region. The critical molecular weight at which the MHS equation deviates from linearity depends on the polymer–solvent combination. For example, the exponent a for the poly (methyl methacrylate)–benzene system at 30°C changes from 0.76 to 0.5 at a molecular weight of about 10^5 [40].

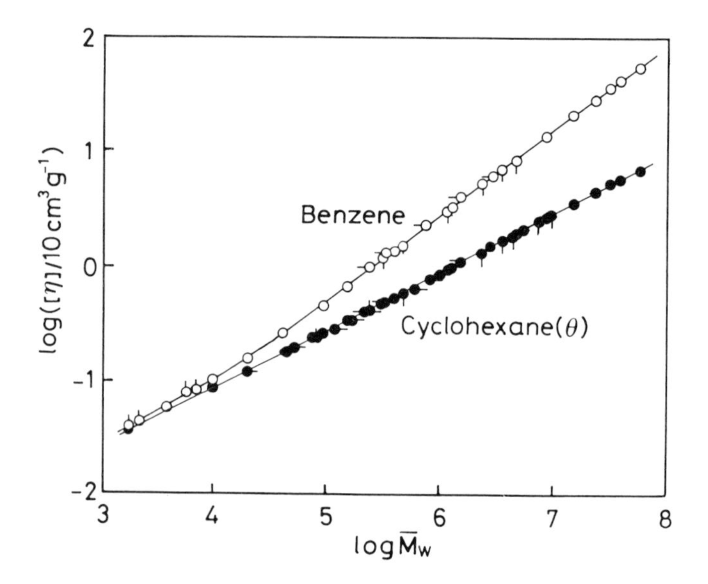

Figure 5.13 Log–log plots of limiting viscosity number [η] for atactic polystyrene solutions against the weight-average molecular weight M_w, ○, Benzene at 25 and 30°C; ●, cyclohexane at 34.5°C (Θ temperature). From ref. 38, reprinted by permission of John Wiley & Sons Inc.

The [η] value of a branched polymer is smaller than that of the corresponding linear polymer. The MHS equation of a segmented polyurethane, synthesised from methylenebis (4-phenyl isocyanate), polycaprolactone and 2-aminoethanol in *N*, *N*-dimethylacetamide shows upward curvature at $M_w \approx 10^5$, owing to an increase in the degree of branching per molecule in the region of $M_w \geqslant 10^5$ [41]. Therefore, it should be noted that M_v of the branched polymer cannot be unconditionally determined by using the MHS equation established for the linear polymer.

One of the factors governing the reliability of the MHS equation is the uniformity of the molecular weight distribution of the polymer samples used to establish it. The literature data for the parameters in the MHS equation for a given polymer–solvent system, such as the systems consisting of cellulose or its derivatives plus solvent, do not always agree. For example, the values of the exponent a in the MHS equation have scattered from 0.67 to 1 for cellulose trinitrate–acetone, 0.75 to ~1 for cellulose diacetate–acetone, and 0.65 to 1.0 for the cellulose triacetate (CTA)–trichloromethane system [42]. This scattering of the exponent a for given cellulose derivative solutions, especially for CTA–solvent systems, can be explained to a first approximation by the erroneous use of M_n for samples with a broad molecular weight distribution. In particular, the systematic molecular weight dependence of the molecular weight distribution of the samples yields incorrectly high a values.

5.6.2 *Measurement of viscosity*

The viscosity of polymer solutions, η, is usually measured by capillary or rotational viscometers. The capillary-type viscometer has the following advantages: high accuracy, ease of operation, speed and low cost of instrumentation. The rotational-type viscometer is frequently used in cases where the mass of the sample available is limited and the viscosity over a wide range of shear rates is required.

Typical capillary viscometers in common use are illustrated in Figure 5.14(a) (original Ostwald type), (b) (modified Ostwald type) and (c) (Ubbelohde type) [43]. The η value of a solution flowing through a capillary of length L' with diameter d' is proportional to the fourth power of d' as follows:

$$\eta = \pi \Delta P d'^4 t/(8QL') \tag{5.42}$$

where ΔP is the pressure difference between the head of the fluid and the end of the capillary and Q is the flux of the solution passing through the capillary in time t. Eqn (5.42), known as the Hagen [44]–Poiseuille [45] law, was derived by solving the Navier–Stokes equation under the assumptions of slow and steady laminar flow of an incompressible Newtonian solution under gravitational force. When ΔP is replaced by $\rho\, g'\, h$, where ρ is the density of the fluid, g' the gravitational acceleration and h the height of the fluid, eqn (5.42) is rewritten as:

$$\eta = \pi\rho\, g' h d'^4 t/(8QL') \tag{5.43}$$

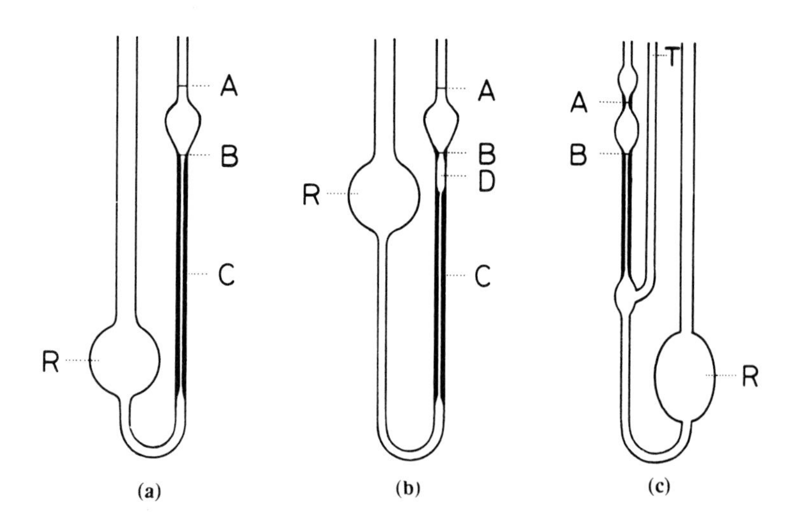

Figure 5.14 Typical suspension-type capillary viscometers. (a) Ostwald type; (b) modified Ostwald type; (c) Ubbelohde type. From ref. 43, reprinted by permission of John Wiley & Sons Inc.

In the actual capillary-type viscometer, all of the assumptions are not always satisfied. Significant deviations from the Hagen–Poiseuille law arise from the additional resistance at the end of the capillary, where the flow diverges or undergoes sudden enlargement (the end effect) and turbulent flow occurs (the Couette correction). In the capillary, the viscous fluid falls with a definite velocity so that the pressure difference, ΔP, is not equal to the static pressure (velocity correction). Moreover, at both menisci of the solution in the viscometer, the effect of surface tension on η should be taken into account (the surface-tension correction). A theoretical equation relating η to flow time t considering the foregoing corrections has also been proposed [46]. As excellently reviewed in refs 47 and 48, various types of capillary viscometers have been designed to minimise the above-mentioned corrections. For the purpose of measuring η under several shear rates, the capillary viscometers designed by Schurz and Immergut [49] and Hermans and Hermans [50] are helpful.

In order to determine the molecular weight the absolute value of the viscosity of the solution is not necessary. It is sufficient to measure the viscosity ratio η_r. In dilute polymer solution ρ can be regarded as being equal to that of the solvent ρ_0. Then η_r is expressed from eqn (5.43) as:

$$\eta_r = \eta/\eta_0 = t/t_0 \qquad (5.44)$$

where t is the flow time of the solution and t_0 that of the solvent. Capillary viscometers, as illustrated in Figure 5.14, make it possible to measure the time for the fluid to fall from the upper mark at A to the lower one at B.

An automatic viscometer which automatically measures the flow time is also commercially available. The models AVS-310 and AVS-400 for transparent liquids and AVS-440 for transparent and coloured liquids are manufactured by Schott Geräte GmbH. They claim that the model AVS-440 is designed to measure flow times ranging from 0.01 to 9999.99 s, by detecting the difference of thermal conductivity between liquid and atmosphere. On all three types of viscometers the optional cleaning assembly (type AVS-24) can be attached. Viscometers that automate the operation of blowing the solution into the measuring bulb, performing dilution, cleaning and drying as well as counting the flow time are marketed by San Denshi Co. (model AVS-2 as shown in Figure 5.15 [51]) and by Shibayama Scientific Instruments Co. (model SS-300LC-1CH). In the model AVS-2 the measuring unit is connected by cable to the control unit. The measuring unit is composed of the viscometric tubes, photosensor, stabilised temperature bath, temperature control unit, and solenoid valves. The control panel is composed of the photo-sensor control circuit, digital counter, controller, keyboard, display, push-button switches and the power source. Injection of the sample liquid into the viscometric tubes and washing of the inside of the tubes are performed by using a negative

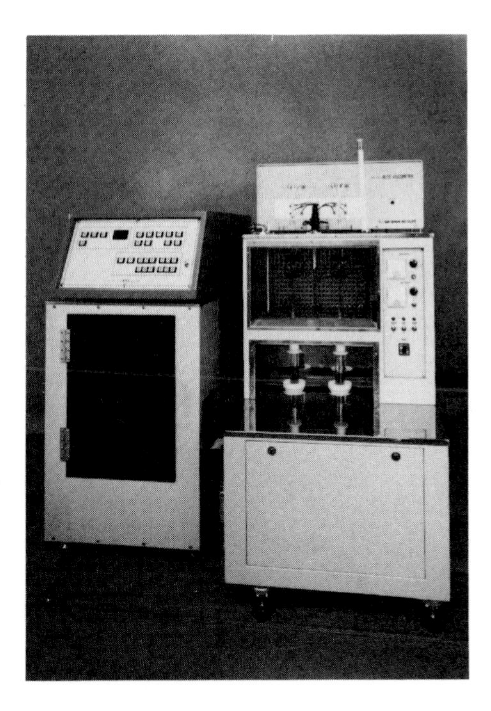

Figure 5.15 The automatic viscometer, model AVS-2 (San Denshi Co.).

pressure generated by the vacuum pump. For viscosity measurement, each viscometric tube is first connected to a bottle containing the sample solution. When the solenoid valves are actuated by commands from the control unit, measurement is automatically started.

Another automatic capillary viscometer which directly detects the pressure difference ΔP (eqn (5.42)) has recently come on the market. The Relative Viscometer manufactured by the Viscotek Co. includes two capillaries (capillary 1 and 2) connected in series, with the sample injection valve located between the two capillaries. Differential pressure transducers (DPT) are connected in parallel to each capillary. The sample is injected into capillary 2 and a pressure change is detected by the DPT. The viscosity ratio η_r is determined as the ratio of the pressures divided by the instrument constant.

5.6.3 *Analysis of viscosity data*

The η_{sp} and η_r values of a dilute solution are represented by a polynomial approximation (neglecting c^3 and higher terms in eqn (5.38)) as follows:

$$\eta_{sp}/c = [\eta] + k'[\eta]^2 c \quad \text{(Huggins' equation)} \tag{5.45}$$

$$(\ln \eta_r)/c = [\eta] + k''[\eta]^2 c \quad \text{(Kraemer's equation)} \tag{5.46}$$

where the coefficients k' (referred to as the Huggins constant) and k'' are characteristic of the polymer–polymer interaction in the solvent. The probability of polymer molecules contacting another is larger in a poor solvent, resulting in larger k'. For polymers in a Flory Θ-solvent k' is about 0.5 or slightly higher, and in a good solvent k' frequently lies in the range 0.2–0.5. A plot of η_{sp}/c versus c is a Huggins plot [52], and a plot of $(\ln \eta_r)/c$ versus c is a Kraemer plot [53]. The constant k'' is related to k' by:

$$k'' = k' - \frac{1}{2} \tag{5.47}$$

Figure 5.16 shows Huggins and Kraemer plots of nylon-6,6 fractions in 98% aqueous H_2SO_4 at 25°C [54]. Both plots of the samples with M_w higher than 3.6×10^5 show slight curvature in the region of $c \geqslant 2\%$, and $[\eta]$ values estimated by the Huggins and Kraemer plots for each sample coincide with each other to within an accuracy of 1%. The slope of the Kraemer plot is always less than that of the Huggins plot, so the relative error inherent to the extrapolation procedure to determine $[\eta]$ is smaller in the Kraemer plot.

Numerous experimental η versus c relations other than Huggins' and Kraemer's have been proposed [43]. They are special cases or equivalent to the Huggins equation (eqn (5.45)).

In very dilute solutions, negative coefficients of $d(\eta_{sp}/c)/dc$ were observed for some polymer solutions (for example, poly(vinyl chloride) [55] and polystyrene in various solvents [56], and isotactic polypropylene in decalin [57].

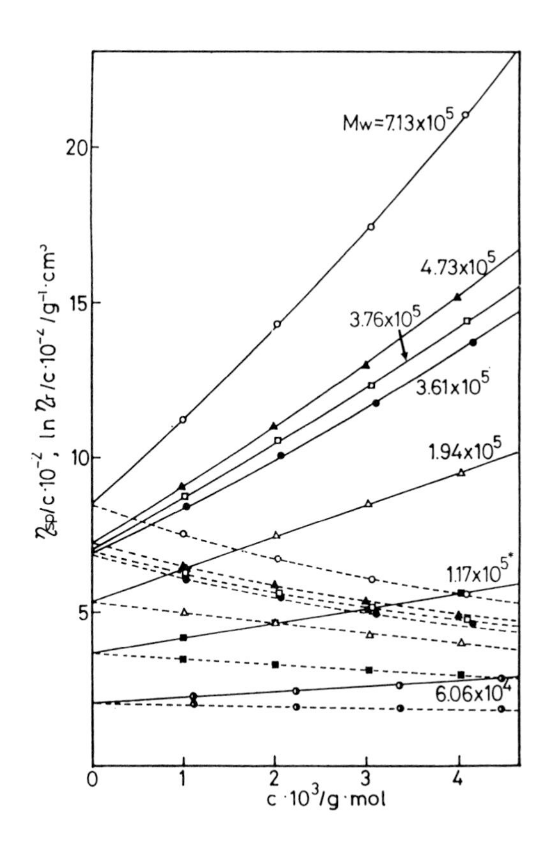

Figure 5.16 Plots of η_{sp}/c versus c and $\ln \eta_r/c$ of nylon-6,6 in aqueous 98% H_2SO_4 at 25°C. From ref. 54.

Interpretations have been given to these extraordinary phenomena, such as adsorption of the polymer molecules on to the surface of the capillary and greater thermodynamic expansion of the polymer molecules in very dilute solution than in the moderately dilute state [58].

References

1. M. Gordon, J. Goldsbrough, B. W. Ready and K. Derham (1973). In *Industrial Polymers*, J. H. S. Green and R. Dietz (eds), Transcripta Books, London, p. 45; K. W. Derham, J. Goldsbrough and M. Gordon (1974). *Pure Appl. Chem.*, **38**, 97; M. Gordon and B. W. Ready (1978). US Patent 4131369 (Dec. 26, 1978); and J. W. Kennedy, M. Gordon and G. Alvare (1975). *Polimery (Warsaw)*, **20**, 463.
2. H. Mark (1938). *Der feste Körper*, Hirgel, Leipzig, p. 103.
3. R. Houwink (1940). *J. Prakt. Chem.*, **155**, 241; and **157**, 15.
4. I. Sakurada (1941). *Kasenkouenshyu*, **6**, 177.
5. K. Kamide (1989). Colligative properties, in *Comprehensive Polymer Science*, Vol. 1, C. Booth and C. Price (eds), Pergamon Press, Oxford, Chapter 4.

6. P. J. Flory (1941). *J. Chem. Phys.*, **9**, 990.
7. M. Huggins (1941). *J. Chem. Phys.*, **9**, 440.
8. K. Kamide and M. Sanada (1967). *Chem. High Polym. (Tokyo)*, **24**, 751.
9. K. Kamide, K. Sugamiya and C. Nakayama (1970). *Makromol. Chem.*, **132**, 75.
10. K. Kamide, K. Sugamiya and C. Nakayama (1970). *Markromol. Chem.*, **133**, 101.
11. K. Kamide and I. Fujishiro (1971). *Makromol. Chem.*, **147**, 261.
12. See, for example, D. F. Rushman (1959). In *Techniques of Polymer Characterization*, P. W. Allen (ed.), Butterworths Scientific Pub., London, Chapter 4.
13. C. A. Glover (1973). In *Polymer Molecular Weight Methods*, E. Ezrin (ed.), American Chemical Society, Washington DC, Chapter 1.
14. K. Kamide, T. Terakawa and H. Uchiki (1976). *Makromol. Chem.*, **177**, 1447.
15. R. E. Dohner, A. H. Wachter and W. Simon (1967). *Helv. Chim. Acta*, **50**, 2193.
16. I. Kucharikova (1973). *J. Appl. Polym. Sci.*, **17**, 269.
17. H. E. Adams, E. Ahad, M. S. Chang, D. B. Davis, D. M. French, H. J. Hyer, R. D. Law, R. J. J. Simkins, J. E. Stuchbury and M. Tremblay (1973). *J. Appl. Polym. Sci.*, **17**, 269.
18. K. Kamide, T. Terakawa and S. Matsuda (1983). *Brit. Polym. J.*, **15**, 91.
19. G. V. Schulz (1942). *Z. Phys. Chem. (Leipzig)*, **A176**(1942)317; and **B52**, 1.
20. R. M. Fuoss and D. J. Mead (1943). *J. Phys. Chem.*, **47**, 59.
21. B. H. Zimm and I. Myerson (1946). *J. Amer. Chem. Soc.*, **68**, 911.
22. I. Jullander (1945). Ark. Kemi, Mineral. Geol., **21A**, 142.
23. N. C. Billigham (1977). *Molar Mass Measurements in Polymer Science*, Kogan Page, London, Chapter 3.
24. T. Fujita Japanese Open Patent H1-280251.
25. G. F. Price (1959). In *Techniques of Polymer Characterization*, P. W. Allen (ed.), Butterworths Scientific Pub., London, Chapter 4.
26. B. H. Zimm, R. S. Stein and P. Doty (1945). Polymer Bull. **1**, 90.
27. M. B. Hugglin (ed.) (1972). *Light Scattering from Polymer Solutions*, Academic Press, London, Chapters 1, 4, 5 and 7.
28. B. H. Zimm (1948). *J. Chem. Phys.*, **16**, 1093; and **16**, 1099.
29. N. C. Billigham (1977). *Molar Mass Measurements in Polymer Science*, Kogan Page, London, Chapter 5.
30. G. C. Berry (1966). *J. Chem. Phys.*, **44**, 4550.
31. P. Debye (1947). *J. Phys. Coll. Chem.*, **51**, 18.
32. B. A. Brice and M. Halwer (1951). *J. Opt. Soc. Am.*, **41**, 1033.
33. P. Debye (1946). *J. Appl. Phys.*, **17**, 392.
34. J. P. Kratochvil, G. Dezĕlić, M. Kerker and E. Matijević (1962). *J. Polym. Sci.*, **57**, 59.
35. E. R. Pike, W. R. M. Pomeroy and J. M. Vaugham (1975). *J. Chem. Phys.*, **62**, 3188.
36. K. Kamide, Y. Miyazaki and H. Kobayashi (1985). *Polym. J.*, **17**, 607.
37. H. Staudinger and H. Freudenberger (1930). *Ber. Dtsch. Chem. Ges.*, **63**, 2331; H. Staudinger and R. Nozu (1930). *Ber. Dtsch. Chem. Ges.*, **63**, 721; and H. Staudinger (1961). *Arbeitserrinerungen*, Dr. Alfred Huthig, Heidelberg.
38. Y. Einaga, Y. Miyaki and H. Fujita (1979). *J. Polym. Sci., Polym. Phys. Ed.*, **17**, 2103.
39. M. Kurata, T. Tsunashima, M. Iwama and K. Kamada (1975). In *Polymer Handbook*, 2nd edn, J. Brandrup and E. H. Immergut (eds), Wiley, New York, Chapter 4.
40. A. Dondos and H. Benoit (1977). *Polymer*, **18**, 1161.
41. Y. Miyazaki and K. Kamide (1987). *Jap. J. Polym. Sci. Technol. (Tokyo)*, **44**, 1; and K. Kamide, A. Kiguchi and Y. Miyazaki (1986). *Polymer J.*, **18**, 919.
42. K. Kamide, Y. Miyazaki and Y. Abe (1979). *Makromol. Chem.*, **180**, 2801.
43. K. Kamide and M. Saito (1989). Viscometric determination of molecular weight in *Determination of Molecular Weight*, A. Cooper (ed.), John Wiley & Sons, Inc., New York, Chapter 8.
44. E. Hagenbach (1860). *Prog. Ann.*, **109**, 385.
45. J. Poiseuille, (1946). *Mém. Savants Étrangers*, **9**, 433.
46. H. Mizutani (1957). *Measurements on Degree of Polymerization, Experiments in Polymer Science Series*, No. 6, Society of Polymer Science of Japan (ed.), Kyoritu Pub., Tokyo, Chapter 2.
47. N. C. Billingham (1977). *Molar Mass Measurements in Polymer Science*, Kogan Page, London, Chapter 7.

48. P. F. Onyon (1959). In *Techniques of Polymer Characterization*, P. W. Allen (ed.), Butterworths Scientific Pub., London, Chapter 6.
49. J. Schurz and E. H. Immergut (1952). *J. Polym. Sci.*, **9**, 279.
50. J. Hermans and J. J. Hermans (1958). *Proc. Koninkl. Ned. Akad. Wetenschap.*, **B16**, 321.
51. San Denshi Co. Ltd, brochure of Automatic Viscometer model AVS-2.
52. M. L. Huggins (1942). *J. Amer. Chem. Soc.*, **64**, 2716; and M. L. Huggins (1943). *Ind. Eng. Chem.*, **35**, 980.
53. E. O. Kraemer (1938). *Ind. Eng. Chem.*, **30**, 1200.
54. K. Kamide, M. Saito and M. Hattori, unpublished.
55. M. Takeda and T. Tsuruta (1952). *Bull. Chem. Soc. Jpn.*, **25**, 80.
56. D. J. Streeker and R. F. Boyer (1954). J. Polym. Sci., **14**, 5.
57. K. Kamide (1964). *Chem. High Polym. (Tokyo)*, **21**, 152.
58. M. Kaneko and N. Kuwamura (1957). *High Polym. Jpn.*, **6**, 324.

Appendix: Manufacturers' address list

VPO

Corona Electronics Co	3517 Higashi-Ishikawa, Katsuta, Ibaraki 312, Japan
Dr. H. Knauer Wissenschaftliche Geräte KG	Heuchelheimer Str. 9, 6380 Bad Homburg, Germany
Jupiter Instrument Co. Inc	407 Commerce Way, A3, Jupiter, FL 33458, USA
Gonotec	Eisenacher Str. 56, 1000 Berlin 62, Germany

Membrane osmometers and membranes

Jupiter Instrument Co. Inc.	As above
Dr. H. Knauer Wissenschaftliche Geräte KG	As above
Gonotec	As above
Sartorius-Membranefilter GmbH	34 Göttingen, Germany
Schleicher and Schuell Inc.	Keene, NH 03431, USA

Light-scattering and refractometers

Otsuka Electronics Co. Ltd	3-26-3 Shoodaitajika, Hirakata, Osaka 573, Japan
Polymer Laboratories Ltd	Essex Road, Church Stretton, Shropshire SY6 6AX, UK
Wyatt Technology Corp.	802 East Cota Street, PO Box 3003, Santa Barbara, CA 93130, USA
Brookhayen Instruments	750 Blue Point Road, Holtsville, NY 11742, USA
C. N. Wood Mfg. Co.	New Town, PA 18940, USA

Automatic viscometers

Schott Geräte GmbH	Im Langgewann 5, Postfach 1130, D-6238 Hofheim a. Ta., Germany
San Denshi Ind. Co.	Takanawa Kaneo Bldg. 3-25-22, Takanawa, Minato-ku, Tokyo 108, Japan
Shibayama Scientific Co	3-11-8 Minami-Otsuka, Toshima, Tokyo 108, Japan
Viscotek	1032 Russell Drive, Porter, TX 77365, USA

6 Chromatographic methods

A. J. HANDLEY

6.1 Introduction

Since the early 1940s and the pioneering work of Martin and Synge, chromatographic techniques have played a major role in helping the analyst to solve problems.

It was not, however, until the early 1960s, with the commercial development of size-exclusion chromatography, that polymer chemists began to explore the potential separating and characterising power of such techniques.

In this chapter the usefulness of the many different chromatographic techniques is highlighted, i.e. the more mature and commercially established techniques such as size-exclusion chromatography (SEC), high pressure liquid chromatography (HPLC), gas chromatography (GC) and thin-layer chromatography (TLC), the newer techniques such as field flow fractionation (FFF), supercritical fluid chromatography (SFC) and hydrodynamic chromatography (HDC), and the coupled technologies, where the chromatographer is looking to harness the appreciable identifying power of the larger spectroscopic techniques such as mass spectroscopy (MS), fourier transform infrared (FTIR) and nuclear magnetic resonance (NMR).

First, it should be said that polymers are a very diverse and complex group of materials. They can manifest themselves in many shapes and forms—they can be viscous colourless liquids, powders, coloured granules, cast or extruded sheet, transparent or translucent products. The type of characterisation asked of the chromatographic technique thus can be equally complex and varied, from the straightforward determination of the molecular size of the polymer to the breakdown, identification and quantification of a fully formulated material.

How then is it decided which chromatographic technique is suitable to the problem? Molecular weight is a key parameter in polymer chemistry, and as such this parameter can be used to select the type of chromatographic technique suitable to tackle the problem. Figure 6.1 illustrates the type and range of the separators available to the polymer chemist. In general, for low-molecular-weight characterisation, gas chromatography–high pressure liquid chromatography and supercritical fluid chromatography can be used; for higher molecular weights, size-exclusion, thin-layer and thermal field flow fractionation come into their own; and finally for the large molecules and particles,

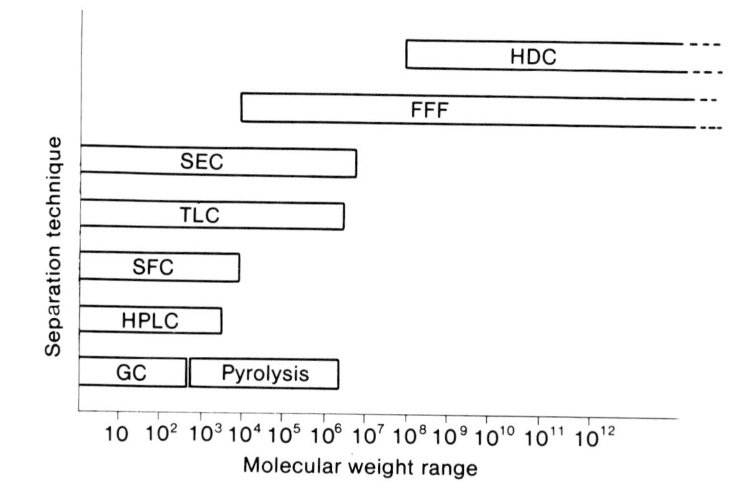

Figure 6.1 Typical working range of chromatographic techniques.

sedimentation field flow fractionation and hydrodynamic chromatography provide gentler separation modes.

Building on this simple selector model, the various techniques, the mechanism of separation, the key technologies used and their applications in polymer characterisation are described.

6.2 Size-exclusion chromatography (SEC) or gel-permeation chromatography (GPC)

This technique is possibly the most widely used chromatographic technique in polymer analysis. It is capable of characterising very high-molecular-weight film and fibre polymers up to 10^6 molecular weight, thermoset resins of molecular weight 10^3 and 10^4, oligomers or pre-polymers in which only a few monomer units are joined, and finally simple molecules present in commercial polymers in the form of additives. SEC thus has the ability for separation and quantitation, but it also offers considerable potential in its use for determining molecular size (weight) and molecular-weight distributions.

6.2.1 Chromatographic process and instrumentation

Size-exclusion chromatography in the main uses conventional liquid chromatographic equipment (Figure 6.2). In SEC a dilute polymer solution is injected into the solvent stream, which then flows through a column or series of columns, packed with material of narrow particle size and controlled pore size (nominally cross-linked styrene/divinyl benzene gels or silica), the pore size being comparable to the size of the molecules to be separated. The solvent molecules pass through and around the packing media, carrying the polymer molecules with them, where possible. The smaller molecules are able to pass through most of the pores and so have a relatively long flow path through the column. The larger polymer molecules are excluded from all but the largest of pores and hence have a shorter flow path, and elute first. The process relies on there being no interactions between the sample and the packing material.

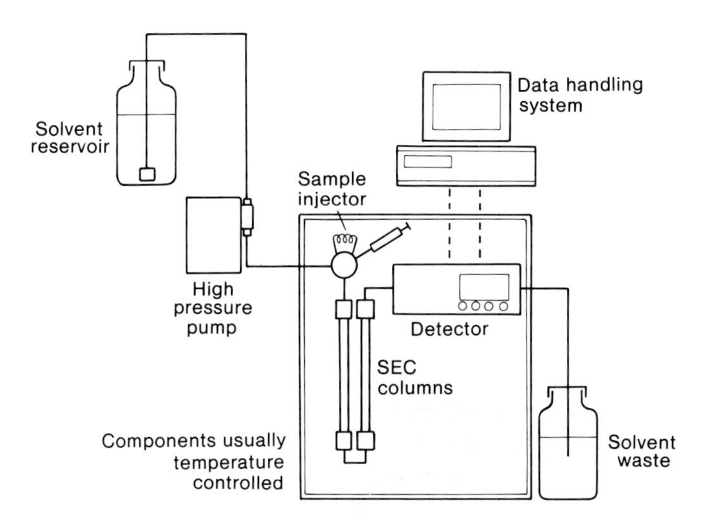

Figure 6.2 Schematic diagram of typical SEC apparatus.

A more detailed discussion on the separation process is given in Janca's paper [1]. As the polymer molecules elute from the column, they are detected by using a suitable concentration detector, to produce an elution volume curve (concentration versus time). A distribution can then be produced after calibration with narrow dispersed polymer standards, usually polystyrene. Figure 6.3 shows a typical molar-mass or molecular-weight distribution produced by a commercial data system. This provides an ideal fingerprint of the polymer and is extremely useful for comparative purposes to show sample trends, polymer growth and degradation and batch-to-batch variations. By suitable data manipulation, molecular-weight averages can also be calculated. Its main advantage over other molar-mass techniques is its speed of analysis. Data can be obtained in a relatively short time. It can be readily automated, and, when used with modern liquid chromatographic equipment such as autosamplers and high-precision pumps, can provide high-quality reproducible data.

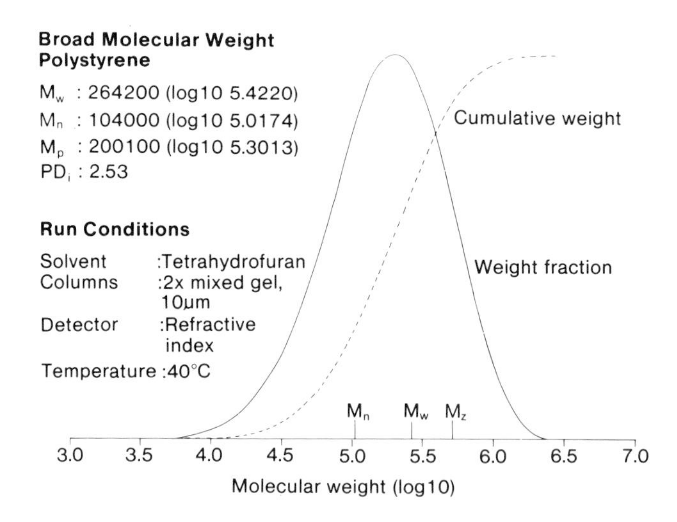

Figure 6.3 SEC molecular-weight determination curve.

6.2.2 *Key components in SEC analysis*

6.2.2.1 The solvent. The solvent should be chosen to best dissolve the sample, to be compatible with the column packing and to permit detection. The most commonly used eluents in SEC are tetrahydrofuran (for polymers that dissolve at room temperature), *o*-dichlorobenzene and trichlorobenzene at 130°C and 150°C (for crystalline polyolefins) and phenolic solvents at 100°C (for condensation polymers, such as polyamides and polyesters). For the more polar polymers, dimethylformamide and aqueous eluents may be employed, but care is required in avoiding interactions between the polymer (sample) and the gel (packing). Flow rates of 1 ml/min are typical for SEC analysis.

6.2.2.2 The column. The heart of the SEC system is the column, packed with a porous gel to effect separation. In non-aqueous or organic SEC, the most common packing materials are polystyrene–divinylbenzene gels, which are highly cross-linked, macroporous and spherical in nature. Silicas have also been used in this area. For aqueous SEC, silanised silicons or hydrophilic polymer-based packings with polyhydroxyl and polyacrylamide surfaces are used. Unlike conventional chromatography (HPLC), where a single column is used, SEC involves the use of a number of columns, suitably joined together with low dead-volume unions. A series of differing pore-size columns can be joined together to cover the desired molecular-mass range. Alternatively, mixed-bed packings can be used, which, as the name implies, contain a balanced mix of pore sizes and cover in one column a wide molecular-mass range.

Modern SEC columns are capable of achieving efficiencies up to 80 000 plates per metre and are available in 3, 5, 10 and 20 μm particle sizes. Analytical columns are typically 300 mm × 7.5 mm i.d. and preparative columns 300 or 600 mm × 25 mm i.d. The strategy of using multiple columns ultimately results in longer analysis times. An SEC separation normally takes 20–30 min.

6.2.2.3 The detector

6.2.2.3.1 Mass-concentration detectors. The traditional detector used in SEC is the differential refractometer (DRI). It is considered a universal concentration detector because almost all polymer–solvent mixtures have a significant refractive-index difference. The detector offers a linear response over a wide range of concentrations, and, with modern cell design, minimum band broadening allied to the capability of being used at high temperatures (up to 150°C). It relies on detecting a difference in the refractive index of the eluent stream, as the solute emerges from the column, with respect to a reference solvent stream. These differences can be both positive and negative. Because refractive indexes are temperature-dependent, these detectors are more sensitive to changes in ambient temperatures than most other detectors.

The ultraviolet absorption (UV) detector is also commonly used when the polymer or additive contains chromophores which will absorb in the UV region. Like the DRI, it offers a wide linear range, but can only be used with non-UV-absorbing solvents. It can only be operated at moderate temperature, but performs well with mixed solvent systems. Variable-wavelength detectors typically cover the range 190 nm–370 nm. UV detectors in series with other detectors can be used to measure the change in copolymer composition as a function of molar mass [2]. The most recent developments in spectroscopic detection have resulted in instruments based on photodiode array detection. These detectors can provide the full absorption spectrum from 200 to 700 nm every 0.1 s. These detectors can thus provide a continuous measurement of the full spectral absorbance with time, and as such can be used to monitor chemical changes in complex polymer systems. Photoresist mixtures containing phenol–formaldehyde resins have been characterised by using such detector systems [3].

Another specific concentration detector is the infrared (IR) flow-through spectrometer. This detector monitors a specific absorbing group in the solute or polymer. Like the UV detector, the IR detector is relatively insensitive to fluctuations around a set temperature and is therefore particularly suitable for temperatures far above ambient conditions. Indeed, one of the uses for IR detection is in the molar-mass measurement of polyolefins at temperatures exceeding 140°C. Infrared detection is also useful in measuring polymer stoichiometry as a function of molar mass (by monitoring specific functional groups in a copolymer [4]). The detector, however, is solvent-limited in terms of its universal use, as a suitable workable absorbance window has to be found to detect the polymer in the presence of a solvent which itself can possess a strong infrared spectrum.

6.2.2.3.2 Molecular-mass detectors. Size-exclusion molar-mass calibration is a complicated matter. Calibration curves differ for different polymer types, and for most commercial polymers, direct molar-mass calibration is not possible because of the lack of suitable known molecular-weight standards of sufficiently narrow mass distribution, and of the same chemical structure. The alternative is to determine the polymers' molar mass in the SEC eluent, *in situ* by the use of an on-line molar-mass-sensitive detector. Two such detectors are now commercially available, the light-scattering detector and the recently introduced viscosity detector[5–8].

SEC with on-line light-scattering detection (SEC/LS)

Such detectors are connected in-line prior to the concentration detector (e.g. differential refractive index, UV or IR) in the SEC system. By combining the outputs from both detectors, absolute measurements of molecular weight and direct molecular-mass distributions, independent of column calibrations or assumptions concerning the hydrodynamic size to molecular weight relationships, can thus be obtained rapidly. The analysis may be performed in aqueous or organic media at temperatures ranging from 0°C to 165°C.

Two types of detector have been used. The low-angle laser light photometer (LALLS), uses a high intensity He–Ne laser source, allowing the measurement of very small scattering volume. The low-angle capability allows measurements to be made at angles as low as 2°, thus eliminating the necessity of extrapolating data to zero angles and concentration, as with conventional light-scattering techniques. The cell volume of the detector is only 10 μl, minimising band broadening in the combined system.

The second type of detector, again suitable for on-line use, is the multi-angle laser light scattering photometer (MALLS). This again uses a laser light source and low-volume cell, but instead of collecting data at only one angle, as with LALLS, measurement can be made at fifteen discrete scattering angles. This allows the deduction of not only molecular weight but also size and conformation. Both detectors suffer from lack of sensitivity at low molecular weights, but have still been used extensively for the characterisation of many polymer systems. Homopolymers such as polyethylene [9], polybutadiene [10], polyisoprene [10], poly(methyl methacrylate) [11], polyamides [12] and polyacrylamides [13] have been analysed, and block copolymers for molecular weight and compositional heterogeneity [14], styrene–butyl acrylate emulsion copolymers for molecular weight [15], linear polyethylene for radius of gyration [16] and solution characterisation of branched polymers [17,18] have been studied.

SEC with on-line viscometer detection

On-line viscometer detectors are now becoming extremely common in characterisation laboratories, as they can easily be incorporated into an SEC system and can be operated with a wide variety of solvents and temperatures. The detector, when combined in series with a mass-concentration detector, can be used to generate absolute molar-mass averages, intrinsic viscosity and long-chain branching information.

A number of commercial designs are now available. In all of these instruments the solution viscosity is measured by the pressure drop across a flow-through capillary, and is monitored by a differential-pressure transducer. One-, two- and four-capillary designs have been used [19–21]; the single capillary, unlike the other two, is sensitive to flow-rate fluctuations, and therefore special pulsation dampeners have to be used. As with the SEC/LS systems, data from the twin detectors have to be captured and processed to produce the intrinsic-viscosity, Mark–Houwink (for branching) and molecular-weight distribution plots. The sensitivity of the viscometer is much greater for low-molecular-weight components than that of light-scattering; for high-molecular-weight components, the converse is true. One of the unique features of viscometry with universal calibration is that molecular-weight data can be obtained from chemically heterogeneous polymers. This is possible for light-scattering only with difficulty. It has found use in the characterisation of many organic SEC separations [22–24], evaluation of branched polymers [25], in the determination of M_n of copolymers, blends, stereoisometric, branched and star polymers [26], and for cationic polymers [27].

SEC With on-line viscometer and light-scattering detection

This combination of detectors coupled with size-exclusion separation possibly offers the ultimate capability for characterising molecular weight and branching of polymers. This technology has recently been commercialised with the development of a low-cost light-scattering detector. The right-angle light-scattering detector (RALLS), as its name implies, uses a 90° scattering cell; opaque glass is used to mask the scattered light and minimise background. A laser diode (670 nm) is used as the light source in place of the traditional gas laser, thus producing a low-cost optical source. Scattering at 90° has been found to be the optimum choice because of its higher signal-to-noise ratio over other scattering angles, and it does not require solvent–glass refraction corrections to determine the true scattering angles. The use of a viscometer detector in series with the RALLS allows one to correct the 90° light-scattering data to zero angle, so the practical considerations of using a single measurement at the optimum angle become dominant. This triple detection system RI–VIS–RALLS can provide extensive information on the polymer. With suitable software to assess the data from all three detectors, accurate fast

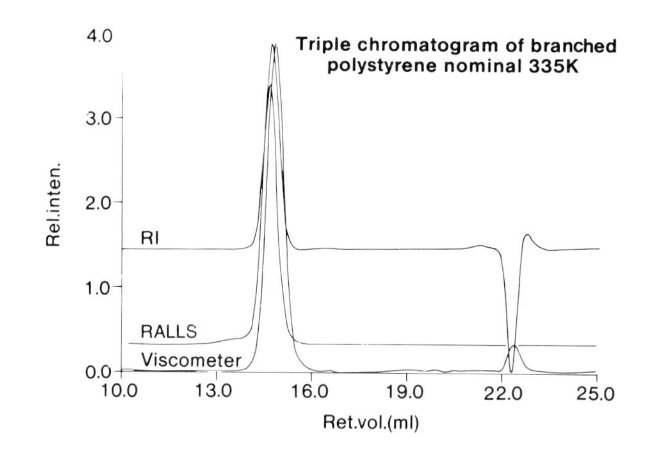

Figure 6.4 SEC with on-line viscometer, refractometer and light-scattering detection. Reproduced by courtesy of Viscotek Inc.

molecular-weight distribution data can be obtained without column calibration for both linear-coil and branched-coil polymers up to 3×10^6 molecular weight. In addition, visual inspection of the various detector traces can yield information on high-molecular-weight tails in a molecular-weight distribution, and aggregation in sample solutions yields information at the high-molecular-weight end of the distribution, where the sensitivity of the RALLS comes into its own (see Figure 6.4).

6.2.3 Compositional separation and detection

As previously stated, polymers are generally highly complex, multi-component materials. They can vary in molecular mass or weight, composition, sequence length, branch length, branch frequency and tacticity. Size-exclusion chromatography is primarily directed at detecting the concentration of each different molecular mass or weight present. This approach is quite straightforward for linear homopolymers. However, if the polymer is a linear copolymer, then the molecules can vary in composition and sequence length as well as molecular weight. The analysis of such systems often needs either a combination of detectors or linking to secondary separation systems.

6.2.3.1 Preparative size-exclusion chromatography

One possible solution to the above problem is to fractionate the polymer using preparative-scale SEC, then examine the fractions with powerful off-line techniques such as mass spectrometry, Fourier transform infrared and nuclear magnetic resonance. This technique can also be used to isolate pure polymer fractions for subsequent use in SEC calibration.

Normal analytical scale SEC equipment can be used, with the essential additions of a high-volume injector and, to reduce operator time, an automatic fraction collector. The columns used contain typical SEC packing materials, but are normally 25 mm internal diameter and 30 or 60 cm in length. Such columns offer up to a tenfold increase in loading over normal 7 mm analytical columns. Flow rates of 10 ml/min of eluent are nominally used. Fractions are taken at specified points in the molecular-size distribution, as monitored on the RI detector at the end of the column. The eluent from the detector is then collected by the fraction collector into designated sample tubes. These fractions are then evaporated to dryness and examined by using other techniques. Since an SEC separation is normally achieved irrespective of the solvent used, volatile solvents can be chosen for SEC analysis to aid fraction isolation. Larger columns (122 cm × 55 mm) are commercially available, but these require more specialised and dedicated preparative instrumentation, as flow rates in excess of 100 ml/min are required, but they obviously offer higher sample loading on the columns. This approach has been used in the study of styrene–methyl acrylate copolymers [28], for poly(vinyl chloride) [29], high-molecular-weight polystyrene, low-molecular-weight epoxy resins [30] and cyclic oligomers of isoprene [31].

6.2.3.2 SEC linked to Fourier-transform infrared

The combination of these two powerful characterisation techniques can provide the ability simultaneously to characterise the molecular distribution and to identify and quantify IR-active functional groups in the distribution. The main problem in linking such technologies is the choice of solvent or eluent used to carry the polymer through the SEC system. Many of the typical SEC solvents have strong absorption bands in wide regions of the infrared spectrum. Developments have centred around the use of low-volume cylindrical internal-reflection flow cells for use in the FTIR system. However, their use is still restricted in many cases to IR-friendly SEC solvents.

Some hope is, however, now being offered with interfaces being designed that remove the solvent prior to infrared analysis. One design recently reported uses a similar approach to that of mass-spectroscopy interfacing, combining the use of a moving belt and a thermospray system. The eluent from the SEC column is deposited on to moving tape via a heated stainless-steel capillary (thermospray). This evaporates the solvent, and the deposited polymer is carried on the tape to a diffuse-reflectance cell.

A second approach, recently commercialised, spray-deposits the eluent from the chromatograph on to a moving collection surface (germanium disc). The solvent is then evaporated in the deposition process, and the polymer is deposited as a dry track. This disc can then be physically taken and scanned in an FTIR spectrophotometer and spectra in the various regions of the chromatogram can be collected. By using the approaches described in this section a number of polymer systems have been examined: ethylene-based polyolefin copolymers [32], low-density polyethylene, poly(vinyl chloride), polycarbonate, unsaturated polyester and polymethyl methacrylate [33], cellulose nitrate [34], linear low-density polyethylene [35] and polymeric components of nitramine propellants [36].

6.2.3.3 SEC linked to nuclear magnetic resonance (NMR) spectroscopy

Of all the possible identification techniques, NMR offers the most in providing information to the polymer chemist. Its link to an SEC system has in the past only been considered in the off-line mode, fractions being collected from a preparative SEC system and subsequently analysed by NMR. On-line linkage has been attempted but has suffered from poor sensitivity and the need for total use of deuterated solvents, thereby making analysis both difficult and costly. Recent developments have now made GPC–NMR a commercial reality. The newer 500 MHz instruments equipped with multiple solvent-suppression software allow the use of typical SEC solvents with small amounts of their deuterated counterparts. Sensitivity, too, has been increased by the use of specially designed ^{1}H probe heads with smaller flow-through cell volumes. The result of these improvements is systems capable of analysing not only structure in both polymers and copolymers but also molecular weight. To date, only a small number of papers have been published using this technology. These include its use in characterising isotatic polymethyl methacrylate [37] and chloral oligomers [38].

6.2.3.4 Orthogonal chromatography

Orthogonal chromatography (OC) or coupled column chromatography (CCC) is a technology which combined multiple separation techniques with selective detection. To accomplish the desired fractionation two SEC systems are coupled together (Figure 6.5) so that the eluent from the first flows through the injection valve of the second instrument. The first instrument is operated as a conventional SEC system; the second is operated to utilise non-size-exclusion mechanisms. At the exit from this second system a selective detector is used. In this combined system, which can be either manually or computer controlled, the polymer solution is injected on to the first column where the copolymer or polymer mix is separated according to molecular size. At a desired time during its elution through the column—say at the start of the molecular-size separation (as monitored by the refractive index detector)—the flow in the system is stopped and the fraction trapped in the injection loop of the second system is transferred to that system. This will contain species of the same molecular size (hydrodynamic volume) but different compositions and/or sequence lengths. This fraction then enters the flow stream of the second system, which can be operated with either a second SEC column but different solvent (usually a polymer solvent/non-solvent mix) or a reactive column (similar to those used in HPLC) with again a solvent mixture. The resulting 'reactive' separation is then monitored using a suitable mass-sensitive detector (refractive-index or evaporative detector; if a solvent gradient is used see section 6.3) or selectively detected using either a variable-wavelength or diode-array detector. This approach has been used in the characterisation of polystyrene–poly(butyl methacrylate) statistical copolymers [39, 40, 42] and polystyrene–acrylonitrile copolymers [41].

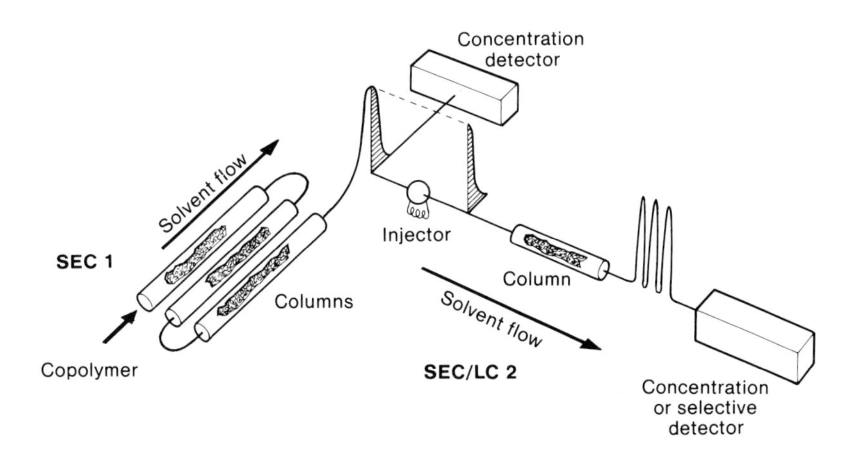

Figure 6.5 Orthogonal chromatography.

6.2.4 The uses of size-exclusion chromatography

SEC in its many forms is capable of providing a vast amount of information on polymers and polymer systems. The molecular weight averages (number, weight and viscosity) can be obtained as described in this chapter. However, the most useful and well-used parameter produced is the molecular-weight distribution curve, which can be used to fingerprint the polymer. Subtle differences in all of the above parameters can affect many of the end-use properties of the polymer such as impact, tensile and adhesive strength, brittleness, drawability, cure time, melt and flow characteristics, solution properties and hardness.

When used in a comparative mode, the SEC molecular-weight distribution can provide data on differences in polymer batch-to-batch variation, growth of a polymer with reaction time, production of high/low-molecular-weight tails in the distribution, degradation of polymers, monitoring of the quality of materials and checking on the purity of monomers, initiators and other starting materials, and comparison of competitors' materials. There are many literature references on the use of SEC in polymer analysis; a useful bibliography can be found in ref. 44. It has found use in many industrial areas: PVC [45], coatings [46], engineering plastics [47], condensation polymers [48], biomedical plastics [49], urethanes [50, 51] and water-soluble polymers [52–55]. As can be seen, the technique offers much both to the researcher and to the polymer manufacturer, with its capability for use in quality control [56] and even on-line analysis [57].

6.3 High-pressure liquid chromatography (HPLC)

High-pressure liquid chromatography is capable of producing much more efficient separation than SEC. As a result, its main use is in the separating, identifying and quantifying of low-molecular-weight species. Polymer additives, starting materials and residual monomers lend themselves to determination by this technique. Recent work also suggests that it can be applied to bulk polymers, in particular copolymers, to gain information on their chemical composition (see section 6.2.3.4).

6.3.1 Chromatographic process and instrumentation

HPLC uses the same instrumentation as shown in Figure 6.2 and, as with SEC, the sample has to be soluble in a suitable solvent. In this technique, the sample is injected into the solvent stream or mobile phase, which, unlike that for SEC, is usually composed of two or more solvents and can vary in composition with time if suitably programmed (gradient elution). The sample flows through the column, which is normally packed with silica or bonded silica commonly 5 μm in particle size, and the separated bands are detected using a concentration detector. This detector is coupled to a data station capable of integrating the signal to produce quantitative data. Unlike SEC, HPLC relies on reaction of the sample with the packing material in the column to effect separation. The two predominant separation modes used for polymer characterisation are normal- and reverse-phase.

6.3.2 Normal-phase chromatography

This refers to separations carried out using a polar stationary phase (packing material) and a non-polar organic mobile phase. The packing material commonly used is silica, which is either irregular or spherical in shape and porous in nature, thereby providing a high surface area for interaction. Bonded silicas can also be used—amino, cyano and diol are the most common. Retention of analytes is primarily dependent on their polar interaction with the polar column surface. The least polar components of the mix to be separated will elute first, followed by the medium-polarity and finally the most polar. Correspondingly, the elution power of the mobile phase increases with its polarity. Hydrocarbon solvents are predominantly used, with smaller amounts of stronger eluents such as dichloromethane, tetrahydrofuran, acetonitrile and methanol. However, care must be taken to ensure that the solvent mix is completely miscible, as such 'modifiers' are only soluble at low concentrations. The adsorption mode has been used extensively for compositional separations of di- and tri-block copolymers, stabilisers, thermoplastic elastomers and adhesives.

6.3.3 Reverse-phase chromatography

In this mode, the interactions are reversed compared with normal-phase separation. Non-polar stationary phases (prepared by bonding alkyl substituents to a silica surface) and a polar mobile phase are used. The most common stationary phase is octadecyl silane (ODS) which is produced by chemically bonding $-(CH_2)_{17}CH_3$ groups to silica. This makes the surface hydrophobic in nature, and non-polar in comparison with silica. C_8-, C_6- and phenyl-bonded phases are also available. The retention of the analyte in this mode depends on the degree to which it is partitioned into the stationary phase, and is largely determined by the hydrophobic interaction with the mobile phase. Both the polarity and the molecular size of the analyte will affect the separation, and generally large non-polar molecules are the most retained (see Figure 6.6). The order of elution for similar-sized molecules is the reverse of that of a normal-phase separation. Mobile phases used include water, methanol, acetonitrile, tetrahydrofuran and dioxan. Buffers can also be used if ionisable compounds are present. This mode is particularly suited to the separation of homologues and low-polarity compounds, and is used extensively for polymer

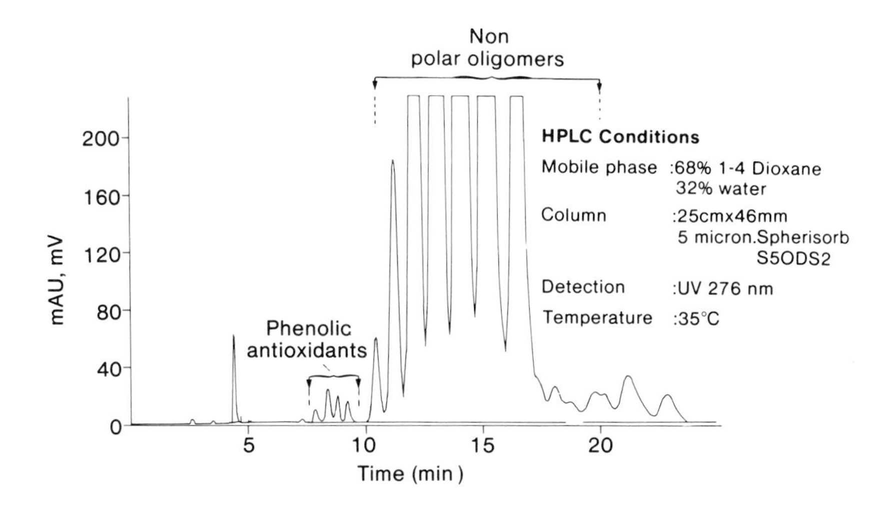

Figure 6.6 HPLC analysis of phenolic anti-oxidants in low-molecular-weight acrylic resin.

additive analysis. It should, however, be noted that often these materials will have to be separated from the bulk polymer prior to the analysis, unless the polymer is of sufficiently low molecular weight or has a high solubility in the mobile phase.

6.3.4 Key components in HPLC analysis

6.3.4.1 The pumping systems. These need to be of sufficient accuracy (flow rate RSD < 0.7) in order to produce good reproducible separations. In HPLC it is quite common to employ mixed solvent systems (isocratic elution). Such solvent compositions can be achieved by using a gradient pumping system and merely dialling up the desired solvent ratios from the solvent reservoirs used in the system. Alternatively, when it has proved impossible to effect a good separation by isocratic elution, gradient elution is often used. Here, the pump can be programmed to change solvent composition with time during the analytical run. Gradient elution is a powerful tool for method development.

6.3.4.2 Detectors

6.3.4.2.1 Ultraviolet detectors. These are the most common detectors used in HPLC, offering good stability and sensitivity. They are best suited to the detection of aromatics; however, by careful choice of solvent, they can be used at wavelengths down to 210 nm. The generic development of diode array detectors (see section 6.2.2.3.1) has now given the technique a powerful quantifying/identifying detector.

6.3.4.2.2 Refractive-index detectors. These detectors are used when the analyte of interest cannot be detected by UV. Their use, however, is somewhat limited as they lack both sensitivity and stability and their very design makes them impossible to use with gradient elution and difficult to use with mixed (isocratic) solvents.

6.3.4.2.3 Evaporative detectors. These detectors represent a viable alternative to refractive-index detectors for HPLC analysis. They are designed such that they can handle both isocratic and gradient elution. In these detectors, the eluent from the HPLC column is nebulised by a gas stream. The vapour then goes into a heated tube or column which allows solvent vapourisation. The non-volatile sample stays as a mist, which goes through a light beam. The scattered light is collected by a photomultiplier located at 120° from the light source, and the observed signal is thus proportional to the concentration. It should, however, be noted that, since the intensity of the scattered light depends on particle diameter, the response of the detector is influenced by the changing particle-size distribution as a function of sample size. This detector has been used in the characterisation of triglycerides [58] and surfactants [59].

6.3.4.2.4 Mass spectrometry

The use of mass spectroscopy and HPLC offers great potential for the quantification and identification of polymer additives. However, its adoption has been fraught with practical problems, created by the very nature of the eluent from the chromatograph. One possible solution to this problem is the moving-belt interface, in which the eluent is sprayed on to a circulating belt and the solvent evaporated in stages, so that finally only the analyte is carried into the mass spectrometer. This approach has been successful in the separation and identification of some polymer additives [60]. However, it can only be used for CI and EI mass spectroscopy, and the interface can only be operated with mobile phases containing less than 50% water. Newer interfaces such as thermospray and particle beam offer potential in this area.

6.3.5 The uses of high-pressure liquid chromatography

High pressure liquid chromatography has found extensive use in the quantitative determination of polymer additives. Antioxidants [61], polymeric amine light stabilisers in polyolefins [62], Tinuvin and Irganox stabilisers [63], plasticisers [64–66], dicarboxylic acids [67] and dialkyltin compounds [68] have all been analysed using this technique. Moving up the molecular-weight range, epoxy [69, 70] and phenol–formaldehyde resins [71], polycarbonate oligomers [72], oligoethylenes, oligostyrenes and oligoethylene glycol non-phenyl ethers [73] have also been separated. Finally, HPLC has also found uses in the compositional analysis of acrylonitrile–butadiene copolymers [74], styrene–methacrylate copolymers [75] and block copolymers [76], acrylic copolymers [77], and even in the separation of stereoisomers of PMMA [78].

6.4 Thin-layer chromatography (TLC)

This is a somewhat underestimated technique. Its principal advantages are that it is simple and quick to carry out, and usually does not involve expensive equipment. It has been used in the separation of many polymer systems.

6.4.1 Chromatographic process and instrumentation

Unlike the situation in most other chromatographic techniques, the stationary phase is not packed into a column, but is coated as a thin layer on to a glass, plastic or metal plate. The samples in solution are applied as spots or bands on to the bottom edge of the plate, which is then placed in a closed tank containing the mobile phase (usually a mixed solvent system), and the solvent is drawn up the plate by capillary action thereby effecting the separation. Two basic separation modes are used in polymer analysis.

6.4.2 *Adsorption TLC*

In this mode, a silica stationary phase is used, along with organic mobile phases such as dichloroethane, chloroform, diethyl ether, methyl ethyl ketone, acetone, tetrahydrofuran and dioxan. The various macromolecular segments of the sample interact with the adsorption centres of the pore surface of the silica stationary phase, and hence a separation is effected. Polymer chains with different chemical structures can be separated from each other, and hence information on the compositional heterogeneity of copolymers can be obtained. The technique has also proved useful in the separation of stereoisomeric macromolecules according to ends, and in looking at polymer architecture (block structure and branching). Low-molecular-weight homopolymers can be characterised by molar mass, and polymer additive separation can also be achieved. Usually more sophisticated mobile phases need to be used in these applications.

6.4.3 *Precipitation TLC*

This mode is used for the separation of higher-molecular-weight polymers according to molar mass. The process of PTLC is related to the separation of a polymer solution into a dilute phase and a concentrated gel phase precipitated on the surface of the stationary phase. The dilute phase is transported with the mobile phase flow, and thus separation is effected. In order to carry out PTLC, polymer adsorption should be suppressed. Mixed mobile phases are usually used, consisting of a highly polar precipitant and a less polar solvent. Whilst SEC seems to be a more suitable technology for such problems, TLC using a two-dimensional separation (the sample is further separated at 90° to the first elution) means it is possible to determine simultaneously the molecular-weight distribution and the compositional inhomogeneity.

6.4.4 *Key components of TLC*

6.4.4.1 TLC plate. The majority of TLC experiments are now carried out on ready-made plates, consisting of the absorbent (silica, alumina or alkyl-bonded silica), the support and an inert binder. Conventional TLC plates are normally 20 cm × 20 cm in size, have an absorbent layer of 100–250 μm thickness, and are coated with 20 μm particle-size material which can be obtained in differing pore sizes (important to polymer separation). Typically 1–5 μl of sample are applied. More recently, high-performance TLC plates have been developed which are smaller in size (10 cm × 10 cm) and are coated with smaller particle-size absorbants (5–15 μm) to increase separating power. When such plates are used, sample loadings can be as little as 0.1–0.2 μl.

6.4.4.2 Sample application. Samples to be analysed by TLC must be applied to the layer so as not to degrade the efficiency of the layer. The sample aliquot must not be large enough to cause volume or mass overload of the layer. If quantitative results are needed the sample can be applied via either known-volume capillaries or microsyringes. Various mechanical devices employing dosimeters, microsyringes or spray-on techniques are also now available.

6.4.4.3 Chromatographic development. Several approaches are available for layer development, depending on whether capillary or forced-flow conditions are used. TLC in its simplest form uses a glass chamber into which is placed the solvent, and separation is achieved by ascending development. Horizontal development chambers are now, however, becoming more popular, thus allowing development of two sets of samples simultaneously. More expensive pressure systems are also finding use in this area to speed up development time. In such systems a mechanical pump is used to force the solvent through the layer.

6.4.4.4 Detection. The spots or bands separated can be detected by spraying the chromatogram with an appropriate visualising reagent, exposing the TLC plate to a reagent vapour, or observing luminance extinction on TLC plates impregnated with fluorescent indicator. If quantitative data are required, scanning densitometers are now commercially available. Measurements are commonly made in the reflectance mode, and occasionally in the transmission mode, by moving the plate on a scanning stage through the measuring beam shaped into the form of a rectangle by adjustable slits. The signal is then detected by using either a photomultiplier tube or a photodiode.

Another approach to detection involves the use of flame-ionisation detection (commonly used in gas chromatography). This detector requires that the separation is carried out not on a TLC plate but on a custom-designed quartz rod, coated with a thin layer of silica. After chromatographic development, the rod is placed in the instrument, where it is advanced at a constant speed through the hydrogen flame of the flame-ionisation detector. Organic substances separated on the thin layer are ionised by the flame, and the ions generate an electric current with an intensity proportional to the amount of each organic substance entering the flame, thereby allowing quantitative determinations [79].

6.4.5 The uses of thin-layer chromatography

TLC has been used in the study of many homopolymers: polystyrene, poly(methyl methacrylate), poly(ethylene oxide), polyisoprene, poly(vinyl acetate), poly(vinyl chloride) and polybutadiene. Their molecular weight, molecular-weight distributions, microstructure (stereo-regularity, isomerism and the content of polar end groups), isotope composition and branching have been studied. For copolymer characterisation (e.g. purity and compositional inhomogeneity), random copolymers such as styrene–methacrylate, and block copolymers such as styrene–butadiene, styrene–methyl methacrylate and styrene–ethylene oxide have been separated. A good review article on polymers and oligomer separators is ref. 80. The technique can also be used in the separation and identification of additives in polymer systems [81, 82].

6.5 Gas chromatography (GC)

Gas chromatography by description does not seem to be a suitable technique for polymer analysis. It is, however, used extensively in the determination of the more volatile, thermally stable polymer additives and monomers, and its ease of linking to both mass spectrometry and infrared spectroscopy makes it an elegant technique for separation, identification and quantification.

One area which is finding increasing use is the monitoring of emissions of hazardous materials during polymer processing (extrusion, moulding and coating) and storage. Determination of such materials often requires specialised analysis and involves particular GC sampling techniques such as 'head-space analysis' and 'thermal desorption'. The polymer itself can also be studied by pyrolysis–GC to yield information about its identity (when coupled to a mass spectrometer), its structure and its purity. In addition, inverse gas chromatography can give physical-property information.

6.5.1 Chromatographic process and instrumentation

The basic instrumentation for gas chromatography is shown in Figure 6.7. The separation is carried out using a gaseous mobile phase which transports the sample through a column containing the liquid stationary phase. The retention and hence separation of the sample depends on the degree of interaction

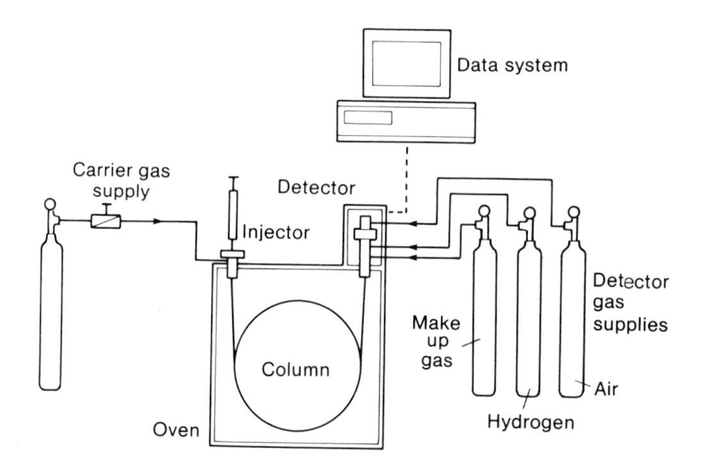

Figure 6.7 GC with flame-ionisation detection.

with the liquid phase and its volatility. The column is normally heated, with operating temperatures of 50–300°C being used. The technique depends on samples having a significant vapour pressure and being thermally stable at the temperatures used for analysis.

6.5.2 Key components in GC analysis

6.5.2.1 Column. A high proportion of GC analysis is now carried out on capillary columns because of their high separation efficiencies. These are made of tubing with an inner diameter of about 30–500 μm fabricated from glass, preferably fused silica material and of length of 1–100 m. The inner wall of the capillary is coated with the stationary phase, anchored by chemical bonding. Film thicknesses of 0.05–10 μm are used in practice. In narrow-bore capillary columns with inner diameters of 30–50 μm and lengths of 10–100 m, 5000 to 1 000 000 theoretical plates can be attained. Solid coating (SCOT columns) can also be bonded on to capillaries. The choice of stationary phase for separation is based on the nature of the components to be separated (i.e. one analyses non-polar samples on a polar phase and polar samples on a non-polar phase), their volatility and their thermal stability. A wide variety of phases are available, all with defined temperature limits; these include polysiloxanes, cyano- and trifluoro-siloxanes, and polyethylene glycols.

6.5.2.2 Detectors. A wide variety of detectors is available, the most common being the flame-ionisation detector (see section 6.4.4.4). This is capable of detecting all organic compounds down to 10–100 pg. For more selective detection, when the matrix being analysed is very complex, other detectors can be used. These include electron-capture detectors (ECD), which are often used for organohalogens, and nitrogen–phosphorus detectors (NPD). Both of these are sensitive down to picogram levels.

6.5.3 Auxiliary techniques in gas chromatography

6.5.3.1 Head-space analysis. As previously mentioned, polymers often contain substances of medium volatility such as monomers and polymerisation solvents. In addition, when heated polymers release volatiles as a result of the thermal degradation of the polymer itself or of the formulation additives used. Characterisation of such volatiles is now becoming increasingly important for regulatory requirements. Head-space gas chromatography as a technique offers considerable potential in this area. Head-space analysis can be carried out in either the static or the dynamic mode and can be readily automated. In static head-space analysis, the substance to be analysed is placed in a sealed vessel, where the material under examination comes into equilibrium with its vapours at a predetermined temperature. When equilibrium has been reached, an amount of the gas phase above the matrix is withdrawn and injected into the gas chromatograph for analysis. In the dynamic head-space technique, the sample to be analysed is placed in a vial and stripped by a flow of inert gas. This gaseous effluent passes through a trapping material (e.g. Tenax), where volatiles are retained. They are subsequently thermally desorbed into the gas chromatograph. Generally, when the gaseous effluent passes through the sample, the method is called 'purge and trap', while if the gaseous effluent passes over the sample the method is called 'dynamic head-space'.

6.5.3.2 Pyrolysis–gas chromatography. Pyrolysis is a useful tool for analysing and characterising polymers and polymer systems. Analytical pyrolysis involves the thermal cleavage of large molecules into small molecules at temperatures above 400°C. A measured sample is heated in an inert atmosphere to temperatures that cause thermal degradation. The resulting fragments (pyrolysates) are more volatile than the starting sample and can be separated by gas chromatography, yielding a characteristic fingerprint or pyrogram. Most commercially available pyrolysers provide highly reproducible thermal treatment of the sample. Two types of pyrolysers are available. The resistively heated pyrolyser, uses a sample probe of either platinum ribbon (for liquid samples) or a platinum coil, into which solid samples, contained in a quartz tube, can be placed. These pyrolyser probes can then be inserted into the injection port of a gas chromatograph, where they can be heated up to temperatures of 1400°C. A second type of pyrolyser is available, known as the Curie-point pyrolyser. This design can provide fast reproducible temperature ramps by heating the wire or foil in the probe ferromagnetically to its Curie point. The specific blend of metals comprising the wire or foil dictates the Curie point and therefore the final temperature of pyrolysis. The information obtained from pyrolysis–gas chromatography has been greatly increased by the use of modern high-resolution capillary columns to separate the volatile fractions and mass spectrometry to identify the peaks eluting from the chromatograph (Figure 6.8).

6.5.3.3 Inverse gas chromatography. The term 'inverse' refers to the fact that the method is used to examine the stationary phase. Here, the injected vapour sample is referred to as the probe.

This is a useful technique for characterising synthetic polymers, copolymers and polymer blends. It is an extension of conventional gas chromatography in which the non-volatile material (polymer) to be investigated is immobilised within a gas-chromatographic column. This phase is then investigated by monitoring the passage of volatile probe molecules of known properties as they are carried through the column via an inert gas. In a relatively short time, the elution data can yield important information on the glass and melting transition temperatures, degree of crystallinity, solubility and interaction parameters, surface areas and adsorption isotherms [83]. Standard GLC equip-

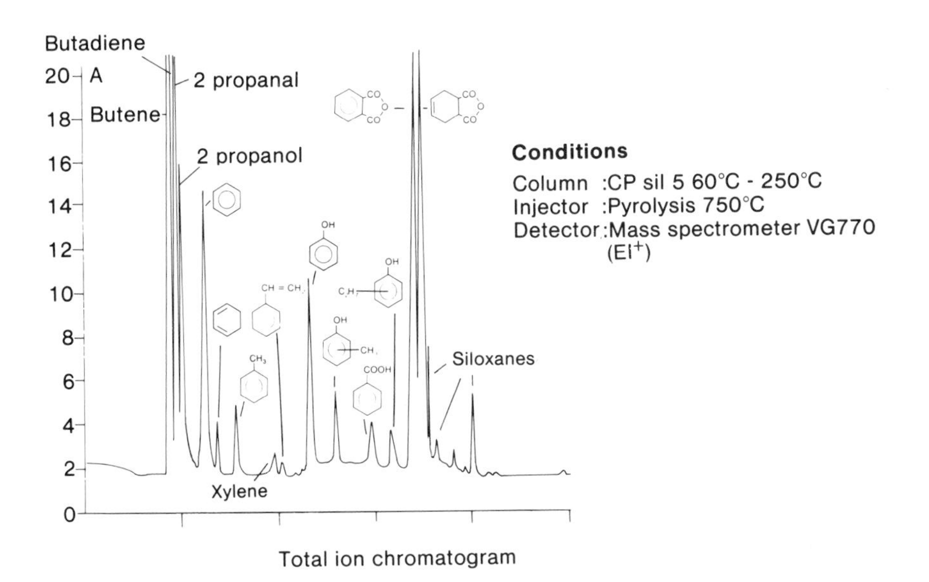

Figure 6.8 Pyrolysis GC/MS of a polyester sample.

ment is used with inert carrier gases (helium and hydrogen) which have to be controlled and measured accurately.

The column is the most important part of the system, as its function is to encourage repetitive partitioning of each solute molecule between the gas and the liquid or solid phase. Packed columns are commonly used with the polymer or blend to be analysed coated on to a non-reactive substrate (Chromasorb W) which is then packed into the stainless-steel GC column. More recently capillary columns have been employed. Here the polymer is dynamically or statically coated as a uniform film on to the walls of the fused silica capillary. Inverse gas chromatography is a very flexible analytical technique; however, full commercialisation of the technique has yet to be realised.

6.5.4 The uses of gas chromatography

Gas chromatography in its simplest form has been used in the separation and quantification of polymer additives (Irganox, Tinuvin, BHA) [85], plasticisers such as phthalates [86–88], blowing agents in polystyrene [89] and cationic surfactants [90]. The technique has also been used to determine monomers in acrylic coatings [91], polyurethanes [92], polymer emulsions [93], PMMA [94] and acrylate dispersions [95]. When combined with more sophisticated sampling procedures (thermal desorption/head-space), the technique has found extensive use in the monitoring of volatiles in plastic packaging materials [96], cable insulation [97], rubbers [98], automotive paints [99] and more specifically in phenol–formaldehyde resins [100]. Polymers have been separated using pyrolysis–GC. This approach, which yields an identifiable pyrogram suitable of being further characterised by mass spectrometry, has been successful in the separation of acrylic plastics [101], phenol–formaldehyde polycondensate [102], vulcanised elastomers [103], silane coupling agents [104], polyquinones [105], polyester, polyamide and polyethylene fibres [106], styrene–butadiene rubbers [107] and in the classification of PVC tape [108].

The use of inverse gas chromatography to produce physical data on polymer systems is increasing. Polymer–solvent interactions have been studied [109–111]; crystallinity and surface area of PMMA [112], melting-point depression of PVDF and PMMA [113], gel point, glass temperature and cross-linking in epoxy resins [114], polymer solubility (of polystyrene, polybutadiene and styrene–butadiene copolymers [115]), heat of adsorption and free energy of adsorption of polymers grafted on to fibres [116], plasticiser compatibility [117, 118] and monomer migration in PVC [119] and general surface characterisation of polymers [120] and coatings [121] have all been examined.

6.6 Supercritical fluid chromatography (SFC)

Supercritical fluid chromatography is a form of chromatography in which the mobile phase is subjected to pressures and temperatures near or above the critical point. Typically one or both parameters (pressure and temperature) extend into the critical region during the chromatographic run. The nature of the supercritical fluid is such that the mobile phase can vary from gas to liquid-like. This results in several advantages for SFC over other separation techniques. The choice of a suitable fluid with a low critical temperature permits operation at temperatures conducive to the analysis of thermally labile polymer additives, and the higher solvating ability of the fluids allows for the analysis of higher-molecular-weight material at higher efficiencies. SFC thus sits between HPLC and GC technologies.

6.6.1 Chromatographic process and instrumentation

The technique uses both HPLC- and GC-like instrumentation. Carbon dioxide is the most commonly used supercritical fluid with a critical temperature of 31°C and critical pressure of 72.86 atm. However, other fluids such as nitrous oxide, sulphur hexafluoride and xenon have been used. The supercritical fluid is pumped through the system by using either a syringe or a reciprocating pump, which is chilled to ensure that the fluid is maintained in the liquid form. The sample for analysis is dissolved in a low-boiling solvent (methylene chloride) and is injected into the system via an HPLC high-pressure injection valve. Separation can be affected by use of either 'LC-like' packed columns or 'GC-like' capillary columns. The critical conditions are achieved by placing the column in an oven and controlling the pressure of the fluid by either a back-pressure regulator or a restrictor, placed either before or after the detector.

6.6.2 Key components in SFC analysis

6.6.2.1 Column. As mentioned above, columns can be either packed or capillary. The majority of the early work in SFC was performed with conventional HPLC columns, 25 cm or 10 cm × 4.6 mm i.d. and packed with either 5 μm silica or bonded silicas (C_{18}, CN or diol). However, narrower-bore columns (2–0.5 mm) with 3 μm packing materials are now being considered. Most packed-column SFC uses carbon dioxide plus a polar modifier such as methanol, acetonitrile, tetrahydrofuran, 1,4-dioxan, methylene chloride or formic acid, typically in the range of 1–20%. Separations are usually effected in the 'normal-phase' mode. This technology has been used successfully in the determination of polymer additives, where good quantitative data can be obtained. More efficient separations can, however, be obtained by the use of

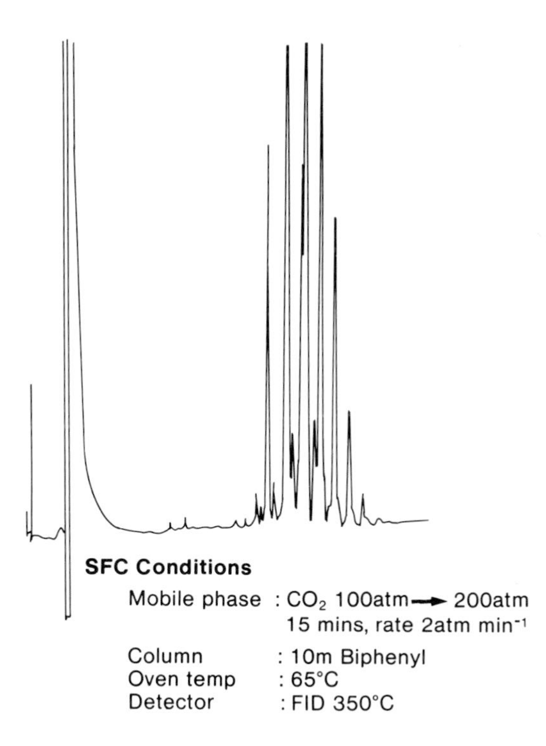

Figure 6.9 Capillary SFC separation. Epoxy acrylate oligomers 5% w/v in $MeCl_2$.

capillary columns. This approach uses GC-like technology, with non-polar, polarizable and polar stationary phases and smaller-diameter columns (50–100 μm) than are normally used in GC. The length of these columns is 10–20 m and film thickness 0.05–0.5 μm. Capillary SFC has proved useful in the separation of low-molecular-weight oligomers and polymers (Figure 6.9).

6.6.2.2 Detectors. The possibility of using both gas- and liquid-chromatography detectors has often been regarded as the major benefit of supercritical fluid chromatography. For example, flame-ionisation detection, the simple and universal GC detector, can now be used with LC-like separations using carbon dioxide. Similarly, nitrogen–phosphorus and electron-capture detectors can be used for selective detection. The use of modifiers in the eluent does, however, reduce the detector options. Here LC-like detectors can be used, such as UV and fluorescence. These detectors are specially equipped with high-pressure flow cells, since in such systems the back-pressure regulator is normally placed after the detector.

6.6.3 Hyphenated systems

The very nature of supercritical fluids means that they lend themselves to easier interfacing to both mass spectrometry and Fourier-transform infrared spectroscopy. SFC–MS has been found to be superior to LC–MS due to its wider applicability and higher sensitivity. The main requirement of such an interface is the ability to handle the gas flows generated from the mobile phase. For capillary SFC, most quadrupole instruments can be used. For packed SFC, which uses higher flow rates, high-capacity interfaces such as thermospray or moving-belt are used. FTIR has also been readily linked to SFC. This can be accomplished in the off-line mode via direct deposition of the chromatographic peaks of interest on to a solid substrate, which is then examined by FTIR. Alternatively, it can be achieved on-line by using a flow cell. In this mode, wavelengths are selected where CO_2 does not absorb or an IR-transparent supercritical fluid such as xenon is used.

6.6.4 Coupled technology: supercritical fluid extraction (SFE)/supercritical fluid chromatography/gas chromatography

The properties of supercritical fluids that make them useful chromatographic mobile phases also make their use as extraction solvents an attractive option for polymer analysis. The low viscosities and high solute diffusivities allow efficient mass transfer during extraction, and the relatively low extraction temperatures reduce the risk of analyte degradation. Thus, the extraction of polymer additives from polymer matrices by using supercritical fluids has many advantages over conventional liquid solvent extraction, with the potential of higher recoveries and shorter analysis times. The further combination of extraction followed by separation by either SFC or GC offers an elegant and fast solution to the analysis of complex polymer systems.

6.6.5 The uses of supercritical fluid chromatography

SFC is establishing itself as a complementary technique to both gas and liquid chromatography. Commercial equipment is now available for both packed and capillary SFC as well as for MS, FTIR and SFE coupling. Whilst capillary SFC to date still offers the most potential in terms of its separating power, several desirable instrument features have yet to be realised (quantitative and discrimination-free injection, column stability and adjustable flow control).

Despite some of these shortcomings, the technique has been used successfully in the separation of many polymer systems. Polystyrene and its oligomers [122–124] and vinyl napthalene [125, 126], vinyl carbazole [125], vinyl biphenyl [127] and epoxy acrylate [128] oligomers have all been separated by SFC, along with polyethylene glycols [129], polypropolene glycols [130] and polysiloxanes [131]. The technique has also been investigated for producing molecular-weight distributions [132] and has been employed for process control in the epoxy-resin area [133]. Polymer additives have been analysed, as mentioned previously. They include phthalate plasticisers [134, 135], antioxidants, UV stabilisers, metal deactivants, slip and antistatic agents [136–139]. SFC linked to mass spectrometry has been reported in the characterisation of non-ionic surfactants [140] and polydimethyl silicone oligomers [141] and SFC–FTIR for the separation and identification of polysiloxane oligomers [142], antioxidants [143] and UV-curing resins [144]. Supercritical fluid chromatography linked to supercritical fluid extraction is still in its infancy, but a number of papers have been published on its use in identifying polymer additives in polyethylene and polypropylene samples [145–147].

6.7 Field flow fractionation (FFF)

Field flow fractionation was developed about the same time as size-exclusion chromatography. However, major instrumentation problems were not solved as quickly for FFF as they were for SEC, and hence it has lagged behind in terms of its commercial exploitation. FFF is a chromatography-like separation method which consists of a family of techniques applicable to the separation and characterisation of macromolecules and colloidal and particulate matter, from a molecular weight of a few thousand up to 100 μm particle diameter.

6.7.1 Chromatographic process and instrumentation

The experimental procedures used are analogous to both SEC and HPLC. A liquid or mobile phase is introduced into a separation column or channel (thin, unpacked and ribbon shaped), samples are injected, separated and eluted from the channel, and the channel effluent is passed into a detector. The output from the detector is then recorded on a data system. In FFF, separation and retention are induced by an external field or gradient applied to the liquid or mobile phase, rather than by partitioning into a stationary phase as with HPLC. The ability to control field strength and type offers a wide separation capability to the technique. Driving forces that have been used include sedimentation, thermal, cross-flow, electrical and magnetic. However, the two most commonly used fields are thermal and sedimentation, whose use has been advanced by their commercial availability.

6.7.2 Sedimentation field flow fractionation (SFFF)

As the name suggests, this utilises sedimentation or centrifugal forces to induce separation. A ribbon-like open channel, approximately 80 cm × 2 cm × 0.01 cm in thickness, and usually made from polymeric material, is clamped to the inside of a centrifuge basket. A liquid stream is brought in and out of the channel via a specially designed rotor. The centrifuge can be mounted either vertically or horizontally, and is controlled to provide accurate field regulation and permit field programming up to 2500 rpm. The mobile phase is pumped through the system using a conventional HPLC pump and similarly the sample is injected by using an HPLC injector. As the technique is used primarily for colloids, emulsions and particles, a turbidimetric UV detector (254 nm) is used to record the separation. The technique is applicable to both small and large particles (0.05–100 μm) and can be used to determine particle size, size distribution and particle density (Figure 6.10). The technique can also be used preparatively, as fractions can easily be collected from the separation channel, thus allowing for further characterisation by techniques such as microscopy.

6.7.3 Thermal field flow fractionation (TFFF)

In TFFF a temperature gradient is applied perpendicular to the channel with separation being induced by thermal diffusion. The channel is formed by clamping two copper bars with highly polished internal surfaces over a sheet of Mylar or Teflon into which the channel space has been cut. The upper bar is equipped with cartridge heaters while water is circulated through the lower bar for cooling. Temperature differentials of up to 80°C, corresponding to gradients in excess of 1000°C cm^{-1} can thus be generated across the channel. As with SFFF, an HPLC injector and pump are used and the separations can be monitored by UV, RI or more recently viscosity and light-scattering detection.

TFFF can be used for the separation and molecular-weight distribution determination of polymers in the molecular weight range 10^4–10^7. As a tech-

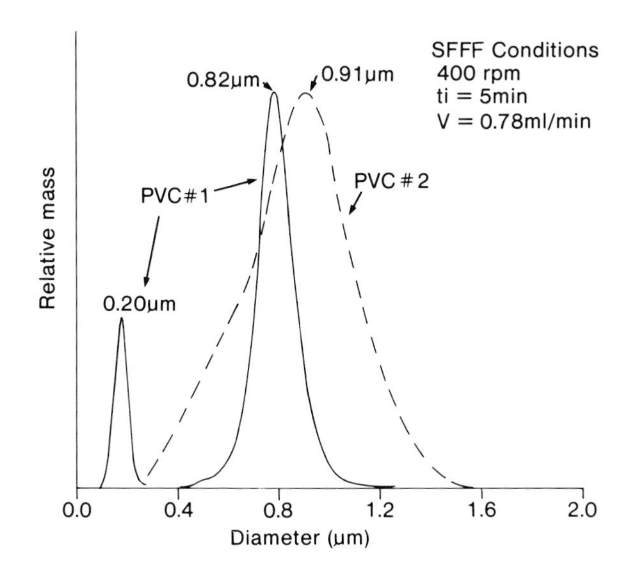

Figure 6.10 SFFF analysis of two PVC powders. Courtesy of Professor J. C. Giddings.

nique it offers higher selectivity and resolving power at high molecular weights than SEC. The TFFF methodology has been found applicable to nearly all polymers, and in addition, since retention depends not only on molecular weight but also on chemical composition, it may be used to determine compositional distributions in copolymers and blends.

6.7.4 The uses of field flow fractionation techniques

Field flow techniques have been reviewed in a number of articles [148–150]. Sedimentation field flow fractionation has found use in the separation of PVC [151, 152], polystyrene [151–153], poly(methyl methacrylate) [153, 154], poly(vinyl toluene) [155] and poly(glycidyl methacrylate) latexes [156] to produce particle-size distributions and particle densities. It has also been applied in polymer-aggregation studies [157], pigment [157] quality control and in the separation of silica particles [158] and its performance has been compared with that of ultracentrifugation [159]. Thermal field flow fractionation has been used successfully in the characterisation of ultra-high-molecular-weight polystyrenes [160, 161], poly(methyl methacrylate), polyisoprene, polysulphane, polycarbonate, nitrocellulose, polybutadiene and polyolefins [162]. In the difficult area of water-soluble polymers, poly(ethylene glycol), poly(ethylene oxide), poly(vinyl pyrrolidone) and poly(styrene sulphonate) have been analysed [163, 164]. In addition, compositional separations have been achieved for polystyrene–poly(methyl methacrylate) mixes [165] and comparisons between TFFF and SEC have been made [166].

6.8 Hydrodynamic chromatography (HDC)

Hydrodynamic chromatography is a chromatographic technique for measuring the particle-size distributions of polymer lattices of approximately 0.1–1 μm and inorganic colloids in the 0.01–40 μm range.

6.8.1 Chromatographic process and instrumentation

The technique is based on the principle that a flowing stream of liquid through a capillary tube generates a laminar-flow velocity gradient. The velocity is greatest at the centre of the tube and decreases towards the wall. On injection of a sample into the flowing stream, the larger particles are exposed to larger average velocities than the smaller particles, which can diffuse close to the capillary wall and into the slower-moving portion of the mobile phase. This differential effect results in the larger particles eluting first, followed by the smaller ones. The resulting chromatogram is similar to that obtained by SEC, but with much lower efficiencies. HDC hardware has much in common with conventional HPLC. It employs a column (packed or open tubular), a liquid mobile phase which is normally an aqueous solution of salts and surfactants, a pump capable of steady, pulseless delivery at moderate pressures, a sample-injection device, a colloid detector (UV, acting as turbidimeter) and a means of data collection and processing. It should be noted that in HDC the solutes or samples are in most cases insoluble in the mobile phase. In a typical experiment a small-molecular marker (usually dichromate) is injected to act as an internal flow-rate marker. The column is the heart of the HDC device. Two types have been used, packed and capillary.

6.8.2 Packed columns

Most of the early HDC work was carried out using packed columns. These were similar in design to HPLC columns, but were packed with homogeneous, non-porous microspheres (ion-exchange or non-functional polymer beads or glass spheres) designed to produce a network of interstitial spaces that would have similar characteristics to a long capillary tube. Columns containing 10–20 μm-diameter materials and up to 1 m in length have been used. The typical separation range of such columns is in the order of 0.01–10 μm.

6.8.3 Capillary HDC

Recent developments in fused-silica capillary technology has led to such columns being used for HDC. The microcapillaries are normally 1–10 μm diameter with column volumes not exceeding 20 nl. The corresponding flow rates required for such systems are thus much lower than those of packed-column HDC. This technology has recently been commercialised with the column taking the form of a removable cartridge containing the microcapillaries. This system is capable of separation in the range 0.015–1.1 μm (Figure 6.11).

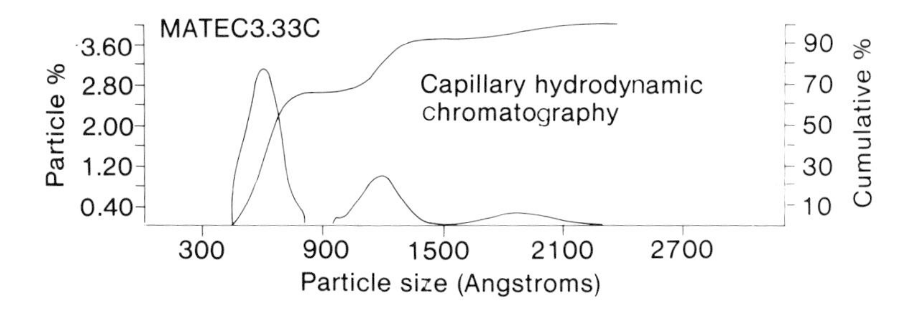

Figure 6.11 High-resolution particle-size distribution of a carboxylated styrene–butadiene latex. Reproduced by permission of Matec Applied Sciences.

6.8.4 The uses of hydrodynamic chromatography

Hydrodynamic chromatography has been used in the separation of polystyrene, butadiene–styrene copolymer, poly(vinyl acetate) [167], carboxylated styrene [168] and carboxylated acrylic copolymer latexes [169]. More specifically, it has been reported for studying particle growth in carboxylated styrene–butadiene systems [170], for polymer dynamics of hydrolysed polyacrylamide [171], to study microgels present in poly(diallyl isophthalate) and diallyl isophthalate–methyl methacrylate copolymer [172], in the off-line monitoring of emulsion polymerisation reactors [173], in the separation of high-molecular-weight polystyrenes and inorganic colloids [174] and to investigate the temperature behaviour of micelles of an isoprene–styrene diblock copolymer [175]. The majority of HDC applications have been carried out with packed-column technology. However, recent developments in the capillary-HDC field have led to a number of applications being published on its use for particle-size determination [176–178].

References

1. J. Janca (1984). In *Steric Exclusion Liquid Chromatography of Polymers*, J. Janca (ed.), Marcel Dekker, New York.
2. Z. Gallot (1980). In *Liquid Chromatography of Polymers and Related Materials II*, J. Cazes and X. Delamare, Marcel Dekker, New York, p. 113.
3. W. A. Dark (1987). *J. Anal. Purif.*, **2**, 62.
4. T. Provder (1975). Analysis of copolymers by GPC, in *Proc. Macromolecular Sci. Div. 58th Canadian Chemical Conference*, Toronto, Canada.
5. C. Kuo, T. Provder, M. E. Koekler and A. F. Kah (1987). In *Detection and Data Analysis in Size Exclusion Chromatography*, T. Provder (ed.), ACS Symposium Series No. 352, American Chemical Society, Washington, DC, p. 130.
6. J. Haney (1985). *Appl. Polym. Sci.*, **30**, 3037.
7. T. A. Chamberlin and H. E. Tuninstra (1988). *J. Appl. Polym. Sci.*, **35**(6), 1667–82.

8. J. Lesec and G. Volet (1990). *J. Appl. Polym. Sci., Appl. Polym. Symp.*, **45** (*Polym. Anal. Charact.*, **2**), 177–89.
9. V. Grinshpun, K. F. O'Driscoll and A. Rudin (1984). *J. Appl. Polym. Sci.*, **29**, 1071–77.
10. N. Hadjichristidis, X. Zhongde and L. J. Fetters (1982). *J. Polym. Sci.*, **20**, 743–750.
11. E. J. Siochi, J. M. DeSimone, A. M. Hellstern, J. E. McGrath and J. C. Ward (1990). *Macromolecules*, **23**, 4896.
12. H. Schom, R. Kosfeld and M. Hess (1983). *J. Chromatog.*, **282**, 579–87.
13. C. J. Kim, A. E. Hamielec and A. Benedek (1982). *J. Liquid Chromatog.*, **5**(7), 1277–94.
14. T. Dumbelow (1989). *J. Macromol. Sci. Chem.*, **A26**(1), 125–46.
15. F. B. Malihi, C. Y. Kuo and T. Provder (1984). *J. Appl. Polym. Sci.*, **29**, 925–31.
16. T. Housaki and K. Satoh (1988). *Makromol. Chem. Rapid Commun.*, **9**, 257–59.
17. A. E. Hamielec, A. C. Quano and L. L. Neberzah (1978). *J. Liq. Chromatog.*, **1**(4), 527–54.
18. D. E. Axelson and W. C. Knapp (1980). *J. Appl. Polym. Sci.*, **25**, 119–23.
19. J. Lesec, D. Lecacheux and J. Marot (1980). *J. Liq. Chromatog.*, **11**, 2571.
20. W. W. Yau, G. A. Smith and J. J. DeStefano (1989). In *Proc. Int. GPC Symp.*, Newton, MA.
21. M. A. Haney (1985). *J. Appl. Polym. Sci.*, **30**, 3037.
22. P. Wang and B. Glassbrenner (1988). Paper presented at *Minnesota Chromatography Forum 27*, Bloomington, MN.
23. P. Young, J. Davis, R. J. Judith and A. C. Chang (1989). Paper presented at the *34th International SAMPE Symposium*, 8–11 May, Reno, NV.
24. See Ref. 5.
25. M. G. Styring, J. E. Armonas and A. E. Hamielec (1987). *J. Liq. Chromatog.*, **10**, 783–804.
26. J. M. Goldwasser (1989). In *Proc. Int. GPC Symp.*, Newton, MA.
27. D. J. Nagy and D. A. Terwilliger (1989). J. *Liq. Chromatog.*, **12**(8), 1431–49.
28. S. Teremachi, A. Hasegawa, Y. Shima, M. Akatsuka and M. Nakajima (1979). *Macromolecules*, **12**, 992–96.
29. S. Hatton, H. Endoh, H. Nakahara, K. Toshio and M. Hamashima (1978). *Polym. J.*, **10**(2), 173–180.
30. J. L. Ekmanis (1987). In *Detection and Data Analysis in Size Exclusion Chromatography*, T. Provder (ed.), ACS Symposium Series No. 352, American Chemical Society, Washington, DC.
31. C. Troeltzch (1986). *J. Prakt. Chem.*, **328**, 454–58.
32. R. Markovich and L. Hazlitt (1991). *Polym. Matters Sci. Eng.*, **65**, 98.
33. J. A. J. Jansen (1990). *Frensenius J. Anal. Chem.*, **337**, 398–402.
34. A. Wirsen (1988). *Makromol. Chem.*, **189**(4), 833–43.
35. K. Nishikida, T. Housaki, M. Morimoto and T. Kinoshita (1990). *J. Chromatogr.* **517**, 209–17.
36. L. J. Mulcahey and L. T. Taylor (1991). *LC/GC Intl.* **4**(2), 34–39.
37. K. Hatada, K. U. Ute, M. Kashiyama and M. Imanrai (1990). *Polym. J.* **22**, 218.
38. K. Ute, M. Kashiyama, K. Oka, K. Hatada and O. Vogl, (1990). *Makromol. Chem. Rapid Commun.*, **11**, 31.
39. S. T. Balke and R. D. Patel (1980). *J. Polym. Sci. B*, **18**, 453.
40. S. T. Balke (1983). *Polymer News*, **9**, 6.
41. G. Glöckner and J. Van den Berg (1987). *Chromatog.*, **384**, 135–44.
42. J. V. Dawkins and A. Montenegro. *Brit. Polym. J.*, **21**, 31–36.
43. W. A. Dark (1987). Automotive challenge and plastics response, in *Proc. Automotive Plastics Regional Technical Conf.*, Dearborn, MI, pp. 210–4, 63 TR, RO, SPE.
44. G. Glöckner (1987). In *Polymer Characterisation by Liquid Chromatography, Journal of Chromatography Library* Vol. 34, Elsevier, Amsterdam, pp. 413–425.
45. W. Chol *et al.* (1990). *Polimo.*, **14**(2), 130–37.
46. S. T. Balke (1989). *J. Appl. Polym. Sci., Appl. Polym. Symp.*, **43**, (*Polym. Anal. Charact.*), 5–38.
47. See Ref. 43.
48. V. V. Guryanova, T. N. Prudskova and A. V. Pavlov (1988). *Int. Polym. Sci. Technol.*, **15**(1), 164–67.
49. E. D. Conrad *et al.* (1985). *Med. Device Diagn. Ind. Part I*, **7**, 124–30.
50. R. E. Sprey (1989). *Elastomerics*, **121**(3), 15–17.
51. Papez (1986). In *Antec 86 Plastics Technology, Proc. 44th Annual Technical Conference*, Boston, 28 April–1 May, pp. 468–72.

52. M. J. Mettillie and R. D. Hester (1989). *Polym. Prep. Amer. Chem. Soc. Div. Polym. Chem.*, **30**(2), 369–70.
53. A. M. Safieddine and R. D. Hester (1989). *Polym. Prep. Amer. Chem. Soc. Div. Polym. Chem.*, **30**(2), 380–81.
54. P. L. Dubin and J. M. Principi (1989). *Anal. Chem.*, **61**(7), 780–81.
55. S. N. E. Ormorodion and A. E. Hamielec (1989). *J. Liq. Chromatogr.*, **12**(13), 2635–60.
56. B. Furth and H. Reidelbauch (1988). *Kunststoffe*, **78**(5), 420–23.
57. C. Deluski, J. Limpert and R. L. Cotter (1987). *Proc. Automotive Plastics Regional Conf.*, Dearborn, MI, pp. 136–139.
58. M. Tsimidou and R. Macrae (1985). *J. Chromatog. Sci.*, **23**, 155–60.
59. G. R. Bear (1988). *J. Chromatog.*, **459**, 91–107.
60. J. D. Vargo and K. L. Olson. *J. Chromatog.*, **353**, 215–24.
61. Yagoubi, Baillet, Pellerin and Baylocq, (1990). *J. Chromatog.*, **522**, 131–41.
62. B. Marcato, C. Fantazzini and F. Sevini (1991). *J. Chromatog.*, **553**, 415–22.
63. K. Jinno and Y. Yokoyama (1991). *J. Chromatog.*, **550**, 325–34.
64. M. R. Khan, C. P. Ong, S. F. Y. Li and H. K. Lee (1990). *J. Chromatog.*, **513**, 360–66.
65. M. Kraxner and F. Barla (1986). *Symp. Biol. Hung.*, **31**, 559–68, Chromatography 84.
66. R. Bodmeier and O. Paeratakul (1991). *J. Liq. Chromatog.*, **14**(2), 365–75.
67. R. Lodkowski, M. Sikora and B. Buszewski (1990). *Chem. Stosow.*, **34**(1–2), 149–52.
68. I. L. Row, Y. L. Liu and C. W. Wang. *J. Chin. Chem. Soc. (Taipei)*, **37**, 203–09.
69. D. W. Eggimann, J. C. Brand and C. K. Elliott (1981). Paper presented at *26th National SAMPE Symposium*, 28–30 April .
70. S. A. Mestan and C. E. M. Morris (1984). *J. Macromol. Sci. C*, **24**(1), 117–72.
71. V. Prussler, K. Slais and J. Hanus (1987). *Angew. Makromol. Chem.*, **150**, 179–87.
72. C. Bailly *et al.* (1986). *Polymer*, **27**(5), 776–82.
73. P. Jandera, (1988). *J. Chromatog.*, **449**, 361–89.
74. H. Hosozawa, A. Toyada and H. Sato (1989). In *136th Meeting 1989 Conf. Proc.*, Detroit, MI, 17–20 October, Paper 82.012, American Chemical Society, Rubber Division, Washington, DC.
75. G. Glöckner and A. H. E. Müller (1989). *J. Appl. Polym. Sci.*, **38**(9), 1761–74.
76. S. Mori and Y. Uno (1987). *J. Appl. Polym. Sci.*, **34**(8), 2689–99.
77. B. L. Neff and H. J. Spinelli (1991). *J. Appl. Polym. Sci.*, **42**, 595–600.
78. H. Sato, M. Sasaki and K. Ogino (1989). *Polym. J.* **21**(12), 965–69.
79. R. G. Ackman, C. A. McLeod and A. K. Banerjee (1990). *J. Planar Chromatog.*, **3**, 450–90.
80. E. S. Gankina and B. G. Belenki (1991). *Chromatog. Sci.*, **55**, 807–62.
81. G. Haesen, R. Depaus, H. van Tilbeurgh and B. LeGoff (1983). *LOXF. J. Ind. Irr. Technol.*, **1**(3), 259–80.
82. T. R. Compton (ed.) *Chemical Analysis of Additives in Plastics*, Pergamon Press, Oxford.
83. J. E. Guillet, M. Romansky and G. J. Price (1989). In ACS Symposium Series No. 391, pp. 20–32, American Chemical Society, Washington, DC.
84. G. Simistad, T. Waaler and P. O. Roksvaag (1989). *Acta Pharm. Nordica*, **1**(1).
85. F. David, S. P. Vanderroost and S. Stafford (1991). Hewlett-Packard application note 228–149, *Gas Chromatography*, October, Hewlett-Packard Co.
86. D. J. Russel and B. McDuffie (1983). *Int. J. Environ. Anal. Chem.*, **15**, 165–83.
87. G. Simistad and T. Waaler (1989). *Acta Pharm. Nordica*, **1**(1).
88. S. Komaromi, A. Hollo, C. Gonczi and J. Somfalvi. *Muanyag. ex Gumi*, **23**(11), 336–39.
89. C. Krutchen and W. Wu (1988). In *Antec 88, Proc. 46th Annual Tech. Conf.*, Atlanta, 18–21 April, pp. 704–06.
90. H. Koenig and W. Strobel (1983). *Fresenius Z. Anal. Chem.*, **314**, 143–45.
91. F. Kang, X. Xue and T. Chen (1982). *Gongye*, **66**, 39–42.
92. I. Padovani and L. Trevisan (1987). *Chim. & Ind.*, **69**(4), 30–31.
93. W. Miller and E. L. Harper (1983). *J. Appl. Polym. Sci.*, **28**(11), 3585–88.
94. T. J. Edkins, V. J. Notorgioacome and J. A. Biesenberger (1990). *Polym. Eng. Sci.*, **30**(23), 1500–03.
95. G. Stoev and M. Angelova (1987). *J. High Resolut. Chromatog. Chromatog. Commun.*, **10**(1), 25–31.
96. S. Jacobsen (1984). *J. High Resolut. Chromatog. Chromatog. Commun.*, **7**(4), 185–90.

97. A. A. Mehta and J. F. Johnson (1987). In *Antec 87 Plastics Pioneering the 21st Century, Proc. 45th Annual Technical Conference*, Los Angeles, 4–7 May, pp. 1085–87.
98. Y. Cai, H. Piao and J. Xiang (1984). *Gongye*, **7**, 207–93.
99. K. L. Olson, C. A. Wang, L. L. Fleck and D. F. Lazar (1988). *J. Coat. Technol.*, **60**, 45–50.
100. Y. D. Abrashkevich, V. M. Kilmovich, G. Filonenko and I. V. Davidenko (1987). *Int. Polym. Sci. Technol.*, **14**(2), 73–74.
101. J. Jason Shen and E. Woo (1988). *Liq. Chromatog. Gas Chromatog.*, **6**, 1020.
102. L. Prokai (1987). *J. Anal. Appl. Pyrol.*, **12**, 265–73.
103. D. A. Bulpett, A. J. Deome, J. F. Jones and J. Patt (1988). In *Proc. Spring Meeting (133rd) 1988*, Dallas, 19–22 April, Paper 69, American Chemical Society, Rubber Division, Washington, DC, p. 35.
104. E. Nishio, N. Ikuta, T. Hirashima and J. Koga (1989). Appl. Spectrosc., 43(7), 1159–64.
105. M. Blazso, E. Jakab, T. Szekely, B. Plage and H. Schulten (1989). *J. Polym. Sci. A*, Polym. Chem., **27**, 1027–43.
106. I. R. Hardin and X. Q. Wang (1989). *Text. Chem. Color.*, **21**, 29–32.
107. Y. K. Lee, M. G. Kim and K. J. Whang (1989). *J. Anal. Appl. Pyrol.*, **16**, 173–82.
108. E. R. Williams and T. O. Munson (1988). *J. Forensic Sci.*, **33**, 1163–70.
109. P. Munk, D. Paul, G. Qiang and A. Abdel (1990). *J. Appl. Polym. Sci., Appl. Polym. Symp.*, **45**, (*Polym. Anal. Charact. 2*), 289–316.
110. K. Chee (1990). *Polymer*, **31**, 1711–14.
111. R. C. Castells and G. D. Mazza (1986). *J. Appl. Polym. Sci.*, **32**(7), 5917–31.
112. J. E. Guillet (1988). *Polym. Mater. Sci.*, **58**, 645–49.
113. P. T. Chen and Z. Y. Al-Saigh (1990). *Polym. Prepr. Amer. Chem. Soc. Div. Polym. Chem.*, **31**, 586–87.
114. Wetzel *et al.* (1990). *Plaste Kautsch*, **37**, 219–24, 27.
115. A. Farooque and D. Mohammed (1991). *Polym. Sci.* (*Symp. Proc. Polym. 91*), **2**, 598–602.
116. D. P. Kamden and J. Wood (1991). *Chem. Technol.*, **11**, 57–91.
117. P. G. Demertzis, K. A. Riganakos and K. Akrida-Demertzi (1990). *Eur. Polym. J.* **26**, 137–40.
118. P. G. Demertzis, K. A. Riganakos and K. Akrida-Demertizi (1991). *Polym. Int.*, **25**, 229–36.
119. D. G. Apostolopoulos and G. Seymour (1988). *Pack. Technol. Sci.*, **1**(4), 177–78.
120. P. J. C. Chappell and D. R. Williams (1987). In *Proc. ICCM ECCM Sixth Int. Conf. Compos. Water Second Eur. Conf. Compos. Water*, Vol. 5, 5.346–5.355.
121. H. P. Schreiber (1989). *Adv. Org. Coat. Sci. Technol. Ser.*, **11**, 192–204.
122. F. P. Schimtz, H. Hilgers and E. Klesper (1983). *J. Chromatogr.*, **267**.
123. C. Fujimoto, T. Watenabe and Jinnok (1989). *J. Chromatog. Sci.*, **27**, 325.
124. D. Leyendecker, D. Leyendecker, F. P. Schimtz and E. Klesper (1987). *Chromatographia*, **23**(1), 38–42.
125. F. P. Schimtz, H. Hilgers, B. Lorenschat and E. Klesper (1985). *J. Chromatogr.*, **346**, 69.
126. F. P. Schimtz and H. Hilgers (1986). *Makromol. Chem. Rapid Commun.*, **7**, 59.
127. F. P. Schimtz *et al.* (1986). *J. Chromatogr.*, **371**, 135.
128. M. W. Rayner *et al.* (1989). *J. High Res. Chromatogr.*, **5**, 300–3 .
129. T. L. Chester (1988). In *Supercritical Fluid Extraction and Chromatography—Techniques and Applications*, B. A. Charpentier and M. Sevents (eds), American Chemical Society Symposium Series American Chemical Society, Washington, DC.
130. F. D. Lee and J. D. Henion (1987). *Anal. Chem.*, **59**, 1309–1312.
131. S. L. Pentoney *et al.* (1987). *J. Chromatog. Sci.*, **25**, 9.
132. T. Altares (1970). *Polym. Lett.*, **8**, 761–66.
133. A. Giorgetti, N. Pericles, H. M. Widmer, K. Anton and P. Datwyler (1989). *J. Chromatog. Sci.*, **27**, 318.
134. T. Takeuchi (1988). *Chromatographia*, **25**, 125.
135. M. Saito (1989). *J. High Resolut. Chromatog. Chromatog. Commun.*, **11**, 741–43.
136. T. Greibrok, B. E. Berg, H. R. Hoffmann, H. R. Norli and Q. Ying. *J. Chromatog.*, **505**, 283–91.
137. K. Bartle *et al.* (1990). *Anal. Proc.*, **27**, 239–40.
138. P. J. Arpino *et al.* (1990). *J. High Resolut. Chromatog.*, **13**
139. D. Dilettato *et al.* (1990). *J. High Resolut. Chromatog.*, **14**, 335–42.
140. H. T. Kalinoski and L. O. Hargiss (1990). *J. Chromatog.*, **505**, 199–213.
141. M. A. Morrissey, W. F. Siems and H. H. Hill (1990). *J. Chromatog.*, **505**, 215–25.

142. S. L. Pentoney, K. H. Shafer, P. R. Griffiths and R. Fluoco (1986). *J. High Resolut. Chromatog.*, **9**, 168–71.
143. R. C. Wieboldt, K. D. Kempfert and D. Dalrymple (1990). *Appl. Spectrosc.*, **44**(6), 1028–34.
144. M. W. Raynor *et al.* (1989). *J. Microcol.*
145. Y. Hirata, F. Nakata and J. Horihata (1988). *J. High Resolut. Chromatog.*, **11**, 81.
146. T. W. Ryan, S. G. Yocklovich, J. C. Watkins and E. J. Levy (1990). *J. Chromatog.*, **505**, 273–82.
147. J. M. Levy, R. A. Cavalier, T. N. Bosch, A. F. Rynaski and W. E. Huhak (1989). *J. Chromatog. Sci.*, **27**, 341–46.
148. J. C. Giddings (1988). *Chem. Eng. News*, Oct., 34–45.
149. K. D. Caldwell (1988). *Anal. Chem.*, **60**(17), 959A–970A.
150. J. Janca *et al.* (1990). Appl. Polym. Sci., Appl. Polym. Symp., **45**, (*Polym. Anal. Charact.*) 39–69.
151. H. K. Jones and J. C. Giddings (1989). *Anal. Chem.*, **61**, 741–45.
152. J. C. Giddings, K. D. Caldwell and H. K. Jones (1987). In *Particle Size Distribution and Characterisation*, T. Provder (ed.), ACS Symposium Series No. 332, American Chemical Society, Washington, DC.
153. B. N. Barman and J. C. Giddings (1990). *Polym. Mater. Sci. Eng.*, **62**, 182–90.
154. J. C. Giddings, B. N. Barman and H. J. Li (1989). *J. Colloid Interface Sci.*, **132**, 554–65.
155. T. Hoshina, M. Suzuki, K. Ysukawa and M. Takeuchi (1987). *J. Chromatog.*, **400**, 361–69.
156. J. Chemlik and J. Janca (1986). *J. Liq. Chromatog.*, **9**, 55–66.
157. L. Koch, T. Koch and H. M. Widmer (1990). *J. Chromatog.*, **517**, 395–403.
158. C. R. Yonker, H. K. Jones and D. M. Robertson (1987). *Anal. Chem.*, **59**, 2574–79.
159. J. Li and K. D. Caldwell (1990). *J. Chromatog.*, **517**, 361–76.
160. Y. S. Gao, K. D. Caldwell, M. N. Meyers and J. C. Giddings (1985). *Macromolecules*, **18**, 1272–77.
161. J. C. Giddings, S. Li, S. P. Williams and M. E. Schimpf. *Mackromol. Chem., Rapid Commun.*, **9**, 817–23.
162. J. C. Giddings, V. Kumar, P. S. Williams and M. N. Meyers (1991). In *Polymer Characterisation*, C. D. Craver and T. Provder (eds), Advances in Chemistry Series 227, American Chemical Society, Washington, DC, Chapter 1.
163. J. J. Kirkland and W. W. Yau (1986). *J. Chromatog.*, **353**, 95–107.
164. M. A. Beninasa and J. C. Giddings (1992). *Anal. Chem.*, **64**, 790–98.
165. J. J. Gunderson and J. C. Giddings (1986). *Macromolecules*, **19**, 2618.
166. J. J. Gunderson and J. C. Giddings (1986). *Anal. Chim. Acta.*, **189**, 1–15.
167. A. Rudin and D. Frick (1985). *Polym. Mater. Sci. Eng.*, **53**, 431–35.
168. R. L. Van Gilder and M. A. Langhorst (1985). *Polym. Mater. Sci. Eng.*, **53**, 440–45.
169. T. W. Thornton and L. B. Gilman (1985). *Polym. Mater. Sci. Eng.*, **53**, 426–30.
170. R. L. Van Gilder and M. A. Langhorst (1987). In *Particle Size Distribution*, ACS Symposium Series No. 332, American Chemical Society, Washington, DC, pp. 272–86.
171. D. A. Hoagland and R. K. Prud'homme (1989). *Macromolecules*, **22**, 775–81.
172. V. I. Kolegov, V. N. Potapov, B. N. Kocheryaev and E. I. Varavina (1986). *Vysokomol. Soedin.*, **28**, 391–94 (in Russian).
173. T. Kourti, A. Penlidis, J. F. MacGregor and A. E. Hamielec (1984). In *Quant. Charact. Plast. Rubber, Proc. Symp.*, J. Vlachopoulos (Ed.), McMaster University, Hamilton, Ont., pp. 56–67.
174. G. Stegeman, R. Oostervink, J. C. Kraak, H. Poppe and K. Unger (1990). *J. Chromatog.*, **506**, 547–61.
175. J. Bos, R. Tizssen and M. E. Van Kreveld (1989). *Anal. Chem.*, **61**, 1318–21.
176. J. G. Dosramos and C. A. Silebi (1989). *J. Colloid Interface Sci.*, **133**(2), 302–20.
177. J. G. Dosramos and C. A. Silebi (1990). *J. Colloid Interface Sci.*, **135**(1), 165–77.
178. J. G. Dosramos and C. A. Silebi (1990). In *Proc. American Chemical Society Spring Mtg.*, **199**, April, p. 15.

7 Thermal analysis

R. E. WETTON

7.1 Introduction

It has long been considered that the most useful workhorse techniques, for a modestly equipped laboratory concerned with charcterising polymeric materials, are differential scanning calorimetry (DSC), dynamic mechanical thermal analysis (DMTA), wide and small angle X-ray diffraction (WAX and SAXS), thermogravimetric analysis (TGA) and optical and electron microscopy.

This list has remained unchanged for nearly twenty years. A more ambitious laboratory should also be considering solid-state NMR, but this represents a quantum leap in cost and in interpretive skills.

It will be noticed that three of the five 'essential' techniques are concerned with thermal analysis (TA). In this chapter, all the main TA techniques are reviewed, but in order to keep to a reasonable length, examples will be selected only from the most important application areas. Some additional interesting applications will be presented from largely unpublished work from the Department of Chemistry, Loughborough University of Technology, UK.

By way of introduction, Table 7.1 itemises the TA techniques covered in this chapter, together with their most important application areas.

7.2 Differential scanning calorimetry (DSC)

Classical adiabatic calorimetry, in which precise quanta of heat are applied and the temperature rise of the sample is noted under shielded conditions, is an extremely slow and laborious technique. By referring the temperature of a sample to that of an inert sample experiencing a closely similar heating profile, equivalent information can be obtained in a rapid scanning experiment. This has many advantages in that structures in the polymer are not annealed and changed during fast thermal scans.

In order to qualify for use as a DSC, rather than just a DTA, the difference in temperature between sample and reference must be capable of calibration, such that:

$$\Delta T = K\, \Delta(mC_p)\, R$$

where ΔT = temperature difference between sample and reference, $\Delta(mC_p)$ = difference in total heat capacity between sample and reference, and R = heating rate (°C/min).

Table 7.1 Application areas of main TA techniques

Technique	Applications
TGA (thermogravimetric analysis)	Weight loss versus temperature or time allows studies of plasticiser or solvent loss, degradation, sorption and desorption.
DSC (differential scanning calorimetry)	Heat capacity versus temperature or time allows measurement of heats of fusion, identification of crystalline and liquid crystalline phases, degrees of crystallinity, etc. Glass transition measurement allows characterisation of ageing, blend compatibilities. Heats of reaction allow cure and degradation studies.
TMA (thermomechanical analysis)	Length change versus temperature. Expansion coefficients and their anisotropy. Empirical softening temperatures of waxes, etc.
DMTA (dynamic mechanical thermal analysis)	Characterisation of all motional transitions including those in the glassy, crystalline and liquid crystalline state. Measurement of relaxation spectra. Definition of morphologies in two-phase systems (e.g. rubber toughened plastics). Blend compatibility. Engineering modulus/damping characteristics over complete materials working range.
DETA (dielectric thermal analysis)	Electrical equivalent of DMTA but less generally applicable. Characterises motional processes involving dipole re-orientation. Mobile-phase content (hence crystallinity). Measurement of conducting polymers over complete working range. Definition of electrical quantities—dielectric permittivity and loss.
TSC (thermally stimulated currents)	Information on charge mobility in polymers. Probe into the nature of the glassy state. Experimentally arduous to obtain quantitative results.
STA (simultaneous thermal analysis)	DSC measurements performed on one arm of a thermal balance, thus avoiding ambiguities of sample temperature for combined DSC–TGA. Characterising minor volatile loss and reaction energies. Heats of desorption.
Hyphenated techniques (TGA/mass spec./FTIR) (STA/mass spec./FTIR)	Products evolved during thermal scan swept into FTIR or mass spectrometer for identification on a continuous basis. Characterisation of polymers and copolymers by breakdown products. Detailed understanding of degradation processes.

It is common practice to use closely similar sample and reference crucibles and if one is used, empty, as reference, then

$$\Delta T = K\, mC_p\, R$$

and this is the usual working equation. All DSC instruments must be calibrated (i.e. K determined) either by measurement of a known specific heat-capacity sample (usually sapphire) or by using samples with known heats of fusion. As K is a slow-moving function of temperature, calibration can be performed from melting curves such as that shown for indium (Figure 7.1). In this case:

$$\text{Area under melting curve} = K\, m\Delta H\, R$$

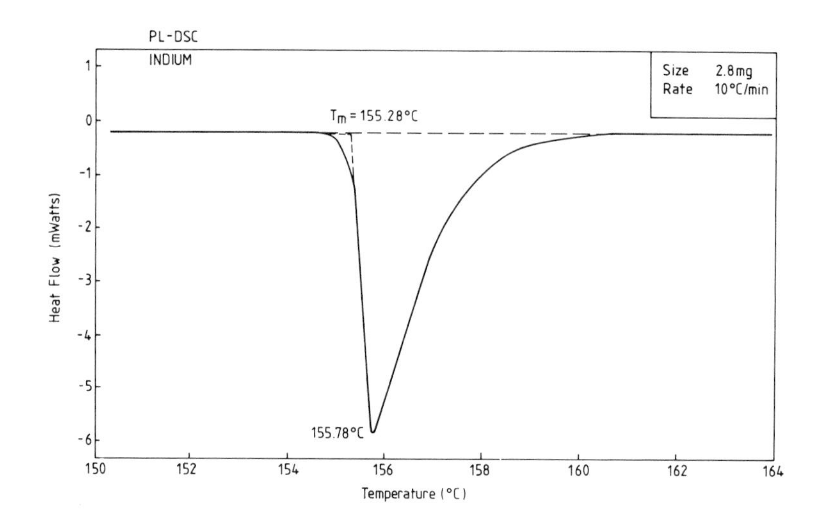

Figure 7.1 DSC trace for sharp melting-point material (indium). Onset construction for T_m is valid in this case.

By using two standard samples with known ΔH and melting points bracketing the temperature range of interest, K is determined at two temperatures and a linear interpolation between them assumed. The reader is referred to other texts for a more detailed discussion [1, 2].

There are two basic methods in use in commercial instrumentation. Figure 7.2(a) shows the power-compensation method employed by Perkin-Elmer. It was this method which first attracted the name 'DSC', because the difference in power required to ramp the sample and the reference at the same rates is measured. In reality, of course, a difference in temperature between sample and reference is required to drive the differential power requirement. This highly elaborate method is then seen to suffer from many of the same problems as the technique originally suggested by Boersma [3] which uses a heat-flow disc to quantify the difference in the heat flow to sample and reference (Figure 7.2(b)). With proper engineering, this heat-flow difference is just proportional again to the temperature difference between sample and reference (ΔT). In

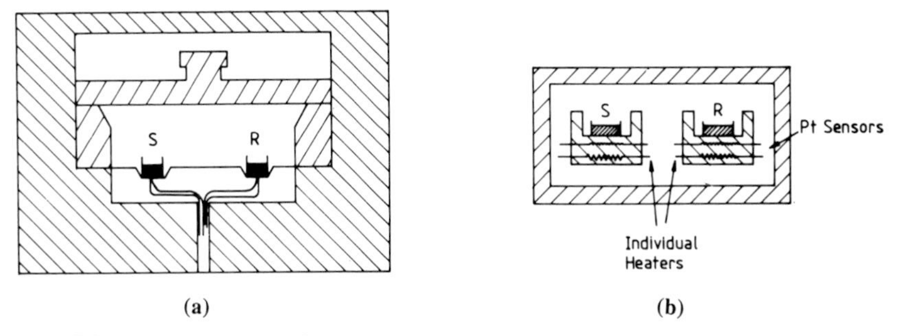

Figure 7.2 Schematic diagram of working principle of (a) heat-flow and (b) electrical-compensation DSC.

cells of the type used by Polymer Laboratories, TA Instruments and others, ΔT can be kept small by appropriate cell design.

Figure 7.3 shows the design of a modern heat-flow cell as manufactured by Polymer Laboratories. In this particular cell, the integral heat exchanger allows direct use of liquid nitrogen to generate programmed cooling curves of the same quality as normal heating curves. In the Perkin-Elmer method, the

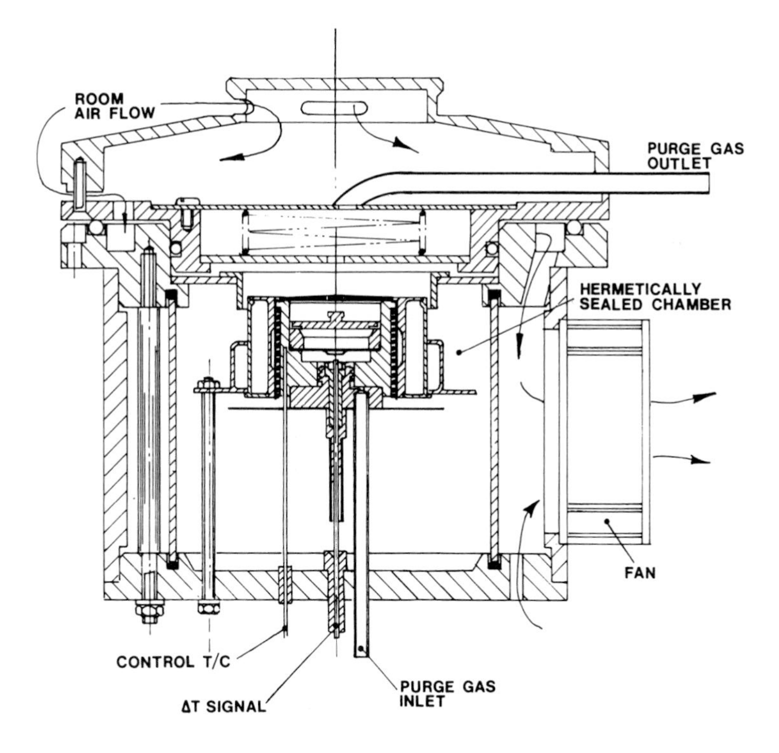

Figure 7.3 Cross-section of PL-DSC showing heat-exchange system for handling liquid N_2 and environmental containment.

size of ΔT to be tolerated (ideally it should be vanishingly small) is more under the operator's control, but only at the expense of varying-quality baselines.

DSC instruments have an inherent response time (timelag) between an event happening in the sample and that event being acknowledged in the DSC signal (ΔT). Unfortunately the sample size, type and quality provide such a major part of this response time that it is not practical to calibrate it out on any type of instrument. The easy availability of DSC traces has led to a complacency over this error, which makes DSC much less of an absolute technique in routine work than most operators realise. A simple practical example can be taken in the definition of melting points from a DSC trace. A sharp-melting crystalline solid should produce an infinitely high sharp spike at the melting point, but in reality a peak with a half width of ~1°C is produced because of the various response-time factors. The best practice in this case is to use the 'onset temperature' via the construction shown in Figure 7.1 (for indium) to pick up the first indication that melting is occurring.

7.2.1 DSC of crystalline polymers

The construction used above for indium is very misleading for semi-crystalline polymers, which have a very broad melting range. Figure 7.3 shows data for high- and low-density polyethylene, and even in the sharper-melting HDPE case the maximum rate of melting defined by the peak maximum is significantly higher than the onset temperature. For polymers the peak maximum is usually taken as the quoted melting point.

For broad-melting polymers, the slow rate of change of the DSC curves allows the peak height at any point to be taken as proportional to the instantaneous rate of melting during a thermal scan. Figure 7.4, by way of example, shows DSC traces for high-molecular-weight poly(tetramethylene oxide) (PTMO) crystallised for different times at the same temperature (293 K). The low-temperature peak is due to uncrystallised material, which crystallises on cooling prior to the thermal scan. This type of feature is very common, and must be properly interpreted, i.e. at 293 K the material melting in the low-temperature peak was in fact still amorphous prior to the 'characterisation' experiment. The major high-temperature peak represents the melting profile (within the errors discussed above) of crystalline lamellae produced at $T_c = 293$ K. The change with prolonged crystallisation time is due to so-called secondary crystallisation, which occurs with distinctly different kinetics after the spherulitic space filling of primary crystallisation is complete. The shift of melting processes to higher temperature signifies increased stability of lamellae. At the melting point:

$$\Delta G_f = \Delta H_f - T_m \Delta S_f = 0$$

for any species present. Thus:

$$T_m = \Delta H_f / \Delta S_f$$

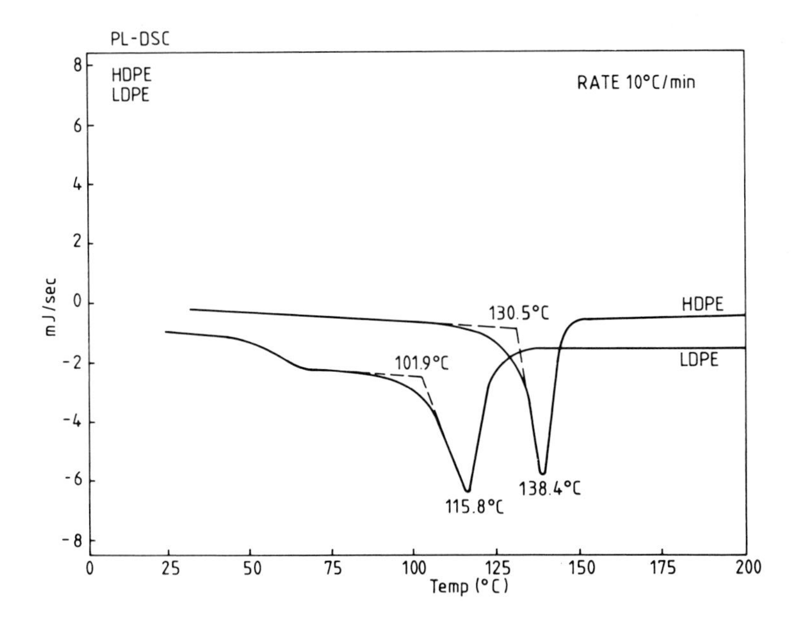

Figure 7.4 Polymer melting processes illustrated by HDPE and LDPE. Even in the sharper-melting HPDE the onset temperature does not represent a meaningful melting point. Peak temperatures are normally used for polymers.

ΔS_f(crystal) is believed to not vary greatly with lamellar size and perfection, but ΔH_f increases during secondary crystallisation due to two factors:

(a) Increase in lamellar thickness with concomitant decreases in the surface free-energy component; and
(b) Chains within existing crystals become better packed, with an increase in density and of ΔH_f(crystal).

Table 7.2 summarises the contribution made from these two processes, derived from very detailed wide- and small-angle X-ray scattering studies [4].

Table 7.2 The effect of secondary crystallisation on the heat of fusion

Crystallisation time (days) ($T_c = 293$K)	ϕ_w (wt fract^n)	ΔH_f (sample) (J/g)	ΔH_f (crystal) (J/g)
1	0.467	71.2	152.5
2	0.473	74.9	158.4
5	0.480	77.2	160.8
14	0.489	82.2	168.1
27	0.494	87.1	176.3
49	0.499	94.1	188.6

It is not usually practicable to derive such detailed results as shown in Table 7.2, and to a first approximation it is assumed that the dominant parameter causing a distribution of melting points is lamellar thickness. Basic thermodynamic arguments lead to the expression [2]:

$$T_m = T_m^\infty (1 - 2\sigma_e/\Delta H\, L)$$

where T_m^∞ is the equilibrium melting point of fully extended chain crystals ($L \to \infty$), σ_e is the fold surface free energy, L is the lamellar thickness, and T_m is the melting point of lamellae of thickness L.

This relation shows how the distribution of melting points in a melting endotherm can be interpreted in terms of a distribution of lamellar thicknesses. In the data in Figure 7.5 this argument even applies to the material crystallised on cooling and giving rise to the lower-temperature melting process. This material has crystallised from the amorphous regions between existing lamellae. The only lamellae capable of forming in this region must be thin, and therefore they can only form at high degrees of supercooling (ΔT). Thickness L_g is given by Lauritzen and Passaglia [5] as:

$$L_g = \frac{2\sigma_e T_m}{\Delta H_c \, \Delta T} + \delta L$$

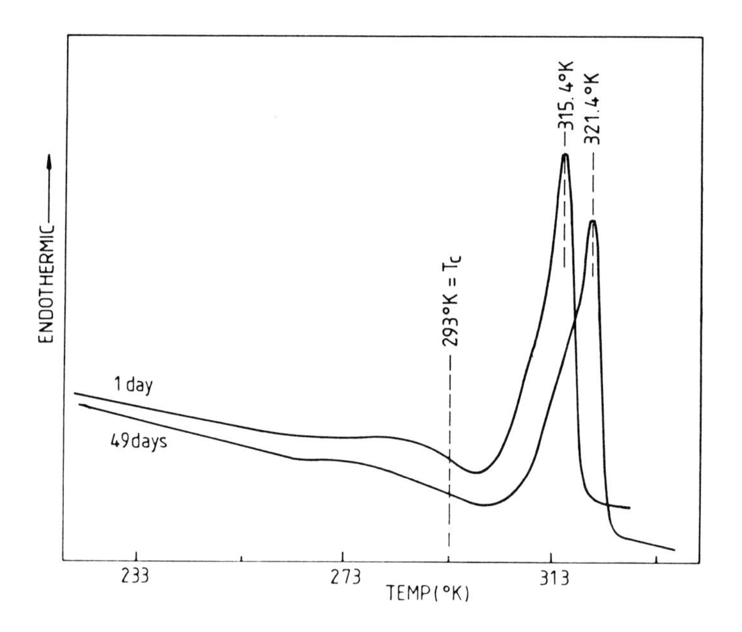

Figure 7.5 Poly (tetramethylene oxide) melting after different times of crystallisation at 293 K. The changes are largely due to lamellar size changes.

where δL is normally ~10% of L_g. Figure 7.6 illustrates the bimodal lamellar structure giving rise to the DSC traces of Figure 7.5.

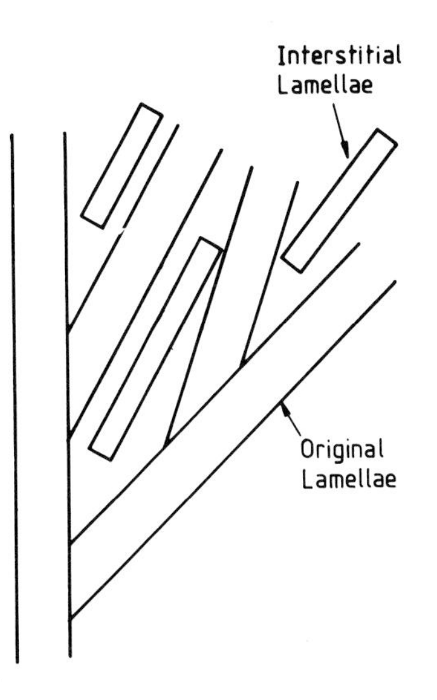

Figure 7.6 Schematic diagram to show the origin of bimodal lamellar-size distribution giving rise to the bimodal melting (see Figure 7.5).

A more normal example of melting endotherms which reflect the lamellar formation conditions can be readily demonstrated in the DSC itself by programmed cooling at higher rates to force the crystallisation process to occur at lower temperatures. The basic reason why this occurs is that crystallisation in the region near T_m is nucleation controlled, and therefore requires a certain induction time before growth can start. The faster the cooling rate, the lower the temperature the amorphous sample achieves before nucleation has occurred. The lamellar thickness is related to the supercooling according to the above relationship. On remelting, a different distribution of lamellae gives rise to the different melting endotherms. Most lamellae can thicken, after they have grown, by chain translation through the crystal. If this did not happen, lamellae would, in theory, melt close to the temperature at which they crystallised.

The overall crystallinity of a polymer can be simply assessed from heats of fusion, and this must be the best quality-control method available. The prime requirement is a knowledge of ΔH_c, the heat of fusion per gram (or kilogram) of crystal phase. Then determination of the heat of fusion per gram of the actual sample ΔH_s allows determination of the weight fraction crystallinity (X_w) via:

$$X_w = \Delta H_s / \Delta H_c$$

In many cases ΔH_c may not be well determined, but for comparative purposes this is relatively unimportant. Values of ΔH_c for the more common polymers are given in Table 7.3, together with T_m.

Table 7.3 Melting points and heats of fusion of polymers

Polymer	T_m(°C)	ΔH_c(KJ/mol)
Polyethylene	141	4.11
Polytetrafluoroethylene	332	4.10
Poly(4-methyl-1-pentene)	250	9.96
Polyisobutylene	44	12.0
Poly-1,4-butadiene, *cis* form	11.5	9.20
Poly-1,4-butadiene, *trans* form	164	3.73
Poly(vinyl alcohol)	265	7.11
Poly(vinylidene fluoride)	210	6.70
Poly(vinyl chloride)-*syndio.*	273	11.0
Polystyrene-*iso*	243	10.0
Polyoxymethylene	184	9.79
Polyoxyethylene	68	8.66
Polyoxytetramethylene	56	14.4
Polyoxypropylene	75	8.40
Poly(methyl methacrylate)-*syndio*	177	9.60
Nylon-6	260	26.0
Nylon-11	220	44.7
Nylon-12	277	48.4
Nylon-6,6	300	57.8
Poly(ethylene terephthalate)	280	26.9
Poly(butylene terephthalate)	245	32.0
Poly(dimethyl siloxane)	–54	2.75

As ΔH_s is obtained from the total area of the melting curves, the absolute errors in the shape of the curve which make melting-point determination problematic are far less crucial. The main practical problem relates to assessing the baseline. Before the peak it relates to the C_p of semi-crystalline polymer, and after the peak to C_p of the melt. The usual approach is to draw a straight line joining the curves at points where the melting is considered to have 'just started' and 'just finished' as indicated in Figure 7.4. A better baseline would be one that varied between the two selected points in proportion to the amount of material melted. A problem peculiar to polymers, because of their broad melting range, is the proper selection of ΔH_c for the temperature required. This is no different from the problem of heats of chemical reaction varying with temperature. The variation of ΔH_c with temperature (invariably an increase) arises because the enthalpy curves of the initial semi-crystalline material and of the melt are not parallel. This variation in ΔH_c is by no means trivial and would need considering, for example, in comparing heats of crystallisation with heats of melting, where the crystallisation could typically be occurring 50 K below T_m.

It should be noted that the crystallinity determined by this method will not agree precisely with that determined by density or X-ray diffraction. Both these techniques determine volume-fraction crystallinity, but more importantly, each technique reflects structural defects and morphology differently.

7.2.2 *DSC of glassy polymers*

Polymers lacking regularity in structure cannot crystallise on cooling and therefore are forced into a glassy state at some low temperature. The lack of structural regularity occurs in many addition polymers because of lack of stereospecificity at the polymerisation stage in monomers, giving an asymmetric carbon in the main chain. Atactic polystyrene, poly(methyl methacrylate) and PVC are common examples. Other irregularity is caused by 1,2- rather 1,4-addition in dienes and, more importantly, by random copolymerisaton or terpolymerisation which prevents formation of single monomer sequences long enough to crystallise.

The main feature of a glassy polymer is the transition from largely brittle glassy behaviour at low temperature to a rubbery or viscoelastic state at high temperatures. This transition has many of the characteristics of a genuine second-order transition, but during normal laboratory timescales must be considered as being kinetically controlled. DSC is a convenient and quick method for determining the temperature change from glass to melt, the so-called 'glass transition temperature' (T_g). It will be seen later that as observations of T_g are concerned with relaxation processes, dynamic mechanical or other techniques that measure molecular motion directly are more appropriate to the study of

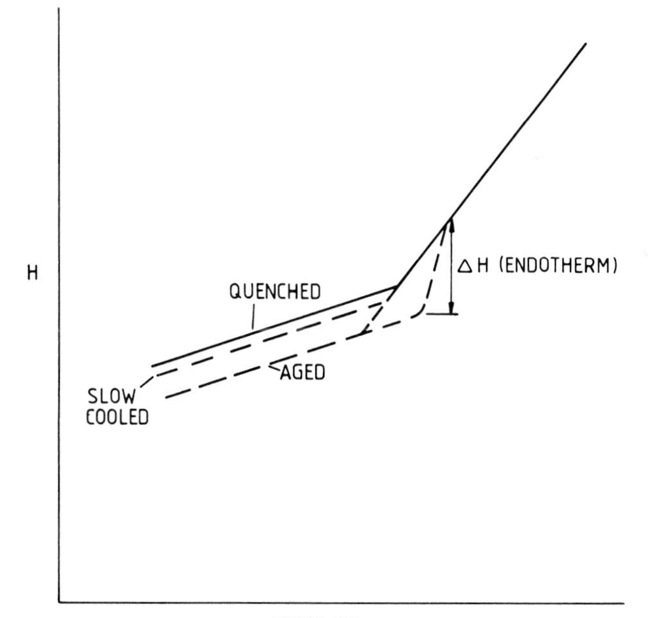

Figure 7.7 Schematic enthalpy changes through the glass-transition region, showing dependence on glass-formation conditions and origin of endothermic relaxation peak in aged glasses.

T_g phenomena. However, DSC can be used to probe the enthalpy content of glasses with respect to the melt state and also changes in the enthalpy content due to ageing.

Polymeric glasses are never at equilibrium; it is thus worth considering the basic changes of enthalpy (or volume) with cooling or heating, as shown schematically in Figure 7.7. If a melt is cooled rapidly, the glass is formed at relatively high free volume, because molecules have had little time to relax their configurational and packing state. A glass of higher than normal volume and enthalpy is produced. Conversely, slow cooling produces a glass of low enthalpy and volume (these changes are actually very small). If a rapidly cooled glass is immediately reheated at the same rate, it will pass back into the melt state at approximately the same T_g as it entered. However, a glass formed by slow cooling or by annealing a fast-cooled glass not too far below T_g will 'superheat' on the upward thermal scan, as molecules trapped in low free-volume regions have acquired long relaxation times.

A typical recovery path taken during a thermal scan of an annealed glass is labelled 'endotherm' in Figure 7.7. The manifestation of these events in DSC heating traces through the T_g is illustrated with respect to data on PVC as a function of ageing at 65°C. The quenched PVC exhibits a simple step function and the annealed PVC shows an endotherm due to the delayed enthalpy

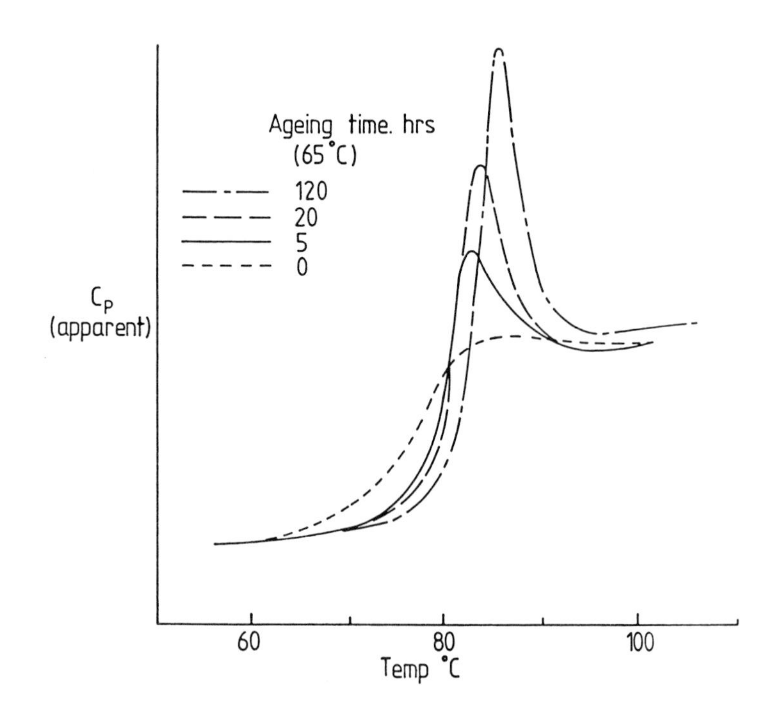

Figure 7.8 PVC aged for the times shown (after quenching) and then rescanned through T_g.

recovery process. Care must be taken not to interpret this as anything to do with a melting transition. These curves are a direct result of the interplay between experimental timescales and molecular relaxation times. It appears from the directly displayed DSC curves that T_g increases on annealing, but Richardson [2] has clearly shown that if enthalpy curves, such as those shown in Figure 7.7, are derived from DSC traces such as those shown for PVC (Figure 7.8) only small changes in T_g are shown. These are in the direction predicted by straight-line extrapolations of the enthalpy curves of Figure 7.7.

Thus the annealed glass enthalpy line cuts the extrapolated liquidus line at a lower T_g, and this agrees with practical extrapolations of enthalpy curves. The actual DSC traces cannot therefore be used directly to define the changes of T_g with ageing. However, as with heats of fusion, the integral changes can be used to characterise the small decreases in enthalpy occurring during ageing. This point is made schematically in Figure 7.9, where the excess area of the endotherm above T_g can be attributed to the lower enthalpy of the aged sample. A practical point that arises from these observations is that an onset construction for T_g, as shown in Figure 7.10, is nearer to the truth than a half-height method.

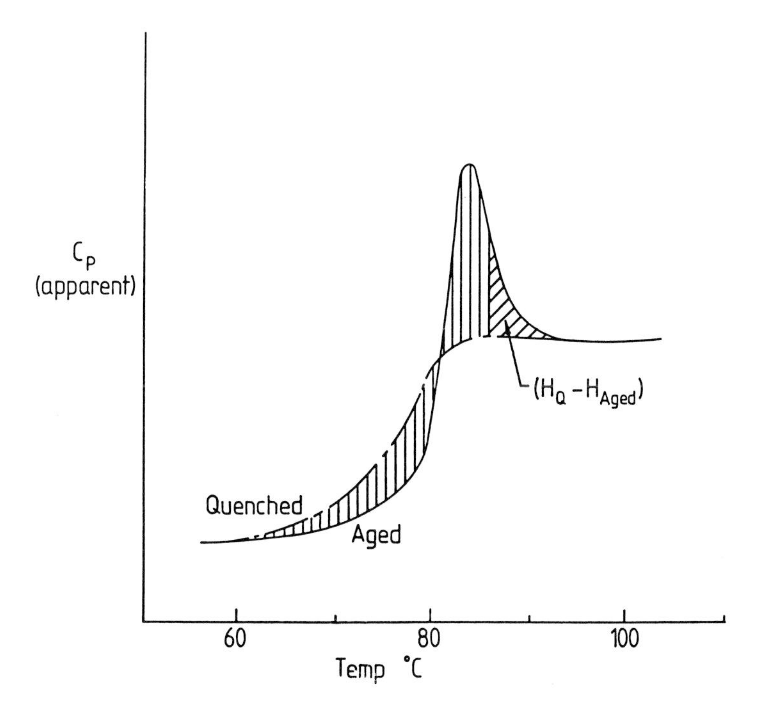

Figure 7.9 Construction to obtain enthalpy difference ($H_Q - H_{\mathrm{Aged}}$) between quenched and aged glasses.

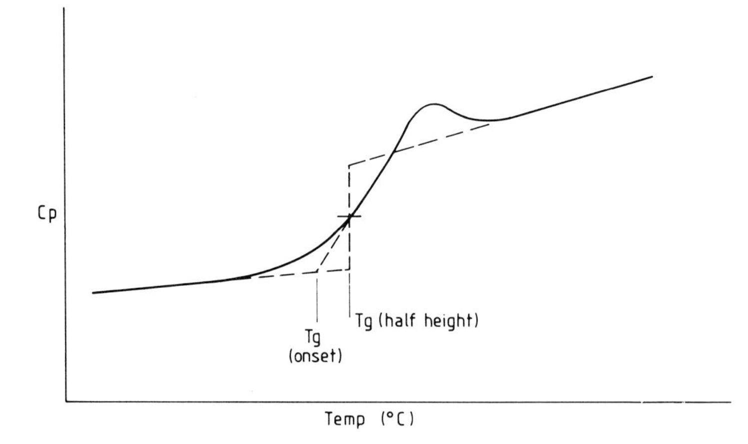

Figure 7.10 T_g definitions from DSC scans.

For many instances within a single laboratory, the precise method used for defining T_g is not important provided that it is always consistent. Problems arise when trying to relate observations to those of other workers. Polymer-blend studies are a case in point where systematic definition of T_g is largely all that is required. Figure 7.11 shows classic work on poly(dimethyl phenylene oxide)–polystyrene blends by MacKnight, Karasz and Frièd [6]. The T_g values defined by the onset and half-height methods give closely similar smooth curves when plotted against composition (Figure 7.12), proving that the system is compatible (single T_g, composition dependent).

7.3 Dynamic mechanical thermal analysis (DMTA)

7.3.1 Theory

The DMTA technique measures molecular motion in polymers, and not heat changes as with DSC. The technique relies on impressing a small sinusoidally varying stress on to the material under test and transducing the strain. The time-dependent response is shown in Figure 7.13. For a completely elastic material the strain is in phase with the stress, and for a purely viscous material the strain lags behind the stress by 90° (out of phase). Polymers are viscoelastic to a greater or lesser extent according to the temperature at which they are measured. It is thus convenient to resolve the behaviour into an in-phase elastic-like component (governed by the storage modulus E' or G') and an out-of-phase viscous-like component (governed by the loss modulus E'' or G''). The lower curves in Figure 7.13 show the stress response broken down into the in-phase and out-of-phase components with respect to the strain response.

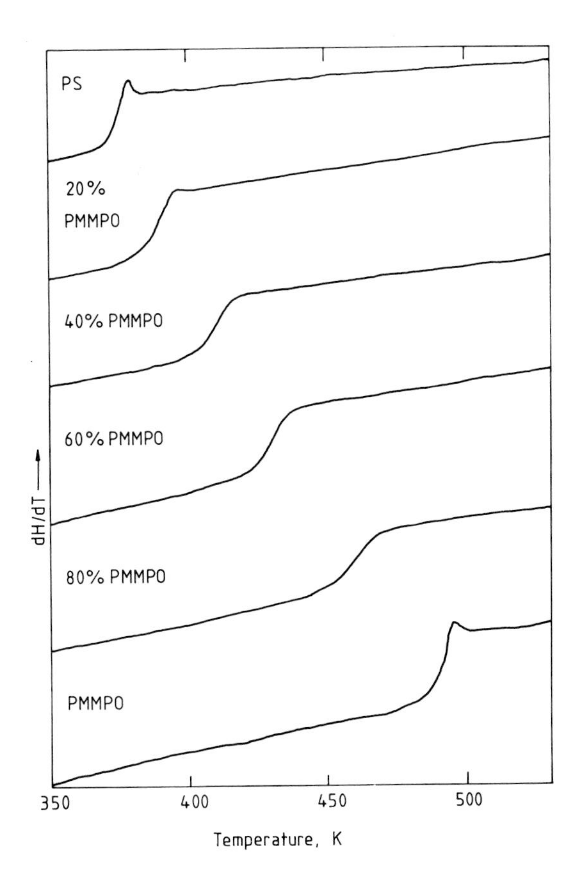

Figure 7.11 C_p traces of polystyrene–poly(dimethylphenylene oxide) blends at compositions stated.

The storage and loss moduli are now defined by:

$$\text{Storage modulus} = \frac{\text{In–phase stress amplitude}}{\text{strain amplitude}} = C/B$$

$$\text{Loss modulus} = \frac{\text{Out–of–phase stress amplitude}}{\text{Strain amplitude}} = D/B$$

These relationships can be summarised in an Argand diagram with the total response governed by the complex modulus (E^*, G^*) as shown in Figure 7.14. The phase angle (lag of strain behind stress) is also clearly defined in the Argand diagram and a convenient dimensionless parameter is the 'loss tangent' ($\tan \delta$):

$$\tan \delta = G''/G' \text{ (in shear)}$$

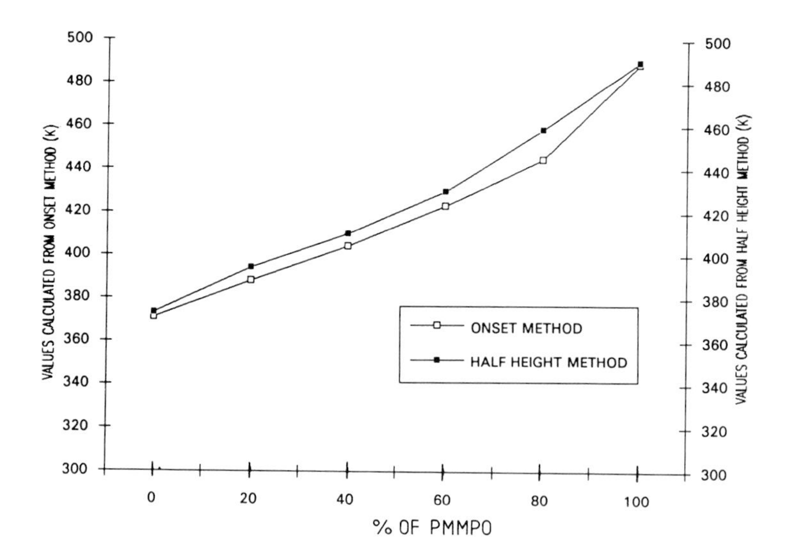

Figure 7.12 Plots of T_g determined by the onset and half-height methods for the blend DSCs shown in Figure 7.11.

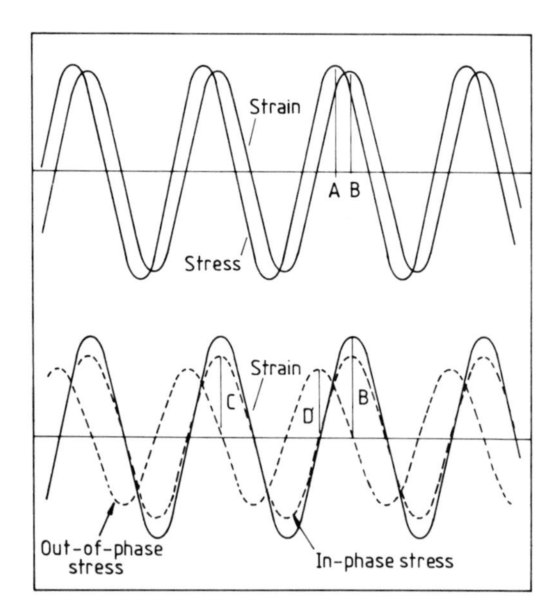

Figure 7.13 Sinusoidally varying stress and strain in a dynamic mechanical experiment. Construction shows definition of $E' = C/B$ and $E'' = D/B$.

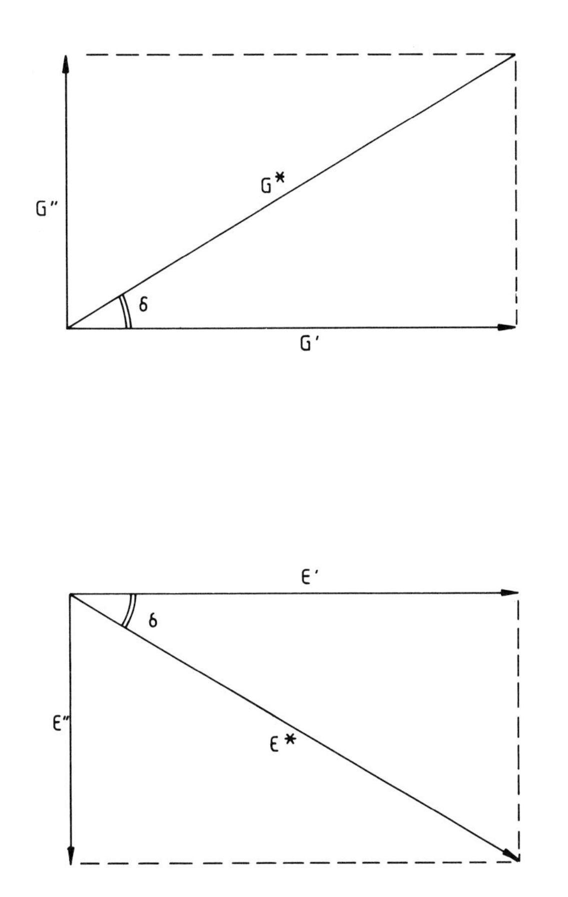

Figure 7.14 Argand diagrams showing relation between dynamic moduli (upper) and between dielectric permittivities (lower).

or:

$$\tan \delta = E''/E' \text{ (for Young's modulus)}$$

This is physically the ratio of energy lost to energy stored per deformation cycle. A peak in $\tan \delta$ occurs when the impressed frequency matches the frequency of molecular relaxation through thermally activated processes. If τ is the average molecular relaxation time at temperature T, then a loss peak will be observed at this temperature if the impressed vibration frequency (f_{max}) satisfies the relation:

$$2\pi f_{max} = \frac{1}{\tau} \text{ at temperature } T$$

τ varies with temperature in an approximately Arrhenius fashion (well above T_g), and thus this simplest relation for relaxation time:

$$\tau = \tau, \exp(\Delta H^* / RT)$$

allows τ to be swept by changing T. The loss peak condition is then:

$$2\pi f = 1/\tau_{max} \text{ at } T_{max}$$

In the context of thermal analysis this is the usual way of observing loss peaks. Temperature is ramped up, usually at a fairly slow rate (1–5°C/min), and the temperatures are observed for maximum loss location at constant impressed frequency, say 1 Hz.

Figure 7.15 illustrates the main features observed during a DMTA scan of a semi-crystalline polymer. Relaxations can occur in crystalline regions as well as amorphous and '*a*' and '*c*' subscripts are used to denote the origin. Relaxations in the amorphous phase are usually the overwhelming events until the melting region is reached. Relaxations in the amorphous phase are labelled with the Greek alphabet (α_a, β_a, γ_a, δ_a, etc.) with decreasing temperature, as are crystalline-phase relaxations if observed. This procedure ensures that the T_g process is always α_a and subsequent glassy state relaxations are β_a, γ_a, etc. Relaxations above T_g in the crystalline phase are usually given the symbols

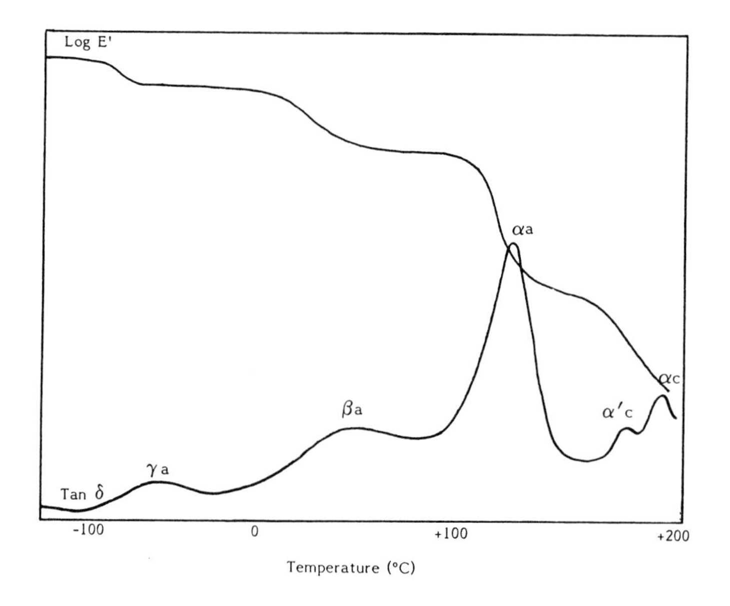

Figure 7.15 Typical transition behaviour in mechanical storage modulus and damping for a semi-crystalline polymer. Subscripts *a* and *c* refer to the amorphous and crystalline phase respectively. Thus α_a is the main T_g process.

α_c', α_c'', etc. in order of decreasing temperature because they are all believed to be due to the melting process. This nomenclature is still awaiting IUPAC approval.

7.3.2 DMTA—experimental considerations

Original pioneering work in relating regions of high loss (so-called loss peaks or relaxations) was carried out very skilfully with free-decay torsion-pendulum or natural-frequency vibrating-reed instruments, where the material under test was the reed or pendulum support.

Modern instruments use direct-phase measurement between the stress and strain sine waves in the convenient experimental range 0.01 Hz to a few hundred hertz. The only problem with measuring at lower frequencies is the time taken to acquire a data point [time $1/f$ (Hz)]. At the high end, fundamental as well as practical problems arise. The practical problems are concerned with instrument resonances, while the fundamental problems relate to propagating waves in samples of similar dimensions to the wavelength. The main deformation modes available in the various instruments available from Rheometrics, Polymer Laboratories and most thermal analysis companies are bending, simple shear, tensile, compression and torsion. The geometry factors for use in these modes are given in Table 7.4, together with the strain conditions. Ideally strain should be held constant throughout a DMTA scan.

Table 7.4 Geometry factors for various DMTA modes

Type of geometry	Form factor (k) and units	Maximum stress	Maximum strain
Uniform strain			
1. Tensile	A/L (m)	$\sigma = f/A$ (*Pa*)	$\varepsilon = x/L$
2. Shear sandwich	$A_1/h_1 + A_2/h_2$ (m)	$\nu = f/(A_1 + A_2)$ (Pa)	$\gamma = x/h$
Non-uniform strain			
3. Cantilever bending (both ends clamped—centre displaced)	$2cd^3/L^3$ (m)	$\sigma = 3fL/4cd^2$ (Pa)	$\varepsilon = 12dx/L^2$
4. Torsion of cylinder	$\pi R^4/2h$ (m^3)	$\nu = 2\delta/\pi R^3$ (Pa)	$\gamma = \alpha R/h$

c = width; A = area; ν = shear stress
d = thickness; L = free length; γ = shear strain
f = force; R = radius; ε = tensile strain
h = height; δ = torque; σ = tensile stress
x = linear displacement; α = angular displacement

The advancement of modern electronics coupled with computer control has now concentrated the problems of DMTA measurements completely at the mechanics end. Selection of optimum geometry for the materials to be measured is thus vitally important. Polymers sag under their own weight and flow away from clamp pressure in the T_g region and above. The 1000-fold stiffness change for amorphous systems through T_g makes the experimental side of DMTA quite difficult and the best instruments have to be judged on the quality of data that they can deliver over the stiffness ranges required of polymeric materials. Moduli for elastomeric foams can be as low as 10^3 Pa and for directional composites as high as 10^{11} Pa in the fibre direction. As the dynamic-modulus variation for polymers is so large, a large dynamic-stiffness range needs to be accessible without change of sample. Thermal scanning is the normal mode of acquiring DMTA data and it is thus desirable, by frequency multiplexing or by simultaneous frequency application, to acquire data at several frequencies during the scan.

Composites are stiff and require long samples to avoid massive end-correction errors. Bending and torsion are the only geometries that can sensibly be used (for E and G, respectively). Three-point knife-edge deformation may be used if the sample is very well cured and does not creep under the permanent loading required for the centre knife edge.

Thermoplastic materials such as polystyrene or polyethylene can be measured conveniently in bending over their whole temperature range in bending or torsion. Thin films and fibres, particularly of semi-crystalline polymers such as PET, can only sensibly be measured in tensile deformation (E). In the elastomeric range, simple shear or torsion can be used to give G for short-length samples, or tensile deformation can be used for E. Viscoelastic liquids (melts) must obviously be suitably contained, and the only options are shear sandwich and parallel-plate torsion, the latter being preferred. Compression modes can be used, but the parameters being accessed are usually obscure, being strong functions of geometry and amplitude. Foams can be measured in simple shear, torsion or fully bonded compression.

An example of a frequency-multiplexed DMTA scan is shown in Figure 7.16. The sample is measured in dual cantilever-bending geometry with small strain amplitude so that the requirements of theory are met. As the temperature is scanned, the on-board processor clocks around the selected frequencies continuously. Each data point is recorded with the instantaneous temperature. The curves shown in Figure 7.16 are generated and software can convert them to three-dimensional plots (see later), and, by interpolation between data points, any frequency-plane or temperature-plane slice can be accessed.

7.3.3 Applications of DMTA to polymers

DMTA is an interesting and rewarding technique in that engineering data are acquired at the same time as molecular information. Figure 7.16 gives an example of engineering data on a composite. This can be further extended by producing a three-dimensional contour map of the data. Such a map is shown for both modulus and damping terms for PVC in Figure 7.17. Here the presence of a β_a process can clearly be seen at low temperatures. Its strength (measured by $\tan\delta$ area or $\log E'$ change) is much smaller than that of the

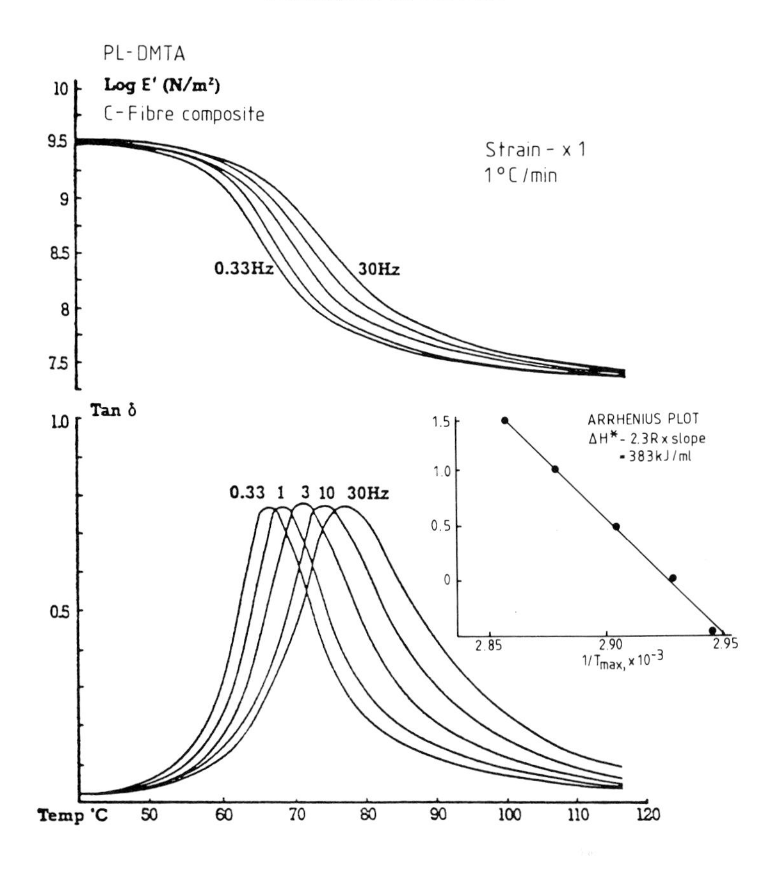

Figure 7.16 Modulus and damping for a carbon-fibre composite measured by the frequency multiplexing method during a single thermal scan. Insert shows the Arrhenius activation-energy plot from the loss peak position.

T_g process. Nevertheless it is of crucial engineering importance, as the strength of a process in the temperature range 0°C to – 60°C tends to convey impact resistance to the polymer [7]. This fact is used in rubber-particle addition to glassy polymers to provide a source of energy absorption. Figure 7.18 shows the α_a relaxation for the dispersed rubber phase is occurring at – 80°C in a polycarbonate–ABS mechanical blend. At low volume fractions of rubber (< 15%), the areas of the loss peak are proportional to the volume fraction of rubber. At higher volume fractions, the rubber particles may no longer be isolated and the detailed morphology will affect the relaxation strength and room-temperature modulus. A useful model in this context is that of Takayanagi [8]. The two phases are considered to be stressed partly in series and partly in parallel, as shown in Figure 7.19. The equation for the modulus of the composite (E_c^*) is:

$$E_C^* = \left[\frac{\phi}{\lambda E_F^* + (1 - \lambda) E_M^*} + \frac{1 - \phi}{E_M^*} \right]^{-1}$$

where E_F^* and E_M^* are the moduli of the matrix and filler, respectively, and λ and ϕ are the series and parallel coupling constants, respectively.

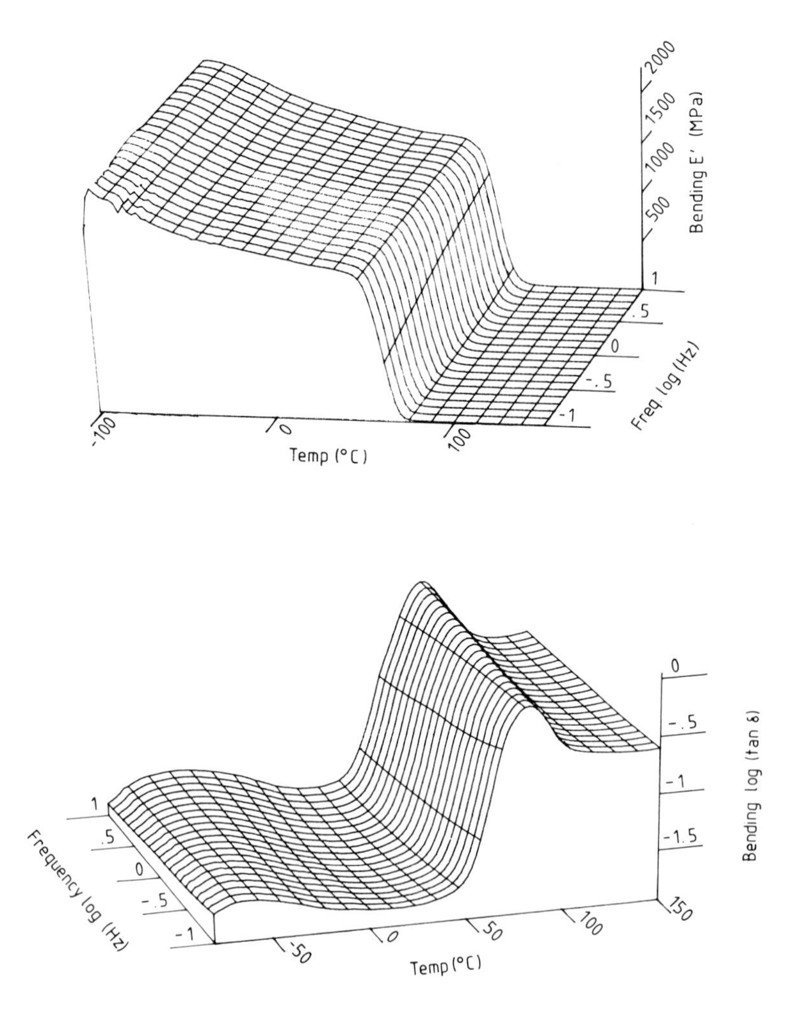

Figure 7.17 Three-dimensional plots of modulus versus T and f and loss versus T and f for PVC.

The material data in Figure 7.18 are in fact an example of an incompatible polymer blend. Each component exhibits a T_g process at the same temperature as it would if unblended. There are considerable advantages in using tan δ peak

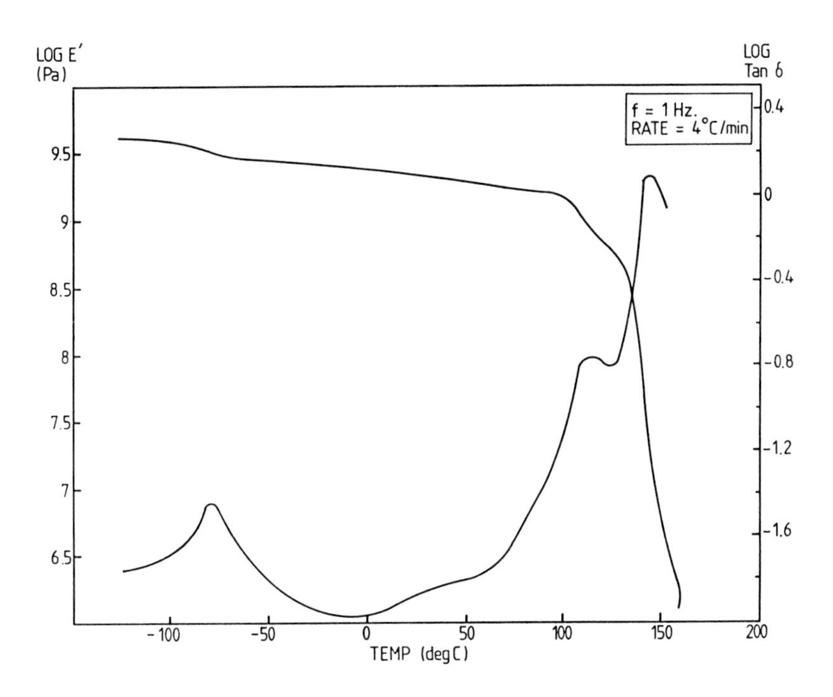

Figure 7.18 Incompatible blend of polycarbonate with ABS. The peaks in order of increasing temperature are due to: (1) α_a rubber phase of ABS (– 80°C); (2) α_a process for styrene–acrylonitrile polymer (110°C); and (3) α_a process for polycarbonate (145°C).

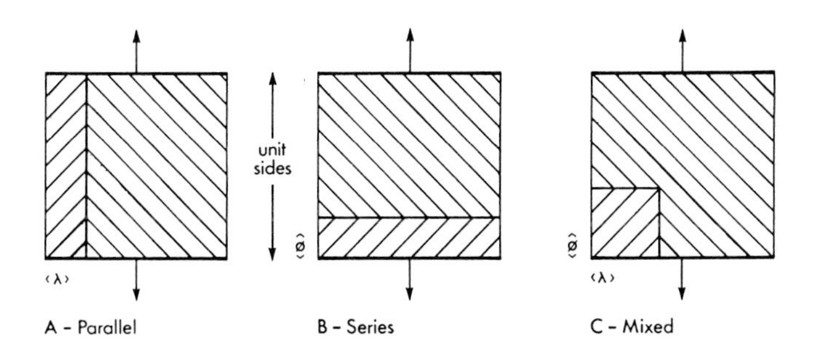

Figure 7.19 Takayanagi models for composite behaviour; λ and ϕ are the series and parallel coupling constants.

locations in this type of study, as G'' and E'' peak locations can be biased by the different influence of morphology on the modulus in different temperature regions. The judgement of compatibility of a blend is made by the presence of a single T_g loss peak that moves between the parent polymer positions with changing composition, just as a single transition is observed in a DSC trace (see Figure 7.11). The DMTA method is approximately 1000 times more sensitive in resolving the strength (not the temperature) of a transition than DSC. The data is Figure 7.20 were obtained by casting the blend concerned from a solvent such as toluene on to a steel shim of thickness 0.2 mm. The resolution of the DMTA gives high precision in the peak position and clearly shows compatibility over the whole compositional range for the poly(glycidyl methacrylate)–polyepichlorohydrin blend.

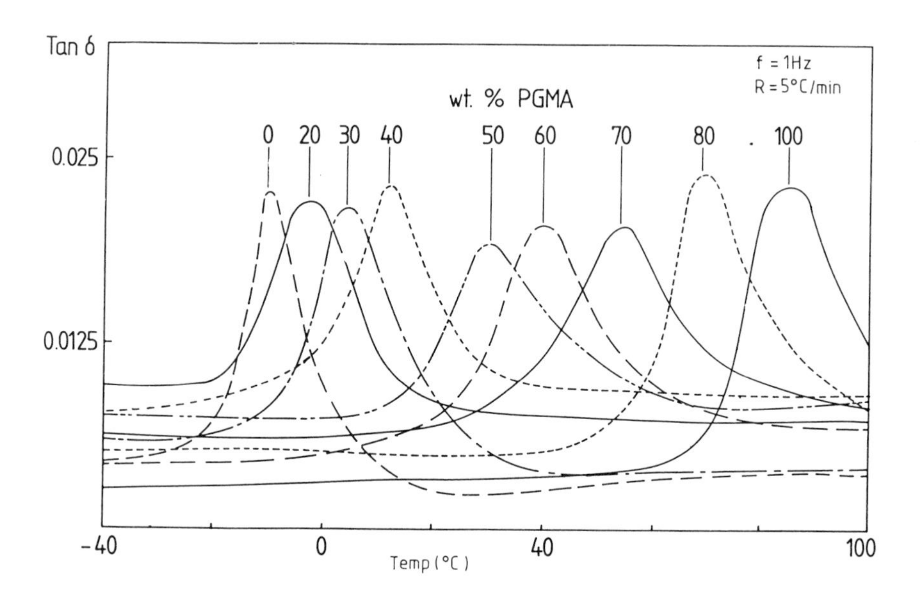

Figure 7.20 Compatible blends of PGMA–PEPC measured in the PL-DMTA as cast film on steel shim.

Ageing of glassy polymers can be studied well by DMTA, which produces information concerning the effect of lowered free volume on modulus, damping and molecular relaxation. Figure 7.21 shows two DMTA scans for polystyrene. The first is for a sample quenched in ice–water and then scanned through T_g at 4 °C/min (1 Hz). The second trace is for a sample similarly quenched and then aged at 80°C for a week prior to cooling and scanning as before. The change in modulus due to ageing is large and is shown (Figure 7.22) plotted against time of ageing at 70°C, 80°C and 90°C. On a linear scale the modulus increases by a factor of two at 80°C, and a feature of ageing is that the sample becomes more brittle. This is in line with the very low tan δ levels observed in the aged glass, which deny absorption of impact energy by internal motion. When a physically aged polymer is taken well above T_g it becomes completely rejuvenated and on quenching follows the unaged

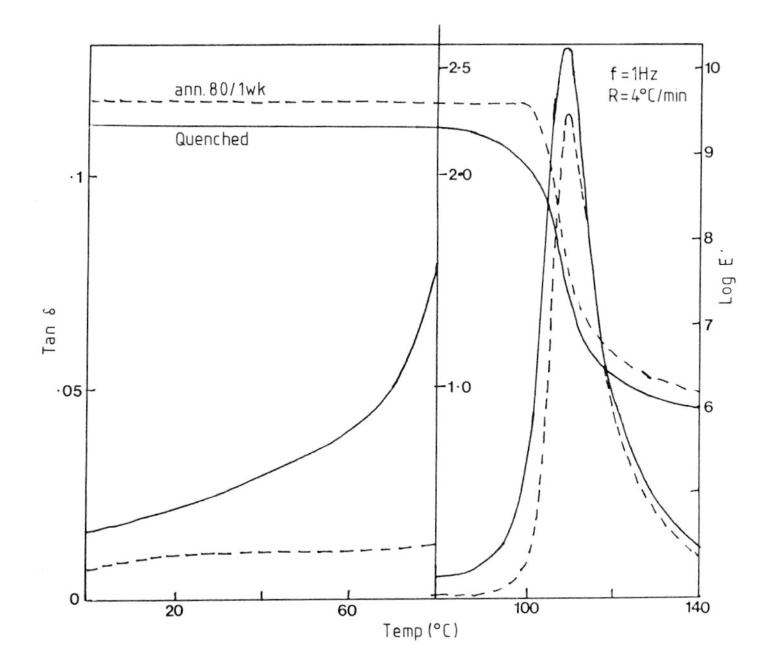

Figure 7.21 Ageing effects in polystyrene studied by DMTA. Freshly quenched polystyrene and a sample aged at 80°C for 1 week are compared.

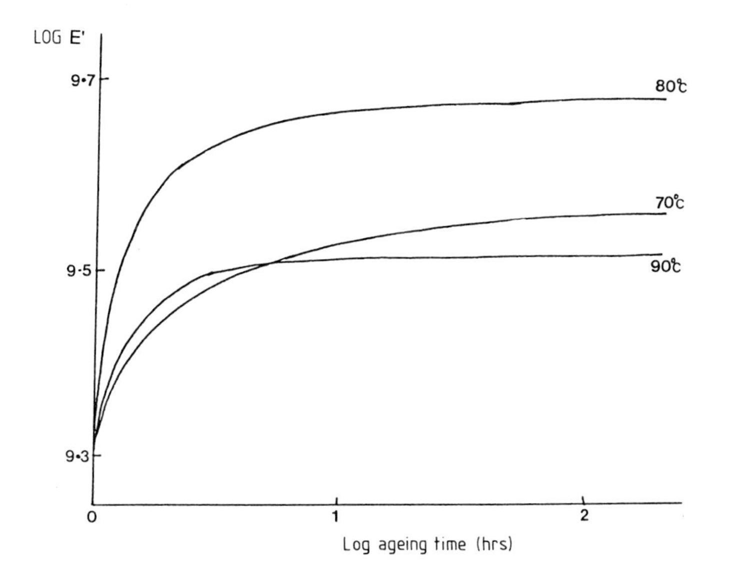

Figure 7.22 Log E' changes for poly(styrene) aged at the temperatures shown after quenching from above T_g.

sample's trace. The reader is referred to Struijk's excellent book [9] for further discussion on ageing phenomena.

Details of relaxation processes such as activation energies and relaxation spectra are obtainable from DMTA. The inset in Figure 7.16 shows an Arrhenius plot for the epoxy composite. The Arrhenius relation:

$$\log f_{\max} = A + \Delta H^*/2.3RT$$

where A is a constant, treats bond-rearrangement processes as if they were a simple reaction, such as optical isomerisation, which is strictly not correct in the T_g region. Figure 7.23 shows the Arrhenius plots for both the α_a and β_a processes in PET. The β_a processes always give straight-line Arrhenius plots over the whole experimentally accessible frequency range. The Arrhenius equation does not, however, strictly apply to the α_a process in the vicinity of T_g, where it acquires curvature, indicating very high activation energies as T_g is approached (see Figure 7.24).

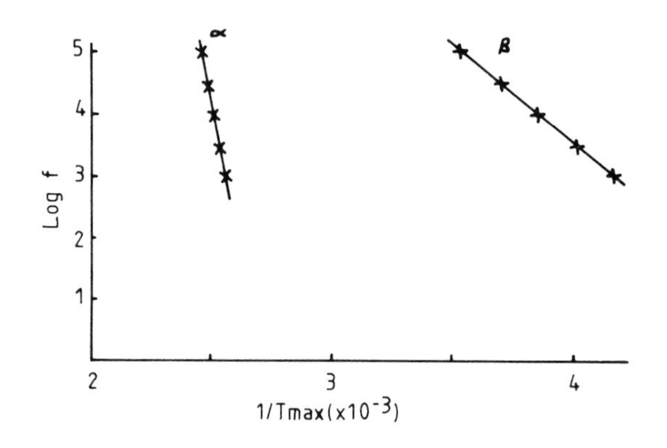

Figure 7.23 Arrhenius plots for the α_a and β_a processes in PET (from Figure 7.32).

Hay and co-workers [10] have used ageing kinetics of glassy polyetherimide to define relaxation times at temperatures below the normally accepted T_g. By combining these with relaxation times from DETA, DMTA and DSC on the same polymer, the composite Arrhenius plot shown in Figure 7.23 was obtained. This clearly shows curvature and enormously high ΔH^* if the Arrhenius expression is applied to the lower-temperature slope. It is clear from Figure 7.23 that the temperature separation of the α_a and β_a peaks (and usually γ_a as well) is greater the lower the measurement frequency. The effect appears larger in the $1/T$ plot than it actually is in linear temperature. The reason why measurements are not regularly performed at a frequency of say 0.01 Hz is that these data will take an absolute minimum of 100 s per point to collect. This data collection would be spread over a temperature range of greater than

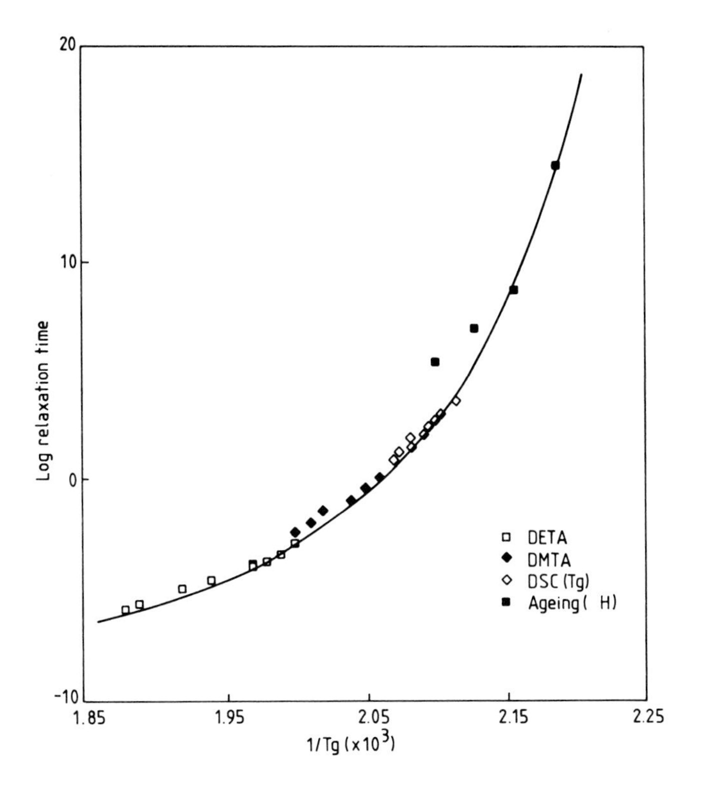

Figure 7.24 Composite 'Arrhenius' plot from different techniques covering the 10^{-5}–10^{15} relaxation time range. Curvature indicates the invalidity of the Arrhenius relation applied to α_a processes.

3°C at a scanning rate of only 2°C/min. Any expected resolution enchancement will be spoilt by lack of temperature precision, unless exceptionally slow rates are used. Lost peak locations for common commercial polymers have been listed elsewhere [11]. .

The reason why β_a follows a simple activation-energy law and α_a does not lies in the origin of the processes. In side-chain polymers some low-temperature relaxations (β_a, γ_a) can occur with little movement of the main chains. Chair–chair flips in cyclohexyl methacrylate are the classic example [12]. Other α_a processes occur in the main chain itself (PET and PVC are examples) without change of main-chain conformation, i.e. without the main chain, even locally, acquiring the rotational and translational freedom that it acquires in the α_a process. Thus, β_a processes are related to such events as in-chain phenyl group libration/rotation, ethylene-sequence large-angle torsional libration, etc. As soon as the main chain is able to translate, the relaxation process becomes co-operative, involving many neighbours. The human analogy would be struggling out of a crowded lift. This type of process becomes catastrophically more difficult the more slowly the neighbours move and actually involves more neighbours, until, at some true T_g, that type of motion becomes impossible and glassy-state conditions prevail.

The so-called WLF (Williams, Landel and Ferry) [13] time–temperature superposition principle and the working relation derived from it give a better

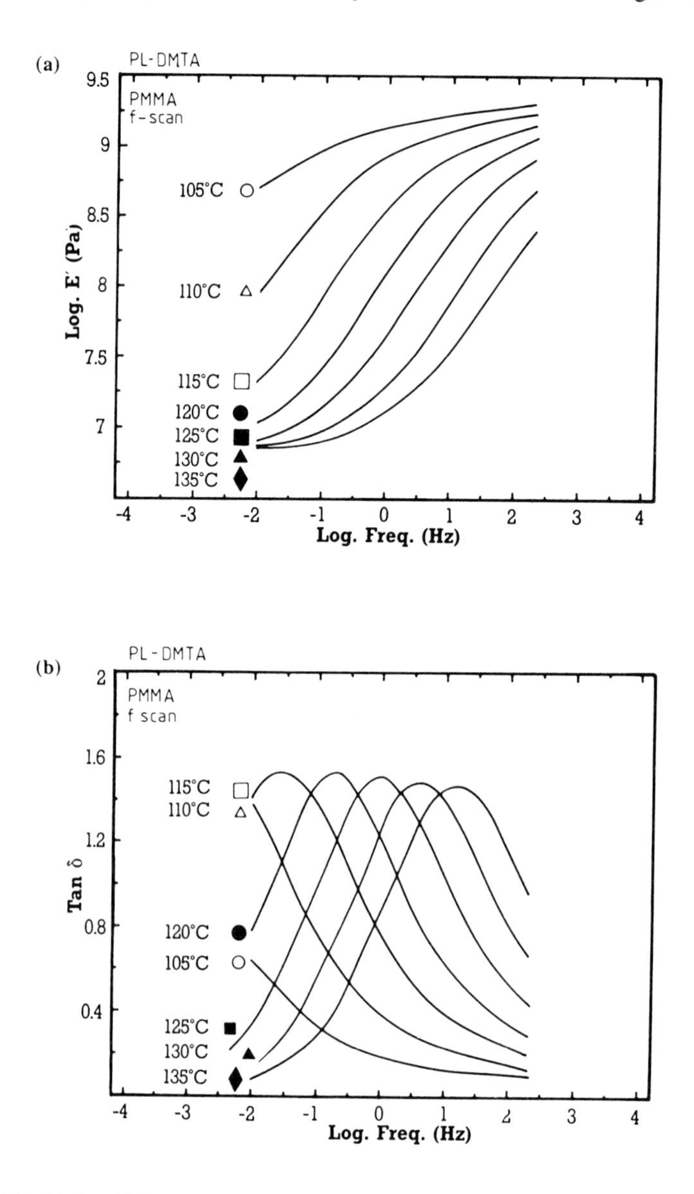

Figure 7.25 (a) Log E' for PMMA measured in the frequency plane at the temperatures shown. (b) Tan δ for PMMA measured in the frequency plane at the temperatures shown.

description of α_a loss peak movement than the Arrhenius relation. Figures 7.25(a) and 7.25(b) show frequency-plane data for poly(methyl methacrylate) (PMMA) at various temperatures in the T_g region. A so-called 'master curve' of viscoelastic response may be constructed by relocating curves to a selected reference temperature by shifting them along the log-frequency axis by $\log f = \log a_T$; $\log a_T$ is given by the WLF equation:

$$\log a_T = \frac{-C_1(T_1 - T_0)}{C_2 + (T_1 - T_0)}$$

This procedure produces the 'master curve' shown in Figure 7.26 for the PMMA at 120°C (data from Figures 7.25(a) and 7.25(b)). It should be noted

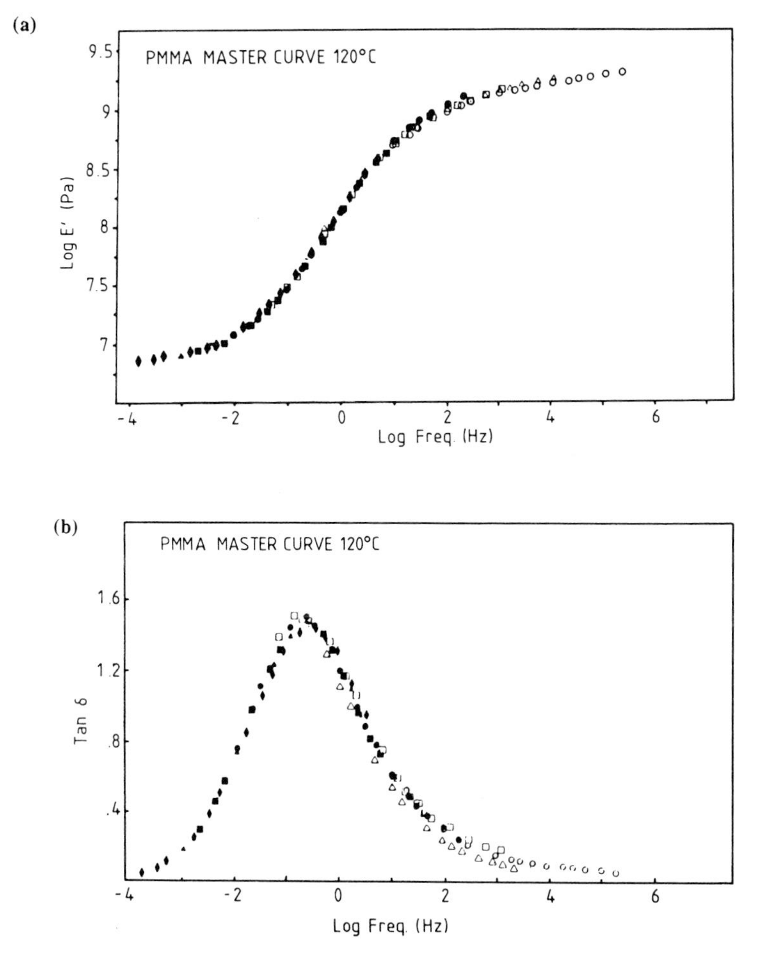

Figure 7.26 (a) Time–temperature master curve of modulus for PMMA at 120°C by shifting data in Figure 7.25(a). (b)Time–temperature master curve for PMMA tan δ, using same shift factors as for (a).

that only one set of data is actually measured at 120°C; the rest of both curves is extrapolation, which will be increasingly unreliable as the data are stretched outside the measurement range. For engineering purposes, two decades of extrapolation would represent the outside limit for any degree of accuracy. It must be pointed out that Williams, Landel and Ferry also advocate the use of a vertical shift to the data. This is designed to correct for density and equilibrium elastic modulus changes with temperature. There is no doubt that some form of vertical shift should be used, but many authors erroneously neglect it, as has been done in Figure 7.26.

Dynamic measurements through the melting region of polymers can only be achieved with difficulty and require parallel-plate, simple-shear, or torsion geometry. Polyethylene shows at least two, if not three α_c' transitions [14, 15] which depend on polymer type (linear, linear low, etc.). Poly(chlorotrifluoroethylene) has also been studied in detail [16]. In reality these data are all related to the lamellar structures formed in these systems. Poly(ethylene oxide) also exhibits loss peaks through the melting region, but their amplitude has been observed to be strain-dependent. The polymer is non-linear viscoelastic, and quantification of the measurements is difficult. This probably applies to most α_c relaxations.

The cure process for composites and the final state of cure are particularly suited to DMTA-type studies, because it is the mechanical properties that are vitally important during production and in final use. Figure 7.27 shows PL-DMTA traces of a carbon-fibre composite precured to different extents. The

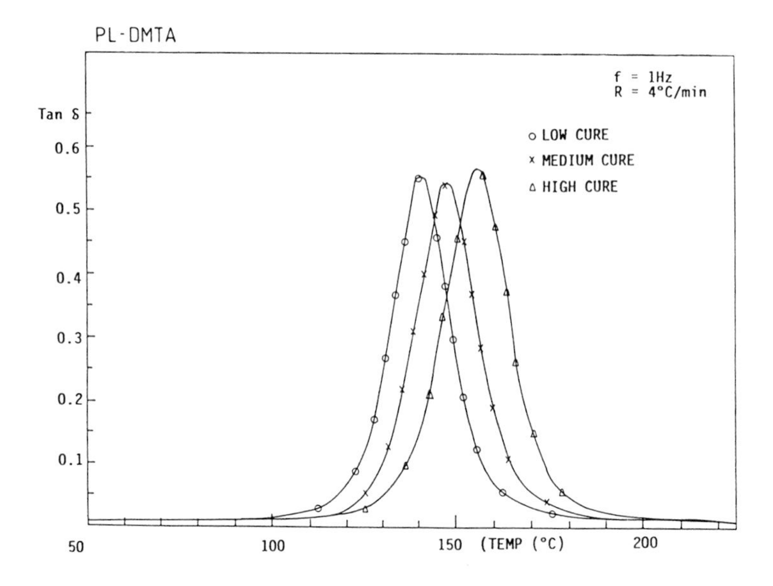

Figure 7.27 DMTA loss peaks used to assess the degree of cure of a carbon-fibre composite.

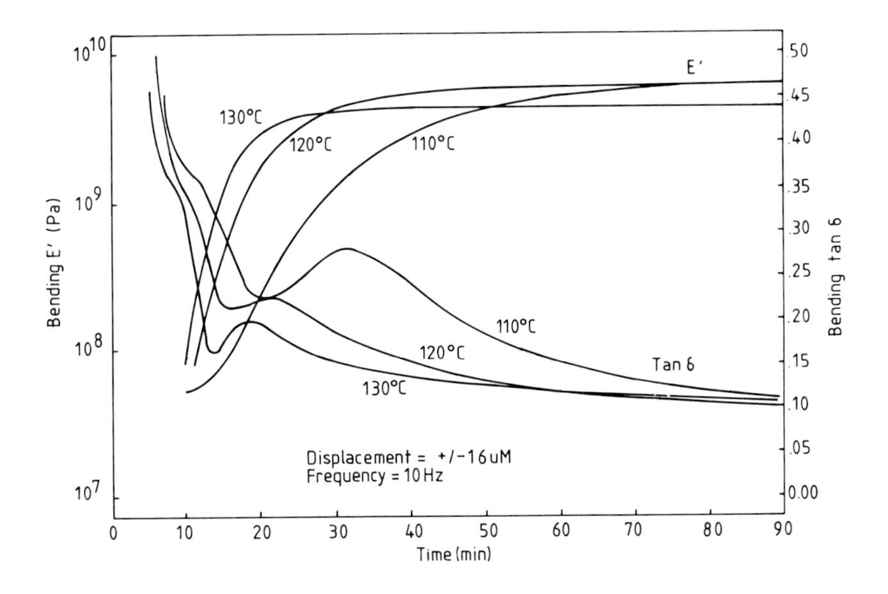

Figure 7.28 Polyester prepreg cure at the isothermal temperatures shown. Modulus and damping changes are shown.

1 Hz loss peak can be used to characterise 'T_g' on real samples; moreover, the loss peak still defines a precise temperature when DSC transitions have become unresolvable from the baseline. The process of cure itself can also be followed by using simple shear or torsion. Figure 7.28 shows the modulus (E') of a polyester prepreg composite during isothermal cure at different temperatures. Short-time data are lost due to the time required to heat from ambient temperature to the isotherm. Table 7.5 compares 'time to cure' by this method with DSC and DETA [17]. The initial modulus levels are rather ill-defined because the system is basically liquid. The two stages resolved at

Table 7.5 Comparative data from three methods

Method[a]	Cure temperature (°C)	Cure time (min)
DSC	110	100
	120	50
	130	30
DMTA	110	120
	120	60
	130	40
DETA	110	180
	120	180
	130	180

[a] DSC value based on per cent conversion, from kinetics analysis; DMTA value from storage modulus E' plateau at 10 Hz; DETA value from permittivity ε' plateau at 20 Hz.

0.1 Hz are gelation and then vitrification. The sample appears to have reached its final cure at 100 Hz before that at low frequencies, because once again it is the kinetics of the molecules which are being probed. These cure processes can well be considered as driving the system into the glassy state by chemical reaction.

7.4 Dielectric thermal analysis (DETA)

7.4.1 Theory

Dielectric measurements are the electrical analogue of dynamic mechanical measurements. The mechanical stress is replaced by an alternating voltage across the sample (a.c. field) and the alternating strain becomes the stored charge (Q) in the simple capacitor. This is always measured as its derivative $\mathrm{d}Q/\mathrm{d}t$ = a.c. current. Samples are typically thin sheets, films or liquids, which can be clamped between parallel-plate electrodes as shown for the PL-DETA, as shown in Figure 7.29. The dielectric data are obtained from phase and amplitude measurements of current and voltage to resolve the components of:

$$\varepsilon^* = \text{Capacitance with sample/capacitance with identical air gap}$$

The Argand diagram (Figure 7.14) shows that this parallels the complex modulus treatment. The $\tan\delta$ values are not equal for the same mechanical and dielectric process.

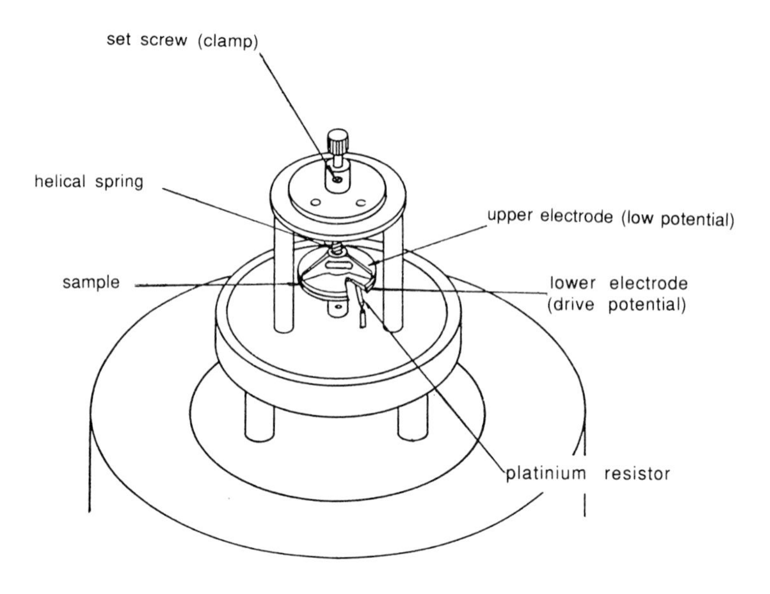

Figure 7.29 DETA cell, showing parallel-plate capacitance clamping of sample.

Dielectric relaxation relies on the presence of dipoles on the moving part of the molecule in order to couple to the electric field and exhibit a dielectric response. Thus, pure unoxidised polyethylene, being non-polar, exhibits a dielectric constant due to atomic and electronic distortions close to the square of the refractive index, as required by theory. It exhibits no frequency-plane relaxation, and the only changes with temperature are due to density changes and melting. A polar polymer such as PET with main-chain dipoles exhibits a DETA scan very similar to a DMTA scan. Figure 7.30 illustrates this. The dielectric permittivity increases with temperature as the sample becomes more polarisable with increasing temperature. In this sense ε′ and

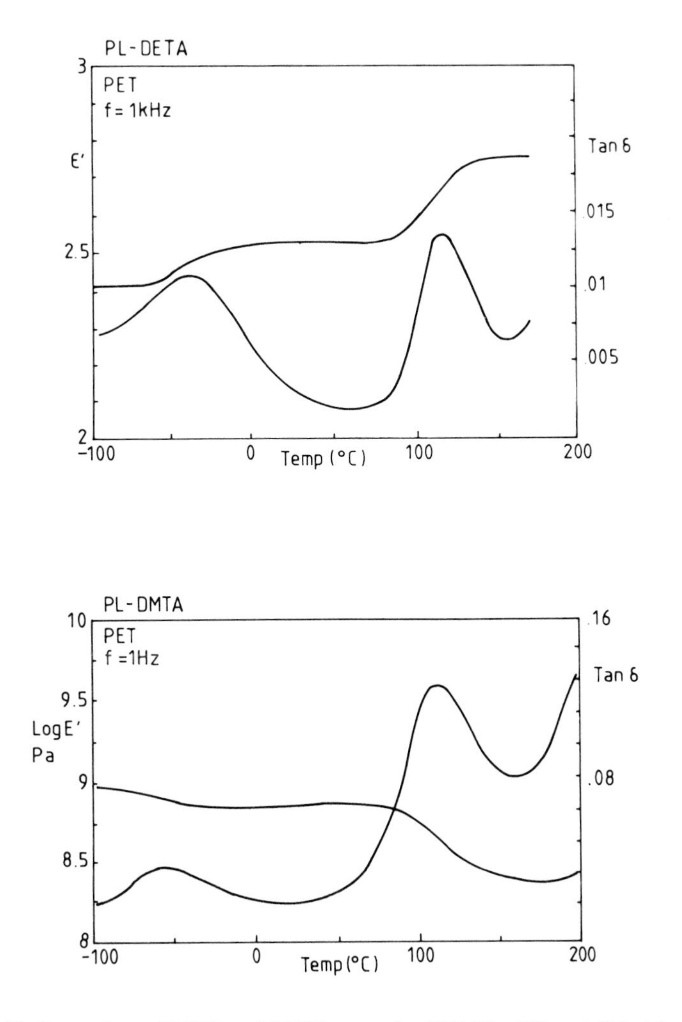

Figure 7.30 Comparison of DETA and DMTA scans for PET. The different dielectric strengths of the processes reflect the extent of dipole orientation in the process.

ε'' behave more like the mechanical compliances J' and J''. The Achilles heel of dielectric measurements is the sensitivity to ionic impurities, which lead to swamping space charges and conduction at low frequencies and high temperature.

It is interesting to compare the temperature of DETA peaks with those of DMTA and with T_g determined by DSC. It is usual to find that, because of the relatively higher frequencies of DETA, the tan δ peaks normally appear at higher temperatures than with DMTA and very much higher than with the relatively static method of DSC. If, however, a proper comparison is made, using the same measurement frequency, typical results are as shown for PVC in Figure 7.31. The DETA loss peak at 100 Hz is approximately 10°C lower than its mechanical counterpart. The reason for this is that although the underlying molecular motions are of course the same—thermal motions of the polymer chain—these are transduced differently by the different techniques. The smallest angular motion of a dipole is reflected strongly in the permittivity. It thus tends towards the 'onset' conditions for the process. The DMTA technique is more sensitive to larger-scale motions, and particularly translation, which can produce significant strain. The spectra from the two methods will thus be different. Both the DMTA and DETA loss peaks, being high-frequency, lie above DSC as normally measured. A rule-of-thumb guide is that a tan δ loss peak at 10^{-4} Hz will agree approximately with a normal DSC T_g.

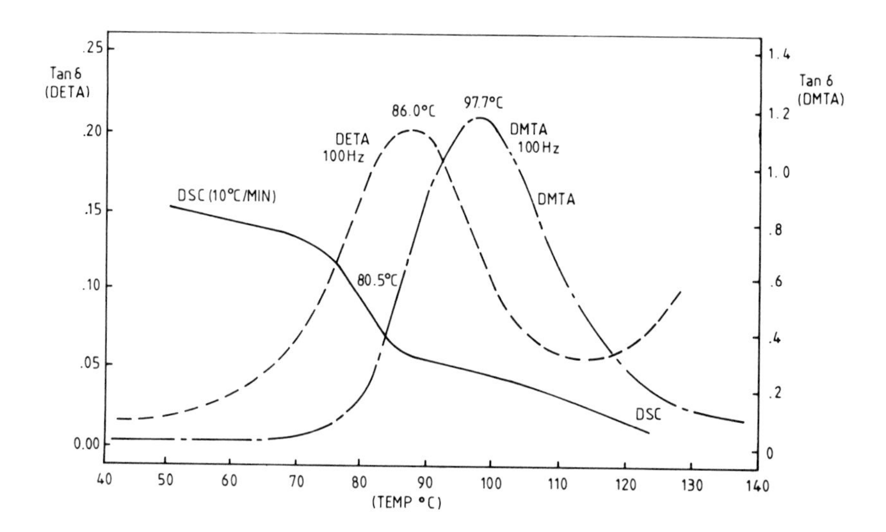

Figure 7.31 Comparison of DSC, DMTA (100 Hz) and DETA (100 Hz) transition-temperature locations.

7.4.2 Applications of DETA to polymers

The data shown in Figure 7.29 indicate that the DETA method can, in principle, provide similar data to the DMTA. This is true at low temperatures, although, because dielectric loss values are low, the DMTA technique wins on sensitivity. Obviously, the strength of various dielectric relaxation peaks compared to their mechanical counterparts allows detailed information on the nature of the molecular motion to be derived. The classic example is PMMA, where the side-group relaxation is stronger in the dielectric case than the main T_g process [18]. The dielectric measuring technique is faster than the DMTA because higher frequencies are involved. Figure 7.32 shows PET data obtained during a 2°C/min thermal scan with frequency multiplexing from 1 to 100 kHz. These data effectively follow the Arrhenius law because at these frequencies the sample is well above T_g.

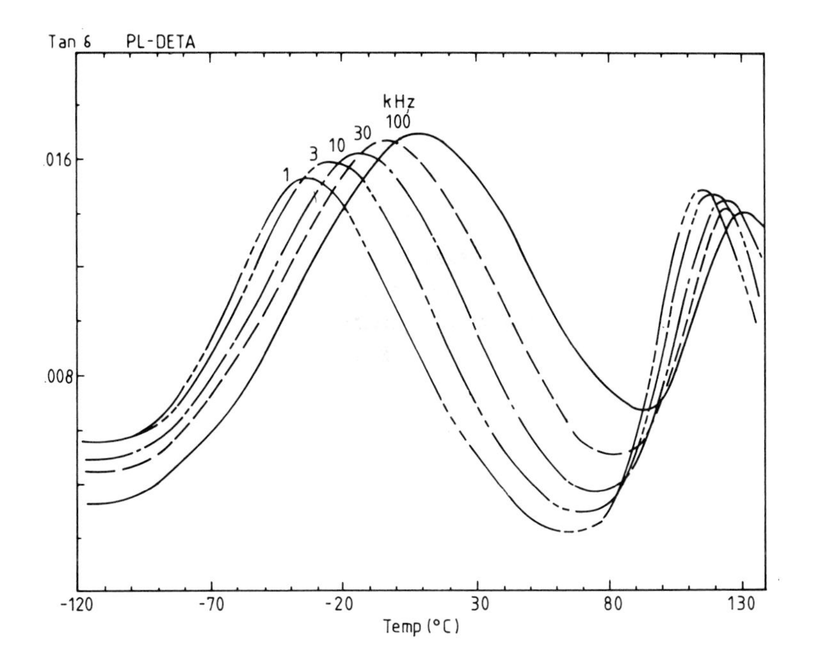

Figure 7.32 Multiplexed dielectric data gathered in a single thermal scan to determine the Arrhenius activation energies.

Crystallinity can be simply and accurately measured by DETA [19] provided that the sample is dipolar and the chains in the crystalline phase do not contribute a relaxation process at the temperature of measurement. The Onsager equation [20] relating dielectric permittivity to the number of orientable dipoles per cm^3 (N) and the dipole moment (μ) is:

$$\frac{(\varepsilon' - n^2(2\varepsilon' + n^2)}{\varepsilon'(n^2 + 2)^2} = \frac{4\pi N\mu^2}{9kT}$$

where n is the refractive index.

Measurement of ε' allows N, the number of orientable dipoles per cm^3 to be determined. In a fully crystalline sample N_c is zero. For a fully amorphous sample, N must be determined (N_a) and then N determined for the semi-crystalline polymer (N_x). The fraction of orientable dipoles is then directly proportional to the volume fraction of amorphous phase (A):

$$A = \frac{N_x - N_c}{N_a - N_c} = \frac{N_x}{N_a}$$

and:

$$X = (1 - N_x/N_a) \times 100\%$$

where X is the required percentage crystallinity. This is a grossly under utilised absolute method.

As a general thermal analysis technique the DETA method needs considerable background knowledge to allow proper interpretation of complex data. Figure 7.33 illustrates this point with an ion-containing polymer complex. This sample is of SBS block copolymer incorporating poly(propylene glycol) in a 15% LiSCN complex to render it more amenable to dielectric heating. The features of the thermal scan are noted on Figure 7.33. By measuring at different frequencies, genuine dipole-motion process and conduction can be distinguished from space-charge effects. When these are caused by two phases they follow the Maxwell–Wagner–Sillars theory [21]. A feature of this process is that they have virtually no dependence of position on frequency, and the loss peak increases dramatically in size with reducing frequency.

7.5 Thermally stimulated currents (TSC)

TSC [22, 23] is essentially a dielectric experiment conducted with static applied electric strain. A sample is placed into the DETA cell and polarised with a high dc voltage at a temperature above the transitions to be studied. The sample is cooled with the voltage still applied to some convenient low temperature (typically −150°C). The cell is then ramped upwards at a reasonably fast rate (~10°C/min), and discharge-current peaks are monitored (by a sensitive picoammeter) with increasing temperature. There is no problem in getting results by this technique; the problem is to have clean enough systems and a good enough understanding of them to interpret them. In this sense the method suffers from more drawbacks than the DETA technique.

Figure 7.34 shows TSC data for a PET film polarised at 150°C. The technique has a timescale imposed by the thermal scan itself, rather like DSC, so

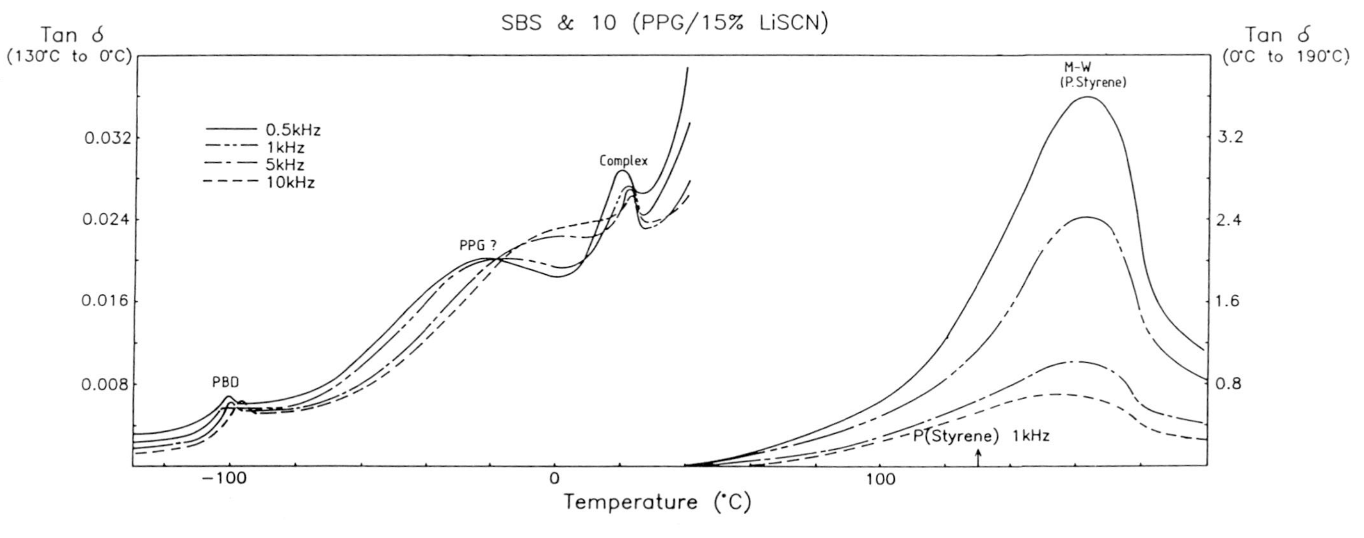

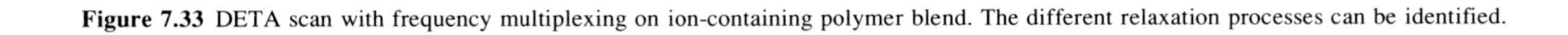

Figure 7.33 DETA scan with frequency multiplexing on ion-containing polymer blend. The different relaxation processes can be identified.

it is a relatively low-frequency technique. The β_a process is only just resolved and the α_a (T_g) process sits as a shoulder on a larger peak which is ascribed to charge build-up and decay at lamellar boundaries. A larger depolarisation peak is always observed as an artifact at the polarisation temperature. This trace should be compared with Figure 7.32. Without this comparison it would be difficult to interpret Figure 7.34. An interesting development of the technique, pioneered by Lacabane [23] and worthy of further consideration, is the 'windowing technique' which takes thermal scans through to different parts of the relaxation peak.

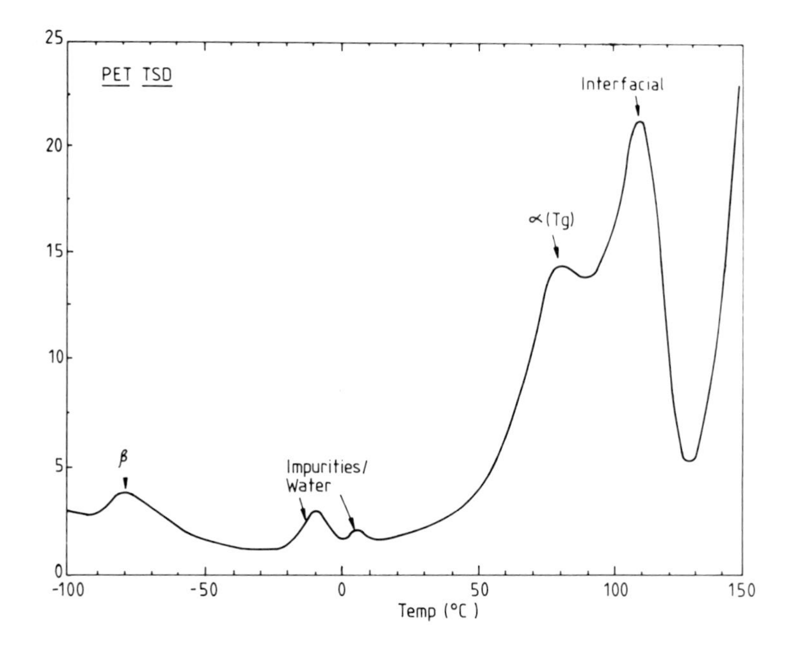

Figure 7.34 Thermally stimulated current experiment performed in the PL-DETA cell. Strong interfacial and poling temperature peaks are present as well as those predicted from dielectric measurements.

7.6 Thermogravimetric analysis (TGA)

TGA is the simplest and oldest thermal analysis technique. It consists of measuring the weight change of a known mass of material as a function of temperature (or time) in a controlled atmosphere. Resolution in weight change on small masses can be better than 1 μg, although buoyancy effects and other problems relating to flowing gases render the accuracy far less than this. Some of the more interesting developments with TGA relate to its combination with FTIR or mass spectrometry, and these will be dealt with under 'hyphenated techniques' (section 7.8). Obviously, weight loss, either isothermal or when measured over a range of heating rates, lends itself to the study of reaction kinetics involving weight change [24]. Measurement of diffusion coefficients is also practicable if well-defined films are available. In this case [25], plots of M_t/M_∞ versus $t^{1/2}$ lead to the calculation of diffusion coefficient of a vapour into a polymer. (M_t is the weight change by time t and M_∞ is the total weight change at long times.)

Space does not permit a long discussion of the well-established technique, but one interesting experiment illustrates the wealth of information accessible. Figure 7.35 shows the determination of the carbon content of a black-filled SBR. The first ramp to 650°C in flowing nitrogen decomposes the polymer, and the breakdown pattern could be studied if desired. The purge is change to

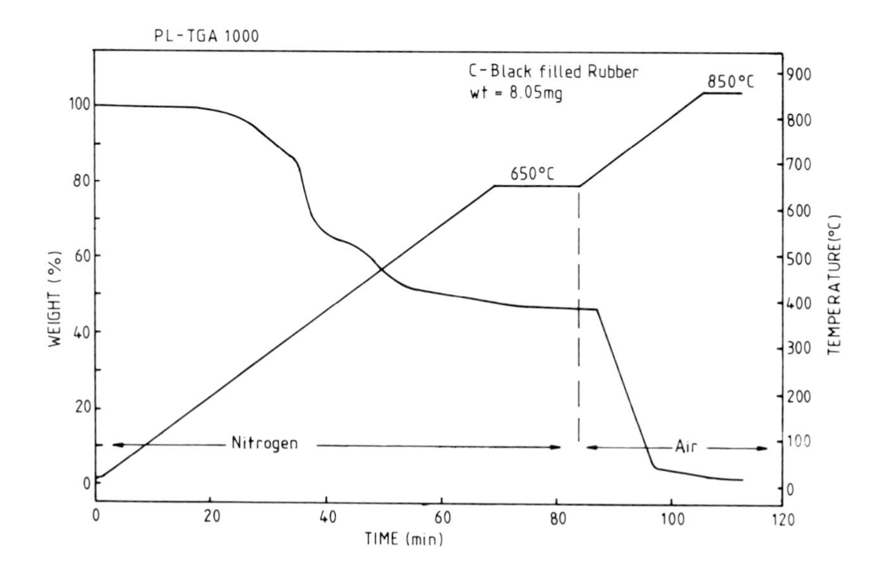

Figure 7.35 TGA used to decompose rubber and combust carbon, hence obtaining carbon content.

air and the ramp is continued to 850°C, where the carbon is converted to gaseous oxides. The small residual ash weight comprises incombustible impurities. The carbon content can be evaluated from either the first or second stage, which checks the self-consistency.

7.7 Simultaneous thermal analysis (STA)

STA is the name coined to signify the simultaneous measurement of weight change (TGA) and heat change (DTA/DSC). This type of instrumentation provides several advantages over the stand-alone techniques. These are summarised as follows:

(a) Removal of uncertainties arising from possible different sample geometries or intimate environmental conditions when measured on separate instruments. Uncertainties due to sample inhomogeneity in such materials as composites and blends are also removed.
(b) Precise one-for-one correlation of weight-change and heat-change observations.
(c) Exact correlation of water content or other volatile content with changes during a thermal scan, i.e. the precise amount of volatile remaining to plasticise T_g or depress T_m is known.
(d) Removal of ambiguity of temperature calibration between instruments.

Details of the measuring head arrangements are given in Figure 7.36. A miniature heat-flux DSC plate is suspended on one arm of a TGA balance. The sensitivity to weight and heat change remains comparable to the stand-alone methods. Lack of predictability of baselines, however, reduces the usefulness of the STA for absolute specific-heat measurements. An example of the STA method is shown for a silicone copolymer in Figure 7.37, where heat events are characterised simultaneously with weight-change events during degradation.

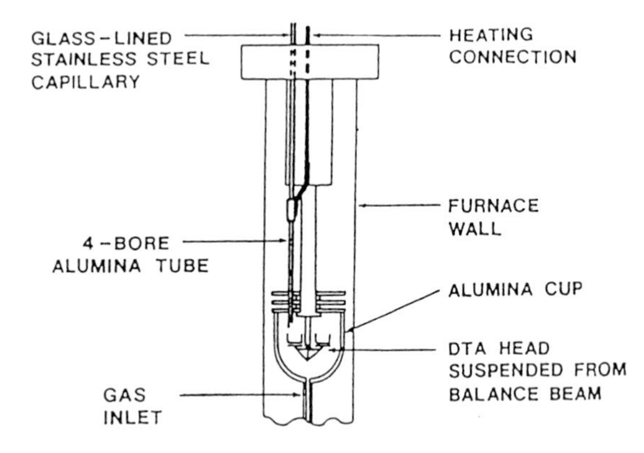

Figure 7.36 PL-STA balance hang-down, showing heat-flow stage. Note the small effective containment volume, which is advantageous when coupling to mass spectrometry or FTIR.

7.8 Hyphenated techniques

It is a logical step to analyse the volatiles evolved during a TGA or STA experiment. The two techniques currently generating most interest are those of mass spectrometry (quadrupole resolution sufficient) and FTIR (one wavenumber resolution sufficient). The volatiles are swept by an inert purge

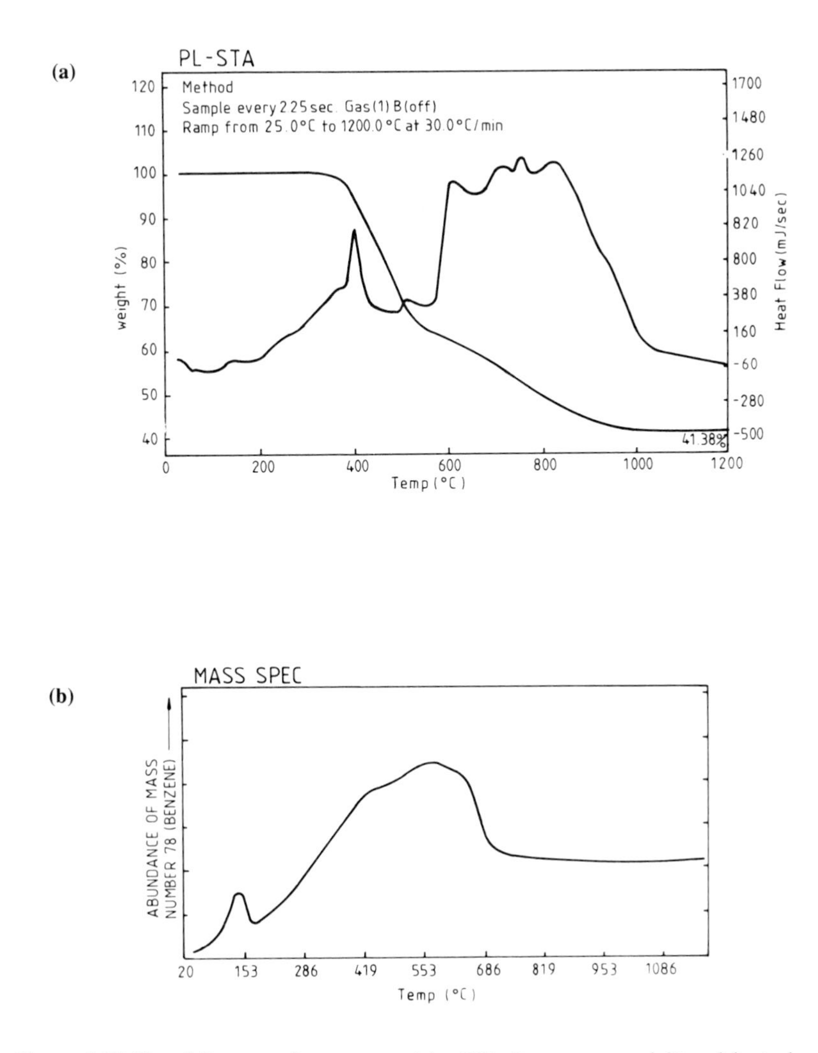

Figure 7.37 Phenylsilane copolymer scanned by STA (to measure weight and heat changes) coupled to mass spectrometry, monitoring benzene evolution amongst other products.

gas (argon) through a heated delivery line to the IR gas cell or mass spectrometer. As the larger degradation products may only be volatile at high temperature, it is important to have a sufficiently heated transfer line and no 'cold spots'. The essential requirement in this instrumentation is to have a low effective oven volume, so that volatiles have small residence times in the oven before being analysed. These keep all the measurements in phase and reduce the occurrence of secondary reactions. Data analysis is usually highly detailed, but one simple example shows the power of a hyphenated technique. Figure 7.37 shows weight and heat-flow data on a random copolymer of dimethylsilane and methylphenylsilane. This is scanned at 30°C/min and rapid, repeated mass spectrometry is used to analyse the evolved gases. The data are subsequently analysed to track the development of different species. In Figure 7.37 the appearance of mass-number 78 (benzene) is shown, and it is seen to be the cause of the major weight change at 400–500°C.

7.9 Overview of thermal analysis methods including thermomechanical analysis (TMA)

In this section transition temperatures for a typical semi-crystalline polymer are compared. Thermomechanical analysis (TMA), which measures the expansion or contraction of small samples with very high resolution, has not been separately reviewed because its use with most thermoplastics is of doubtful worth. The data always contain the sum of at least two effects: thermal expansion and time-dependent creep. The contributions of these terms will vary in an unknown way with heating rate. In addition to this, there are non-linear effects due to the absolute load on the sample. The technique has, therefore, to be applied with care to most thermoplastics.

Figure 7.38 shows the results obtained from a thermal scan at about 5–10°C/min for a semi-crystalline polymer such as PET. To explain the events from top to bottom of Figure 7.38, it is seen that TGA does not respond to T_g, T_m, etc., but will give measurements of volatile loss and degradation at high temperature.

DSC responds to T_g with a sigmoidal step in C_p, which has various shapes depending on the comparison of molecular relaxation kinetics with experimental timescales. It actually always depends on heating rate, but a 'static' T_g is frequently defined by extrapolating T_g at various rates to zero rate—not actually a correct procedure as it varies with log (rate). Secondary processes in the glassy state are not observed. T_m and crystallisation are strong processes in DSC. The heats of fusion and hence crystallinity can be accurately evaluated, but care must be taken in over-quantifying peak shapes and temperatures because of ill-defined instrumental relaxation times.

TMA, if it represented a pure length change of an isotropic sample, would give the expansion coefficient (α), shown as a dashed line in Figure 7.38. The volumetric expansion coefficient $(\mathrm{d}V/\mathrm{d}T)_p$ is a second-order thermodynamic function parallel to C_p. Through T_g, α should therefore show much the same shape curve as C_p. In reality, creep, relaxation of residual moulding, stresses, complex interplay with heating rate, etc., make this technique semi-empirical.

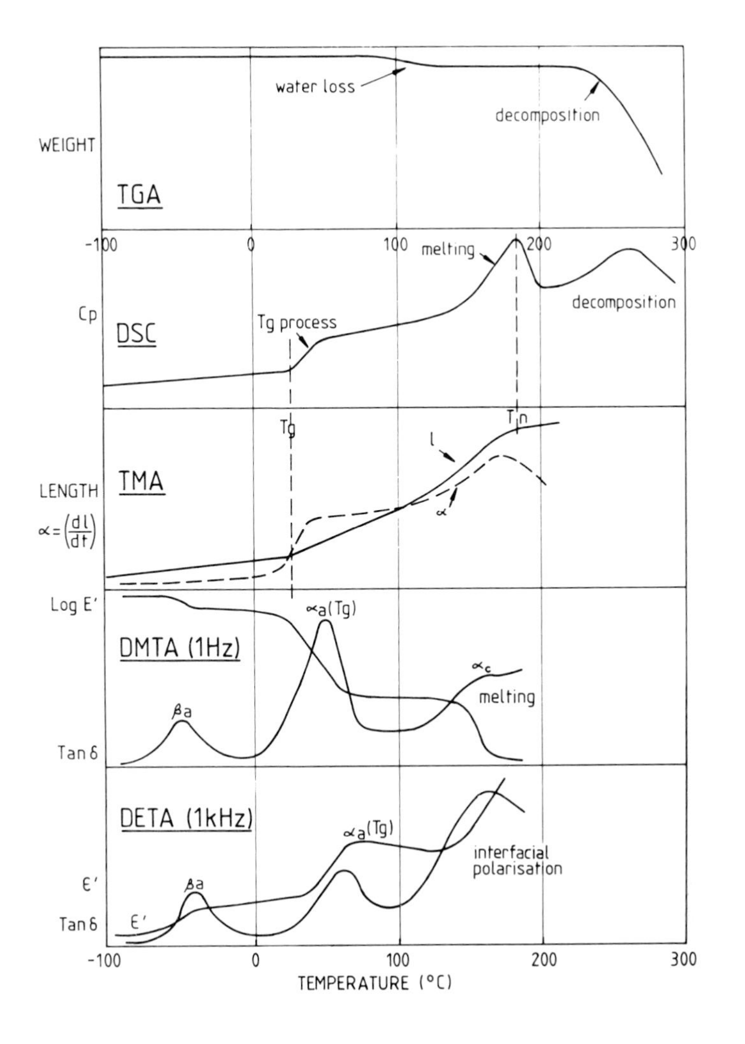

Figure 7.38 Comparison of transitions sensed by the various thermal analysis methods for a typical semi-crystalline polymer.

DMTA gives engineering data on moduli and loss as well as defining all motional transitions. It is not very useful for studying melting or crystallinity. The damping peaks will increase in temperature with the frequency (normally 0.01 to 100 Hz) of vibrational measurement, but do not depend on heating rate. This is where it differs from DSC. DMTA α_a (T_g) loss peaks will always lie above T_g measured by DSC. Secondary relaxations in the glassy state, relating to toughness, are easily studied. It is the technique of choice for polymer-blend studies.

DETA is the electrical analogue of the DMTA method but normally operates at higher frequencies (10 Hz–1 MHz). All dipole-active motional processes can be measured, but data at higher temperatures are usually obscured by polarisation and conductivity. Engineering data on dielectric permittivity are obtained and are useful in chip and electric-insulator design. The closely related technique of TSC is in essence a very-low-frequency dielectric method. Its generally high level of sensitivity makes the data highly confused by impurity and space-charge problems.

Thermal analysis techniques are becoming more user-friendly and of continually wider scope. Most laboratories concerned with polymers should now boast a suite of instruments based on principles outlined in this chapter.

Acknowledgements

The author thanks his colleagues at Polymer Laboratories Ltd for help in the preparation of this chapter, and in particular Dr John Duncan and Dr Paul Larcey for supplying data and helpful discussions.

References

1. B. Wunderlich (1990). *Thermal Analysis*, Academic Press Inc.
2. M. J. Richardson (1978). In *Developments in Polymer Characterization—I*, J. V. Dawkins (ed.), Applied Science Publishers Ltd, London, Chapter 7.
3. S. L. Boersma (1955). *J. Amer. Ceramic Soc.* **38**, 281.
4. F. P. Warner (1975). Ph.D. Thesis, Loughborough University of Technology, UK.
5. E. Lauritzen and J. I. Passaglia (1967). *J. Res. Nat. Bur. Standards*, **A71**, 261.
6. J. R. Fried, F. E. Karasz and W. J. MacKnight (1978). *Macromolecules*, **11**, 150.
7. M. Kisbengi, M. W. Birch, J. M. Hodgkinson and H. G. Williams (1979). *Polymer*, **20**, 1289.
8. M. Takayanagi, H. Harima and Y. Iwata (1963). *Mem. Faculty of Eng., Kyusha Univ.*, **23**, 1.
9. L. C. E. Struijk (1978). *Physical Ageing in Amorphous Polymers*, Elsevier, Amsterdam.
10. F. Biddlestone, A. A. Goodwin, J. N. Hay and G. A. Mouledons (1991). *Polymer*, **32**, 3119.
11. R. E. Wetton (1986). In *Developments in Polymer Characterizations*, J. V. Dawkins (ed.), Elsevier Applied Science, London, Chapter 5.
12. J. Heijboer (1969). *Brit. Polymer J.*, **1**, 3.
13. J. D. Ferry (1961). *Viscoelastic Properties of Polymers*, John Wiley & Sons, N.Y.
14. J. B. Jackson, P. J. Flory, R. Chaing and M. J. Richardson (1963). *Polymer*, **4**, 237.
15. J. A. Sauer and A. E. Woodward (1960). *Rev. Mod. Physics*, **32**, 88.
16. J. D. Hoffman, G. Williams and E. Passaglia (1966). *J. Polymer Sci. C*, **14**, 173.
17. M. Connolly and B. Tobias (1992). *Amer. Lab.*, **24**, 1.
18. N. G. McCrum, B. E. Read and G. Williams (1967). *Anelastic and Dielectric Effects in Polymeric Solids*, John Wiley & Sons.
19. G. S. Fielding-Russell and R. E. Wetton (1967). *Polymer Lett.*, **5**, 761.
20. L. Onsager (1939). *J. Amer. Chem. Soc.*, **58**, 1486.
21. R. E. Sillars (1937). *J. Instn. Elect. Engrs.*, **80**, 378.
22. J. Van Turnhout (1975). *Thermally Stimulated Discharge of Polymer Electrets*, Elsevier, Amsterdam.

23. A. Lamure, N. Hitini, C. Lacabane, M. F. Herdmand and D. Herbage (1986). *IEEE Trans El*, **21**, 443.
24. Ref. 1, p. 404.
25. J. Crank (1956). *Mathematics of Diffusion*, Oxford University Press, Oxford.

8 Small-angle neutron scattering and neutron reflectometry

R. W. RICHARDS

8.1 Introduction

Diffraction methods, interpreted loosely, could be applied to the techniques of wide-angle X-ray scattering, small-angle X-ray scattering, electron diffraction, small-angle neutron scattering, small-angle light scattering and reflectometry. Each of the above techniques is defined by the size range explored and the property that causes the scattering. Both X-ray and electron diffraction methods depend on the electron density of the sample to produce a signal, light scattering is generated by variations of polarisability (and hence refractive index) in the sample, and neutron scattering requires a difference in scattering length density (*vide infra*) for scattering to be observed.

The variation of scattered intensity with scattering angle is the scattering law for the system; the nature and form of this law depends on the range of scattering vector (Q, see below) being probed, and this determines the size range analysable. Hence, for wide-angle X-ray scattering, the length scale probed is of the order of a few angstroms and the information obtainable relates to the crystallinity, local conformation, packing and orientation. For small-angle scattering methods, length scales of 20Å to about 2000Å may be probed, giving access to polymer molecule dimensions and the sizes of aggregated species (micelles, block copolymer domains). Moreover, in this region of very low Q, the extrapolation to low Q is not too hazardous, and thus one can attempt to obtain the osmotic compressibility and hence some information about the thermodynamics of the scattering system. This is clearly of relevance to polymer blends.

In principle, neutron reflectometry can explore an extremely wide range of length scales from about 10Å to thousands of angstroms. However, practical limitations more often restrict this range to about 30–1000Å. Moreover, this length-scale range is only relevant in a direction normal to the surface at which the neutrons are incident. Nonetheless, this apparent limitation in direction provides a unique probe of the nature and composition of the surface, and has been used with much ingenuity in recent years.

Attention will be focused here on the use of small-angle neutron scattering (SANS) in polymer systems and the application of neutron reflectometry (NR) to polymer surfaces and interfaces. SANS has been used to investigate molecular configuration, diffusion, phase separation, block copolymer organisation, the dimensions of network chains and segment density profiles of adsorbed polymers, and to evaluate thermodynamic parameters of polymer blends. It is for these reasons that SANS is dealt with to the exclusion of other diffraction techniques. Even with this limitation, space precludes a discussion of all the areas listed above. This chapter is confined to blends, block copolymers and liquid crystal polymers in the main. Periodically, reviews [1–4] on SANS from polymers have appeared and these should be consulted for earlier work.

Small-angle neutron scattering has been available for some twenty years now. By contrast, neutron reflectometry has only been actively pursued in the last five years. Like SANS, NR has been rapidly applied to many different materials, notably surfactants, and has not been confined to polymers alone. Essentially, NR can be used to measure the density profile and thickness of a surface layer provided that sufficient contrast is available. Applications of NR to polymers have included surface segregation, Langmuir–Blodgett films, interdiffusion and adsorption at the solid–liquid interface, and these will be mentioned here.

This chapter is divided into two large sections, one on small-angle neutron scattering and the second on neutron reflectometry. Within each section the theory of the method will be discussed and some basic aspects of the experimental features set out. Finally, selected applications of each method in polymer science will be described.

8.2 Small-angle neutron scattering

8.2.1 Theory

A rigorous derivation of the theoretical basis of neutron scattering will not be presented here; the reader is referred to the text by Lovesey [5] for this. Only elastic scattering is considered, i.e. where there is no energy transfer between the neutrons and the scattering system. Figure 8.1 shows the SANS experiment in schematic form. A well-collimated neutron beam of intensity I_0 and wave vector k_0 is incident on the sample. Most of the beam is transmitted through the sample but a fraction (0.1–0.5, typically) is scattered with a wave vector k. The scattered beam at an angle 2θ to the direct beam is collected by a detector which subtends a solid angle $\Delta\Omega$ at the sample. The scattered intensity per unit volume at this angle is given by:

$$I = I_0 N (\mathrm{d}\Sigma/\mathrm{d}\Omega)\Delta\Omega T(\varepsilon/r^2) \tag{8.1}$$

where N is the number of nuclei per unit scattering volume, T is the transmission of the sample, ε is the detector efficiency, r is the sample detector distance, and $(\mathrm{d}\Sigma/\mathrm{d}\Omega)$ is the differential scattering cross-section. This latter

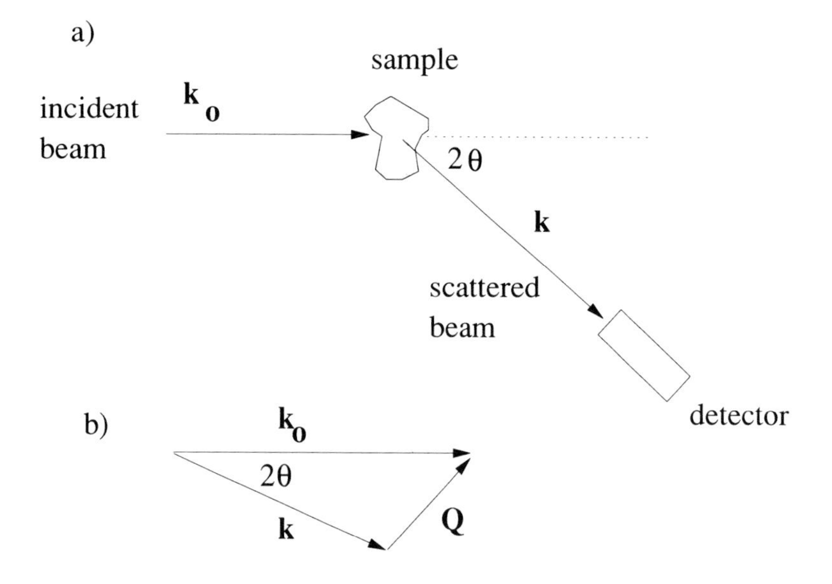

Figure 8.1 (a) Schematic diagram of a small-angle neutron-scattering experiment. (b) Vector diagram of scattering process for an incident wave vector k_0 and scattered wave vector k.

term can be factorised into coherent and incoherent components; only the coherent differential cross-section contains information on the structure of the scattering specimen. The incoherent differential scattering cross-section is featureless and constitutes a flat background on which the coherent cross-section is superimposed. Structural correlations reveal themselves in an angular dependence of the scattered intensity; the fundamental parameter is not the scattering angle but Q the scattering vector, defined in Figure 8.1(b). The coherent differential scattering cross-section can then be written as:

$$\frac{\mathrm{d}\Sigma^{coh}}{\mathrm{d}\Omega}(Q) = \frac{k}{k_0}\left\langle \left| \sum_i b_i \exp(ikR_i) \right|^2 \right\rangle \tag{8.2}$$

where b_i is the coherent scattering length at position vector R_i, for elastic scattering $k = k_0$ and, from Figure 8.1, $Q = |Q| = (4\pi/\lambda)\sin\theta$, and then:

$$\frac{\mathrm{d}\Sigma^{coh}}{\mathrm{d}\Omega}(Q) = \left\langle \left| \sum_i b_i \exp(iQR_i) \right|^2 \right\rangle \tag{8.3}$$

A two-phase model, where the polymer molecule is dissolved or dispersed in a medium denoted by subscript m, is now adopted; furthermore, the coherent scattering length is averaged over the volume of a polymer segment or a medium molecule. Then contribution of the medium to the sum in eqn (8.3) is:

$$\rho_m\left[\int \mathrm{d}R \exp(iQR) - \int_{V_p} \mathrm{d}R \exp(iQR)\right]$$

The first integral is over the whole sample volume and is zero everywhere except at $Q = 0$. The second integral is over the volume occupied by the polymer molecule, and if $\rho(R)$ is the scattering length density variation within the molecule, then:

$$\frac{\mathrm{d}\Sigma^{coh}}{\mathrm{d}\Omega} = \left|\int_{V_p} \mathrm{d}R(\rho(R) - \rho_m)\right|^2 \tag{8.4}$$

using the average scattering length density per polymer segment, then:

$$\frac{\mathrm{d}\Sigma^{coh}}{\mathrm{d}\Omega} = (\rho_p - \rho_m)^2 V_p^2 |F(Q)|^2 \tag{8.5}$$

where $F(Q) = V_p^{-1} \int_{V_p} \exp(iQR)\mathrm{d}R$ is known as the form factor and is defined so that at $Q = 0$, $F(Q) = 1$. Since $|F(Q)|^2$ is the Fourier transform of the density correlation function of the scattering particle, it provides information on the distribution of scattering length density in the particle, i.e. the structure. In principle, the observed scattered intensity could be Fourier-inverted to give the density correlation function directly. However, such a numerical procedure requires data of high quality extending over a wide Q range (ideally from $Q = 0$ to $Q = \infty$); moreover, on Fourier inversion all information on the phase is lost. A variant of this approach has been described by Glatter [6] (the indirect transformation procedure); however, this requires some prior knowledge of the morphology of the scattering particle. This highlights the feature common to all small-angle scattering: interpretation of the data is always dependent on the application of a model. Even in classical intensity light scattering, the evaluation of the radii of gyration from a Zimm plot presumes the adoption of a Gaussian-coil configuration in dilute solution. Other models are applied (e.g. scattering from rod-like molecules) only when additional evidence supports this, for example the observation of depolarised light scattering. Some typical forms for $|F(Q)|^2$ are given below.

The incoherent differential cross-section contributes an isotropic, Q-independent scattered intensity whose magnitude is $(\Sigma_{inc}/4\pi)$, where Σ_{inc} is the macroscopic incoherent scattering cross-section of the scattering species. Consequently eqn (8.1) becomes:

$$I(Q) = I_0 N \frac{\varepsilon}{r^2} \Delta\Omega T\left[(\rho_p - \rho_m)^2 V_p^2 |F(Q)|^2 + \frac{\Sigma_{inc}}{4\pi}\right] \tag{8.6}$$

The collection of parameters $((I_0(\varepsilon/r^2)\Delta\Omega)$ is constant for the particular set-up of the small-angle diffractometer, and its value will be known by calibration with known standards. N and T will depend on the sample, as will the values of ρ_p, ρ_m and Σ_{inc}.

8.2.2 Form factors and scattering laws

The terms 'form factors' and 'scattering laws' have come to be used interchangeably. Strictly, the term 'form factor' pertains to the scattering from an isolated particle; the scattering law is the scattering from an assembly of particles in the sample. For an isolated macromolecule, the most generally applicable form factor is the Debye equation for a Gaussian distribution of scattering segments:

$$\left|F_p(Q)\right|^2 = (2/u^2)\,(\exp(-u) - 1 + u) \qquad (8.7)$$

where $u = Q^2R_g^2$ and R_g is the root mean square radius of gyration of the polymer molecule. A more generally applicable model, which can be used to describe both flexible coil and rod-like molecules, is the Kratky–Porod worm-like chain [7]. This model has persistence length as a parameter of the molecule, where the persistence length, a, is defined as:

$$a = \lim(1/(1 + \cos\theta))$$

where θ is the angle between bonds of length l. Calculation of the form factor for this model requires a calculation of the distribution function, for which there have been many attempts described, using both analytical and Monte Carlo methods. Yamakawa [8], in particular, has devoted much attention to worm-like chains and helical worm-like chains. One expression obtained [9] is:

$$\left|F_p(Q)\right|^2 = \left(\frac{2}{x^2}\right)(\exp(-x) + x - 1) + \left[\frac{4}{15} + \frac{7}{15x} - \left(\frac{11}{15} + \frac{7}{15x}\right)\exp(-x)\right]\frac{a}{2L} \qquad (8.8)$$

where $x = LaQ^2/12$ and L is the rod length.

Figure 8.2 shows the form factors for a worm-like chain and a coil plotted as a Kratky plot. The intersection between the plateau of the Gaussian coil and the rod scattering defines the value of the persistence length, since at this point $Q = 6/\pi a$. If the scattering from a coil-like molecule is extended to a high enough Q value, then this upturn in the plateau level will be observed. In this region of Q, the radiation is probing short sections of the polymer molecule that have the scattering behaviour of an assembly of randomly oriented rods. A word of caution is warranted here since, in the region of Q where this observation is made (typically, 0.5–1.0Å^{-1}), the scattered intensity is usually not significantly greater than the background, and there may be large errors in the excess scattering (sample scattering-background).

Form factors for scattering particles or molecules with a regular morphology (spheres, cylinders, hollow spheres, lamellae) are available in the literature

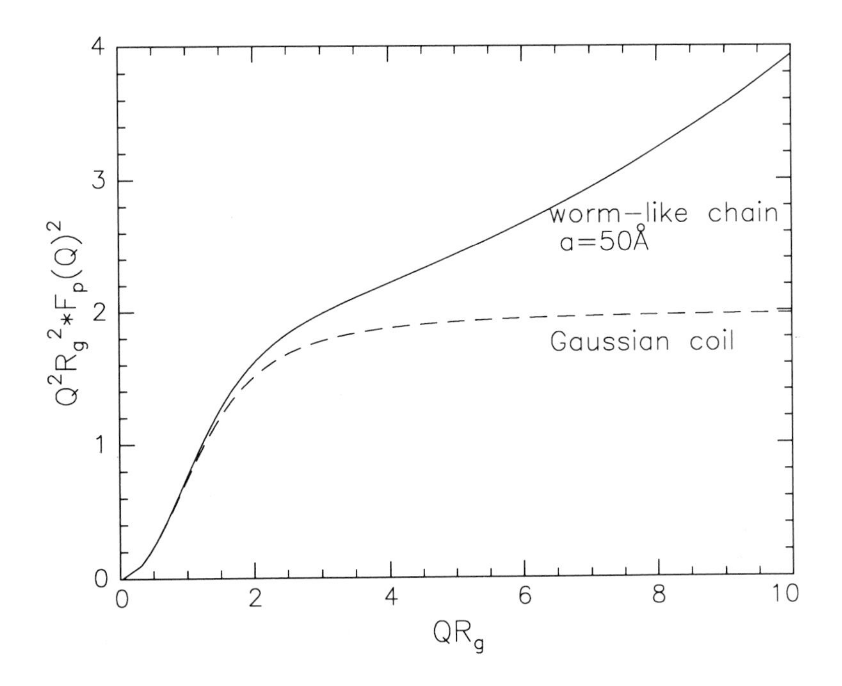

Figure 8.2 Single particle form factors for a Gaussian coil and a rigid rod of the same radius of gyration plotted as Kratky plots.

and the modifications to these form factors consequent on a distribution in size are also given [10–12]. These individual form factors are of direct use when the polymer molecule is at a low concentration in the matrix; however, low concentration necessarily means that the scattered intensity will be small. To improve the scattered signal-to-noise ratio higher concentrations would be favourable. The use of higher concentrations would appear to introduce unnecessary complications due to intermolecular scattering contributing to the total scattering, as well as the intramolecular scattering which is described in the form factors discussed above. Under certain conditions this has been shown not to be the case for SANS, where multiple scattering is negligible for samples of typical thickness of about 1 mm.

In the most general case, that of a deuterium-labelled polymer (subscripts D) together with its hydrogenous equivalent (H) dissolved in a solvent (S), eqn (8.3) can be expanded to:

$$\frac{\mathrm{d}\Sigma}{\mathrm{d}\Omega}^{coh}(Q) = b_D^2 S_{DD}(Q) + b_H^2 S_{HH}(Q) + b_S^2 S_{SS}(Q) + 2b_D b_H S_{DH}(Q) + 2b_D b_S S_{DS}(Q) + 2b_H b_S S_{HS}(Q) \qquad (8.9)$$

where:

$$S_{XY}(Q) = \left| \sum_X \sum_Y \exp(iQR_X)\exp(iQR_Y) \right|$$

The system is assumed to be incompressible, i.e. density fluctuations of H, D and S components are self-cancelling, so that the overall fluctuation in density is zero, in which case:

$$\frac{\mathrm{d}\Sigma^{coh}}{\mathrm{d}\Omega}(Q) = (b_D - b_S)^2 S_{DD}(Q) + (b_H - b_S)^2 S_{HH}(Q) + 2(b_D - b_S)(b_H - b_S)S_{HD}(Q)$$

Averaging the coherent scattering lengths over the volume of the solvent molecule or segment volume (assumed to be equal), then this equation can be written in terms of the scattering length density:

$$\frac{\mathrm{d}\Sigma^{coh}}{\mathrm{d}\Omega}(Q) = (\rho_D - \rho_S)^2 S_{DD}(Q) + (\rho_H - \rho_S)^2 S_{HH}(Q) + 2(\rho_D - \rho_S)(\rho_H - \rho_S) S_{HD}(Q) \qquad (8.10)$$

If the total number of polymer molecules is N, then let $N_D/N = x$ and $N_H/N = (1 - x)$. The scattering laws $S_{XY}(Q)$ in eqn (8.10) have intra- and inter-molecular contributions:

$$S_{DD}(Q) = x\,[\,|F(Q)|^2 + (1 - x)Q(Q)]$$

$$S_{HH}(Q) = 1 - x\,[\,|F(Q)|^2 + (1 - x)Q(Q)]$$

$$S_{HD}(Q) = x(1 - x)Q(Q)$$

where it is assumed that the H and D molecules have the same number of segments and $Q(Q)$ is the intermolecular interference function. If the solvent is removed, then the scattering-length density of the solvent is replaced by the average scattering-length density of the whole sample, i.e. $\rho_S = x\rho_D + (1 - x)\rho_H$, and in this case we have;

$$\frac{\mathrm{d}\Sigma^{coh}}{\mathrm{d}\Omega}(Q) = (\rho_D - \rho_H)^2 x(1 - x)\,|F(Q)|^2 \qquad (8.11)$$

Consequently, from eqn (8.11), the maximum scattering intensity is obtained for $x = 0.5$, i.e. a mixture of equal weights of hydrogenous and deuterated polymer. One is still able to extract the single-molecule form factor and thus obtain the radius of gyration and molecular weight (see below). To use eqn (8.11) successfully, it is necessary that the molecular weights and the molecular-weight distribution of both H and D polymers are exactly similar [13,14].

8.2.3 Scattering lengths, scattering-length density and contrast variation

From sections 8.2.1 and 8.2.3 above, it is evident that apart from the concentration, the intensity of small-angle neutron scattering is determined by the difference $(\rho_p - \rho_m)$, known as the contrast factor. Scattering-length densities are calculated by summing the coherent scattering lengths of the constituent atoms in the scattering unit over the volume of that unit:

$$\rho = \sum_i b_i d N_A / m \tag{8.12}$$

where d is the density of the material, m is the scattering unit molecular weight, and N_A is Avogadro's number. Similarly, the incoherent scattering cross-section is the sum of the atomic incoherent scattering cross-sections averaged over the scattering volume. Values of b and incoherent scattering cross-sections (σ) for isotopic species are tabulated [5]. Unlike X-ray scattering cross-sections, the values of b and σ do not have a regular variation with position of the element in the periodic table; furthermore, values of σ are neutron-wavelength-dependent, and this should be borne in mind when calibration factors are determined using purely incoherent scatterers. Table 8.1 gives values of b and σ for some typical elements occurring in polymers, and Table 8.2 gives scattering-length densities for selected pairs of hydrogenous and deuterated monomers. The large difference in b values for hydrogen and deuterium is particularly fortunate in polymers, since this leads to large values of the contrast factor being obtained when the deuterated and hydrogenous polymers are mixed together. Consequently, by mixing deuterated polymer with hydrogenous polymer, the form factor for a single polymer molecule can be obtained by small-angle neutron scattering and hence the radius of gyration obtained. This was the first use of SANS on polymers. Up to that time there had been considerable argument about the configuration of amorphous polymers in the solid state. The results proved conclusively Flory's hypothesis that such polymers would have unperturbed dimensions. Since those initial experiments SANS has been used to investigate the dimensions of polymer chains in networks [2], to investigate the rotational isomeric state scheme [4] for polymers and to confirm the predictions of scaling theory for semi-dilute and concentrated polymer solutions [2].

The large difference in coherent scattering length of hydrogen and deuterium need not be confined to generating large contrast factors for whole-molecule observation. It can be used selectively to vary the contrast of particular parts of the scattering sample. For example, one block type in a block copolymer can be deuterated to 'highlight' that part of the microphase separated structure. Additionally, Rawiso [15] selectively labelled parts of the styrene monomer and showed that the form factor observed was very dependent on the location of the centre of gravity of the scattering-length density relative to the position of the main-chain backbone. This facility of contrast variation has

Table 8.1 Coherent scattering lengths and incoherent scattering cross-sections for selected nuclei

Isotope	$b(10^{-12}\text{cm})$	$\sigma_m(10^{-24}\text{cm}^2)$
^{1}H	− 0.374	79.7
^{2}H	0.667	2.0
^{12}C	0.665	0
^{16}O	0.580	0
^{19}F	0.565	0
^{28}Si	0.41	0
^{35}Cl	1.17	4.42[a]

[a] Chlorine nuclei have an absorption cross-section of 44.1×10^{-24} cm^2.

been particularly useful in colloidal dispersions, since H_2O has a negative scattering-length density and D_2O a positive, and hence they may be mixed together in different proportions to obtain a range of values of ρ including zero. This possibility is also available for some organic solvents, e.g. cyclohexane.

Table 8.2 Scattering-length densities and macroscopic incoherent scattering cross-sections

Monomer	$\rho(10^{10}\text{cm}^{-2})$	$\Sigma_{inc}(\text{cm}^{-1})$
Ethylene	− 0.342	0.729
Ethylene-d4	8.24	0.165
Styrene	1.42	3.889
Styrene-d8	6.50	0.098
Methyl methacrylate	1.07	4.570
Methyl methacrylate-d8	7.03	0.114
Butadiene	0.426	4.897
Butadiene-d6	6.823	0.123

8.2.4 Experimental

Small-angle neutron diffractometers consist of at least six components: (a) a neutron source; (b) a combination of a neutron guide and a collimation system (both evacuated) which deliver a beam of neutrons to the sample area; (c) the sample environment; (d) an evacuated flight tube after the sample; (e) an area detector on which both scattered and transmitted beams are incident; and (f) a control and data-analysis computer. If the neutron source is a steady-state nuclear reactor, then there will usually be a monochromator in the incident neutron beam before the sample. This may be a velocity selector, as at the Institut Laue–Langevin, which can allow discrete wavelengths to be selected from the Maxwellian spectrum of wavelengths produced by the reactor. The monochromator could also be a crystal lattice selected so that only a particular neutron wavelength is transmitted. For pulsed neutron sources, a 'white' beam of neutrons is incident on the sample and the scattered neutrons are analysed by time-of-flight methods to give the scattered intensity as a function of Q. Diffractometers on steady-state reactors are available at the Institut Laue–

Langevin, Grenoble, France, the HFIR reactor at Oak Ridge National Laboratory, Tennessee, USA, and the NIST reactor, Gaithersburg, Maryland, USA. Pulsed neutron sources with small-angle diffractometers include ISIS at the Rutherford–Appleton Laboratory, UK, LANSCE at Los Alamos, New Mexico, USA, and the Argonne National Laboratory in Illinois, USA. In all cases, an area detector is used since the collected scattered intensity can be radially averaged for isotropic scattering samples, and for anisotropic samples (stretched polymers or flowing systems), the scattering collected in different regions of the detector can be analysed separately.

The sample environment is generally an easily accessible region where a variety of equipment can be mounted in the beam, e.g. heating jackets, flow cells, magnets, stretching devices and automatic sample chargers. Each instrument has its own particular features, and details are obtainable from the parent institutions. Additional details are provided here of the LOQ diffractometer at ISIS.

Neutrons are produced by the spallation process when a pulsed (50 Hz) beam of accelerated protons collides with the uranium atoms of a cooled target. The neutrons are slowed by passage through a moderator of liquid hydrogen (this boosts the population of long-wavelength neutrons), and the fast neutrons of wavelength less than 1.5Å are removed by bending the beam using a super-mirror bender. A problem with pulsed sources is that the slow neutrons (long wavelength) in pulse i can be overtaken by the fast neutrons in pulse $(i + 1)$ (frame overlap). To prevent this, a chopper is placed in the beam, which removes every other pulse, followed by a series of frame-overlap mirrors (nickel on silicon) which reflect neutrons of $\lambda > 12$Å out of the incident beam. Scattered neutrons are detected by a two-dimensional ^{3}He detector, and the use of a semi-transparent beam stop allows the transmission of the sample to be collected at the same time as the scattered intensity if desired. The collected data are normalised to the incident beam flux and corrected for transmission over the spectrum of neutron wavelengths incident on the sample. Absolute cross-sections can be obtained by a calibration run with a specimen of known cross-section. For LOQ this is generally a mixture of deuterated and hydrogenous polystyrenes. Data collection, normalisation and correction are all handled by the instrument's MicroVAX computer.

8.2.5 Applications of small-angle neutron scattering

An exhaustive survey of the uses for SANS in polymer science would be too long for the space available here. Attention is focused on three aspects. Polymer blends provide an example of the application of the random-phase approximation and have also enabled a better appreciation to be obtained of the thermodynamic changes consequent on deuteration. Block copolymers in the homogeneous state are also analysable by using the random-phase approximation, and the theory of the segregation in these systems has progressed rapidly in recent years. Lastly, liquid-crystal polymers are the most recent class of polymers to be examined by SANS. They present new practical problems as well as the opportunity to investigate the influence of preferential alignment.

8.2.5.1 Polymer blends. Although it is well known that the mean-field Flory–Huggins theory of the thermodynamics of polymer systems is not a rigorously accurate description, especially for polymer blends, it is sufficiently valid that its use does not incur serious errors. Furthermore, de Gennes [16] used the mean-field random-phase approximation to obtain the scattering law for a binary polymer blend as:

$$S^{-1}(Q) = [N_1 \phi g_D(R_g^1, Q)]^{-1} + [N_2(1-\phi) g_D(R_g^2, Q)]^{-1} - 2\chi \qquad (8.13)$$

where χ is the Flory–Huggins interaction parameter, N_i is the degree of polymerisation of the polymer i, ϕ is the volume fraction of polymer 1 and $g_D(R_g^i, Q)$ is the Debye function (eqn (8.7)) for polymer i. Warner *et al.* [17] derived an extended version of this equation which incorporated the possibility of having only a fraction of one of the polymers deuterium-labelled. At the time it was still thought that the labelled polymer had to be dilute to remove intermolecular effects. Additionally, to remove any Q-dependent effects, values of χ were extracted from the cross-section at $Q = 0$ obtained by extrapolating data to $Q = 0$. Tomlins and Higgins [18] used this extended approach (but concentrated on mixtures where all of one polymer was deuterated) to determine χ for oligomeric mixtures of polystyrene and polybutadiene (PS–PB) and methoxylated poly(ethylene glycol) and poly(propylene glycol). Molecular weights were between 1000 and 4000 g mol^{-1}, and for the PS/PB system two sets of mixtures were examined, one with deuterated PS and the second with deuterated PB. The temperature range investigated was 60–150°C. Figure 8.3 shows the temperature dependence of χ obtained for different volume fractions of labelled polymer for these two systems. Evidently, χ is not only temperature-dependent but also dependent on the composition of the mixture. To discuss this aspect, Tomlins and Higgins [18] used the interaction parameter defined by Koningsveld [19], $g_{12}(\phi, T)$ which is related to χ by:

$$\chi = (\partial^2 g_{12}/\partial \phi_2^2)_{\phi_1} \phi_1 \phi_2$$

and:

$$g_{12} = \alpha + \frac{\beta_s + \beta_H/T}{1 - \gamma\phi_2} \qquad (8.14)$$

where $\gamma = 1 - (\sigma_2/\sigma_1)$, (σ_2/σ_1) is the ratio of the surface areas of the polymer segments, α and β_S are empirical parameters, and β_H is related to the internal energy per contact of the segments. Empirically, the variation of χ with temperature is written as $\chi = a + b/T$, and by comparing this with eqn (8.14) it can be seen that:

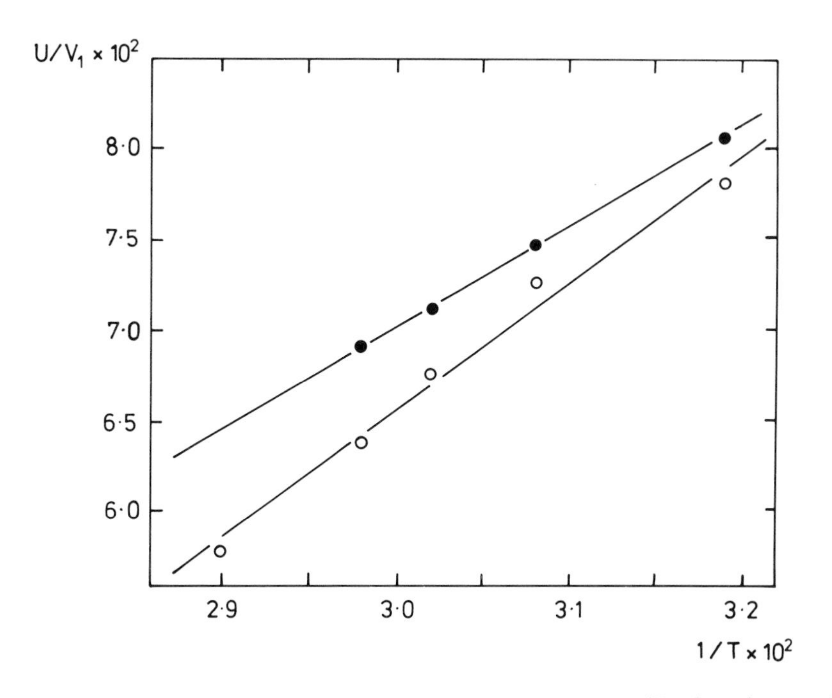

Figure 8.3 Temperature dependence of the interaction parameter (U/V_1) for mixtures of methoxylated poly(ethylene glycol) and methoxylated poly(propylene glycol) of two different compositions. Reproduced with permission from ref. 18.

$$a = \alpha + \beta_S(1-\gamma)/(1-\gamma\phi_2)^3 \tag{8.15}$$

(the entropic part of χ), and:

$$b = \frac{\beta_H(1-\gamma)}{(1-\gamma\phi_2)^3} \tag{8.16}$$

(the enthalpic part of χ).

Values of β_H obtained ranged from 19 K to 57 K for the two types of mixtures concerned, and the error quoted on these values was ± 20%.

The basis for using SANS to obtain the dimensions of polymers in the solid state was that deuteration of the molecule did not significantly alter the chemical and thermodynamic properties. It was evident from some of the earliest SANS experiments on polyethylene that this assumption could not be generally valid since clustering of deuterated polyethylene molecules was observed. Furthermore, Buckingham and Hertschel [20] predicted that mixtures of hydrogenous and deuterated polymers of the same chemical type should have an unfavourable free energy of mixing, due to small differences in the molar volumes of the two forms of the polymer. However, it was believed that such effects would only become significant close to a first-order transition. It was also noticed that the phase boundaries of polymer mixtures containing deuterated components were in very different temperature ranges from their fully hydrogenous counterparts. Despite all this evidence that deuteration does lead to a change in the interactions between the molecules, no attention was paid to the problem for some time. It was the work of Bates which addressed the problem and provided experimental evidence. Using the random phase expression of de Gennes, Bates and Wignall [21] examined mixtures of deuteropolystyrene with hydrogenous polystyrene and showed that when the molecular weights of each polymer were greater than about 10^6, the scattering at 160°C was greater than that predicted for a system at the limit of single-phase stability, and that $\chi = 1.7 \times 10^{-4}$ at this temperature. This system has been examined in detail by Schwahn *et al.* [22], who confirmed the results of Bates and also calculated the phase diagram and the upper critical temperature (130°C). They were also able to show the presence of 'critical slowing down', i.e. the decrease of the diffusion coefficient to very small values in the region between binodal and spinodal curves. This was shown by determining the change in the scattering on quickly changing the temperature of the system but remaining within the homogeneous phase, i.e. following the relaxation of the PSD–PSH system as it approached a new equilibrium. Although the data could not be fitted by an exact form of the Cahn–Hilliard [23] equation, they were able to obtain interdiffusion coefficients and compare the results with the theory due to Binder [24] (Figure 8.4). A fuller examination of the isotope effect on polymer thermodynamics and using SANS data was made by Bates, Dierker and Wignall [25], using mixtures of deuterated and hydrogenous polybutadienes. One mixture consisted of both polymers having a high molecular weight ($\sim 200 \times 10^3$ g mol^{-1}) and the second mixture was a deuterated polybutadiene of high molecular weight ($\sim 200 \times 10^3$ g mol^{-1}) and a hydrogenous polybutadiene of lower molecular weight ($\sim 50 \times 10^3$). This second mixture was homogeneous at 23°C and the SANS data could be fitted by the random-phase expression (eqn (8.13)) and gave a value of 8.7×10^{-4} for χ. By computing the phase diagram ($\chi N_1 f(\phi_1)$) using the mean field expression for the Gibbs free-energy change on mixing per segment, they were able to show that the higher-molecular-weight specimens should be in the phase-separated region, which concurred with the observation that those samples scattered light strongly. Even though the high-molecular-weight system was phase-separated, they were able to extract evidence of extensive interfacial mixing. By noting the deviations from Porod's law ($I(Q) \alpha Q^{-4}$ when $Q >>$ (characteristic dimension)$^{-1}$), a diffuse interfacial region thickness of circa 250 Å was estimated to be present at the boundary between the phase-separated region and the matrix.

In addition to the isotope effects discussed above, the miscibility of polymers will depend on the stereochemical microstructure and the tacticity of the constituent polymers. The influence of microstructure on the thermodynamics

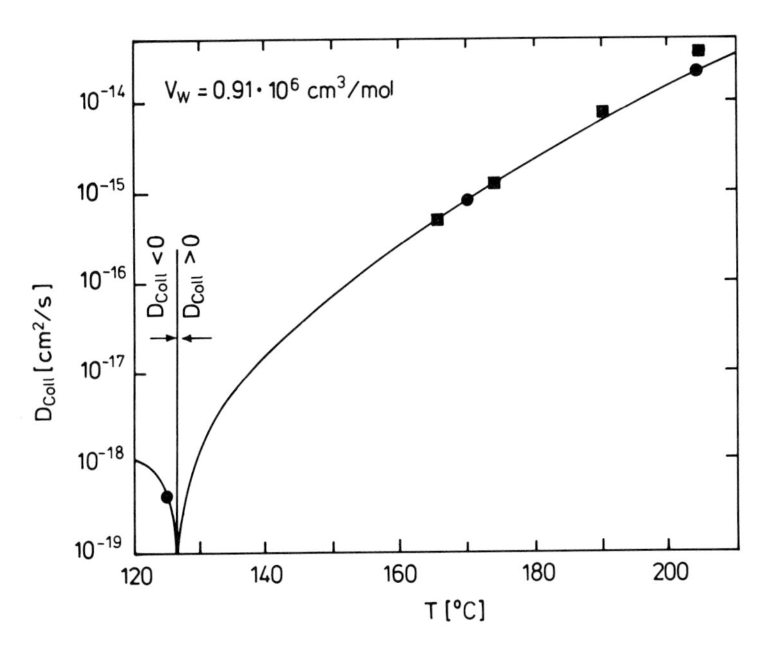

Figure 8.4 Collective diffusion constant for a mixture of polystyrene and poly(vinyl methyl ether). ●, SANS Data; ■, FRES data. Reproduced with permission from ref. 22.

of polymer blends has been investigated by Han and Hashimoto [26–29] in a series of papers on mixtures of polybutadiene and polyisoprene. In addition to using the random-phase expression to obtain χ and $S(Q=0)$, they also used the Ornstein–Zernike form of the equation for the region where $QR_g << 1$:

$$S(Q) = 0.5(\chi_S - \chi)^{-1}/(1 + Q^2\xi^2) \tag{8.17}$$

where χ_S is the value of χ at the spinodal curve for the volume fraction ø, and:

$$\chi_S = [(\phi N_1)^{-1} + (N_2(1-\phi))^{-1}]/2 \tag{8.18}$$

and ξ is the correlation length of the composition fluctuations which eventually result in phase separation (see section 8.2.5.2). A relation for ξ has also been given by de Gennes, with a being the Kuhn statistical step length of the polymer:

$$\xi = (a/6)[\phi(1-\phi)(\chi_S - \chi)]^{-\frac{1}{2}} \tag{8.19}$$

The spinodal curve defines the region of ϕ and T where the mixture is unstable to any fluctuation, and at this boundary the scattering intensity will be infinite, as will the value of ξ. Figure 8.5 shows the variation of ξ^{-2}, $S(Q=0)$ and χ as a function of $1/T$ for a mixture of a hydrogenous and deuterated polybutadiene each with approximately 65% 1,2-addition. The composition dependence of

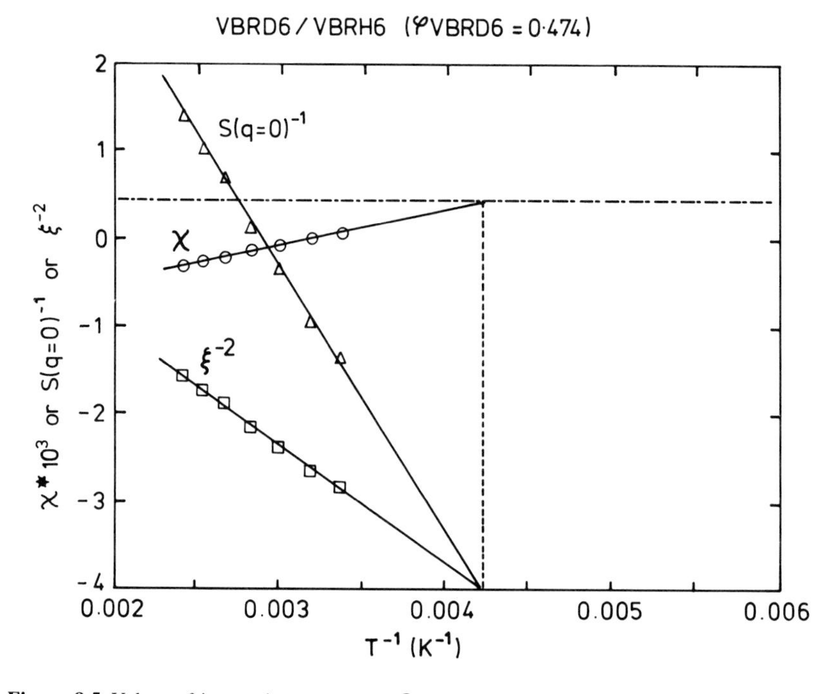

Figure 8.5 Values of interaction parameter (○), intensity at $Q = 0$(△), and the reciprocal of the (correlation length)^2 (□) as a function of reciprocal temperature, for mixtures of deuterated and hydrogenous vinyl butadienes. – · – · –, Value of interaction parameter at spinodal temperature. Reproduced with permission from ref. 29.

χ for this system was essentially negligible; however, at higher temperatures χ became negative. This was attributed to the copolymer effect rather than the existence of specific interactions. The polybutadienes were treated as random copolymers of 1,4- and 1,2-addition units, and there are deuterated and hydrogenous homologues in the mixture, i.e. there are four different types of monomer segment, A, B, C and D. Using the copolymer theory of ten Brinke *et al.* [30], then χ can be written as:

$$\chi = xy\,\chi_{AC} + (1-x)y\,\chi_{BC} + x(1-y)\chi_{AD} + (1-x)(1-y)\chi_{BD} - x(1-x)\chi_{AB} - y(1-y)\chi_{CD}$$

where x is the number fraction of A in the A–B copolymer and y the number fraction of C in the C–D copolymer. Solution of this equation for all of the χ terms requires six sets of data, which the authors did not have. However, by making some simplifying assumptions they were able to reduce the number of χ terms to three, and the variation of these with reciprocal temperature is shown in Figure 8.6. Mixtures of high vinyl-content and high *cis*-content polybutadienes only appeared to be miscible when one component had a low molecular weight.

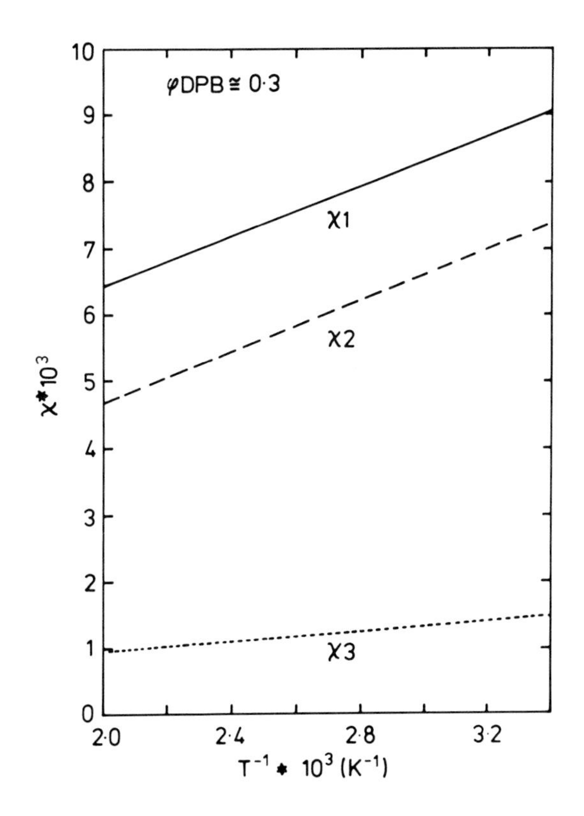

Figure 8.6 Temperature dependence of interaction parameters for mixtures of deuterated and hydrogenous vinyl butadienes. $\chi_1 = \chi_{AB} = \chi_{CD}$, $\chi_2 = \chi_{AD} = \chi_{BC}$, $\chi_3 = \chi_{AC} = \chi_{BD}$. Reproduced with permission from ref. 29.

A similar series of experiments were carried out on mixtures of deuterated polybutadiene (PBD) and hydrogenous polyisoprene (PIH). The PBD had a content of 1,2-units of 12–28% and the PIH had a 3,4-unit content of 7–15%. For a fixed temperature, χ was observed to decrease as the content of 1,2-units in PBD increased and to increase as the content of 3,4-units in the PIH increased. Although the change in χ with microstructure is small ($\sim 10^{-3}$), it was pointed out that the net interaction parameter is χ multiplied by the degree of polymerisation, and hence small changes in χ can still produce significant effects. Furthermore, changes in microstructure influence the entropic part of χ (i.e. a in the empirical expression for χ) more than the enthalpic part (b).

Apart from the influence of temperature on the thermodynamic state of a polymer blend, flow also affects the equilibrium position. Both flow-induced mixing and demixing have been observed. Nakatani *et al.* [31] have published results on polystyrene (PS)–poly(vinyl methyl ether) blends where the PS was fully deuterated. The composition of the blend was close to the critical composition and shear rates of 0.02–3.0 s^{-1} were applied in Couette flow for temperatures of 100–142.5°C. Due to the SANS diffractometer having an area detector, the scattered intensity could be analysed both perpendicular and parallel to the direction of flow. Perpendicular to the direction of flow no influence of shear rate on the scattering was observed; parallel to the direction of flow a decrease in scattered intensity with increasing shear rate was noted. It was pointed out that the random-phase approximation formula could not be applied to polymer mixtures in shear flow; hence, only the fluctuation correlation lengths were extracted from the scattering data and these were used to evaluate the spinodal temperature. Table 8.3 gives the spinodal temperatures obtained as a function of shear rate, and these data clearly show a shear-induced mixing in the direction of flow. Further analysis of these data suggested a rotary motion being applied to the concentration fluctuations; if the rotary motion has a relaxation time greater than the reciprocal of the shear rate, then that fluctuation mode is depressed, this decreases the long-wavelength fluctuations and thus there is a decrease in the scattered intensity at low Q. Application of mode–mode coupling theory to these data suggested that the hydrodynamic mode was the dominating factor of the relaxation rate of the concentration fluctuations of the one-phase mixture under shear flow.

Table 8.3 Spinodal temperatures for polystyrene–poly(vinyl methyl ether) blends

Shear rate s^{-1}	$T_S^{\parallel}$ (°C)	$T_S^{\perp}$ (°C)
0	138.3	140.4
0.026	139.1	139.6
0.067	140.9	139.5
0.13	141.3	141.0
0.26	143.8	141.0
0.67	144.2	140.9
1.30	146.1	142.9

$T_S^{\parallel}$ = spinodal temperature parallel to shear direction.
$T_S^{\perp}$ = spinodal temperature perpendicular to shear direction.
$\overline{M}_W$ PS = 4.4×10^5 g mol^{-1}.
$\overline{M}_W$ PVME = 1.8×10^5 g mol^{-1}.
Weight fraction of PS in mixture = 0.2.

Results have also been published on the polystyrene–poly(vinyl methyl ether) mixture, both as a mixture of homopolymers [32] and where the polystyrene was crosslinked [33].

Lastly, before leaving polymer blends, attention is drawn to the possibility of the interaction parameter χ being dependent on Q, the scattering vector, as well as the composition of the polymer mixture. Evidence for this behaviour has been given by Brereton *et al.* [34] for the mixture of deuteropolystyrene with poly(tetramethyl carbonate) (PTMC). The reciprocal scattering as a function of Q^2 for different temperatures was not a series of parallel lines of data, as predicted from the de Gennes random-phase approximation formula. To accommodate this, the functional form proposed for $\chi(Q)$ was:

$$\chi(Q) = \chi^0(1 - KQ^2 + \ldots)$$

where the parameter K is related to range and form of the potential between segments. Depending on the spatial variation of the potentials for *ii* and *ij* interactions, then it was shown that $\chi(Q)$ could vary from positive to negative over a small range of Q. A value for the parameter K was obtained from the variation of the correlation length, ξ, with the intensity of scattering at $Q = 0$. For the PTMC system, $K \approx -6000\text{Å}^2$; for mixtures of polystyrene with poly(vinyl methyl ether) and poly(α-methylstyrene) $K \approx 0$.

It is perhaps advisable to reiterate here that all of the thermodynamic analysis of SANS data is based on a mean-field theory, essentially that of Flory and Huggins, and this is known to be inadequate. It may be that, if a proper account were taken of the different expansibilities of the polymer mixture components, as in equation-of-state or lattice-fluid theories of polymer thermodynamics, then the dependencies on microstructure and a Q-dependent χ might disappear.

8.2.5.2 Block copolymers. The microphase-separated state of block copolymers is well known, and examination of the molecular-weight and composition dependence of the domain morphology and organisation by both SANS and small-angle X-ray scattering is well documented. These results have been compared with the theory of Helfand [35], which is an example of a strong-segregation theory, and pertains to a regime where $\chi N >> 10$ with N being the total degree of polymerisation of the block copolymer. There exists also a weak segregation regime ($\chi N \approx 1$), where the microdomain morphology is not evident, but the length scales of the composition fluctuation are of the order of the radius of gyration of the copolymer. These fluctuations lead to a peak in the small-angle scattering, and the peak intensity and width are related to the value of χ, the interaction parameter between the two blocks. A recent review of block copolymer thermodynamics [36] provides an excellent summary of both of these theories. Some experiments exploring the weak segregation regime are discussed here.

Leibler [37] produced the seminal theory of block copolymers in the homogeneous disordered state using random phase-approximation theory, the scattering law obtained being:

$$S(Q) = N/(F(u) - 2\chi N) \tag{8.20}$$

where $u = Q^2R_g^2$ and $F(u)$ is a composition-weighted sum of Debye functions for the component blocks of the copolymer. Figure 8.7 shows a schematic diagram of the disordered homogeneous phase and the microphase-separated state, together with the monomer unit density fluctuations. In the weak-segregation regime, the monomer density changes smoothly. In this region the value

Disordered Ordered

Figure 8.7 Schematic diagram of the weak segregation and strong segregation regions of block copolymers.

of χN has a marked influence on the scattered intensity. The scattering is characterised by one peak (unlike the strong-segregation regime, where the high degree of organisation leads to Bragg scattering and many maxima), located at $Q \approx 2/R_g$ for a diblock copolymer, as χN increases (Figure 8.8). Eventually, at the spinodal point the peak maximum is located at infinity, and this condition for a symmetrical diblock copolymer is reached when $\chi N(\equiv (\chi N)_S) = 10.495$. Subsequently, Fredrickson and Helfand refined this value, and Hashimoto *et al.* [38] have derived scattering laws for multiblock copolymers and mixtures with homopolymers, and have included polydispersity effects [39]. The main characteristic is retained, i.e. a single maximum in the scattering whose Q location depends on the exact nature and composition of the copolymer–homopolymer mixture. Provided that there is sufficient electron density between the constituent blocks, then small-angle X-ray scattering may equally well be used to examine block copolymers in the homogeneous disordered state. To use small-angle neutron scattering, one of the blocks must be deuterium-labelled and hence one can make 'isotopic' block copolymers of the monomers of the same chemical type. Bates [40] has reported such a study on a block copolymer with a deuterated 1,4-polybutadiene block and a hydrogenous 1,2-polybutadiene block. Molecular weights used varied between 25 000 and 50 000, and for one particular block copolymer, with a volume fraction of deuterated component of 0.5, the temperature-dependence of the SANS scattering was obtained. From these data and fitting of the Leibler expression, the temperature-dependence of χ was obtained. As Figure 8.8 shows, the scattering maximum increases rapidly as the spinodal temperature is approached, i.e. as $(\chi N)_S$ has a value near to 10.495. Near the spinodal point of the system, the maximum intensity scales as:

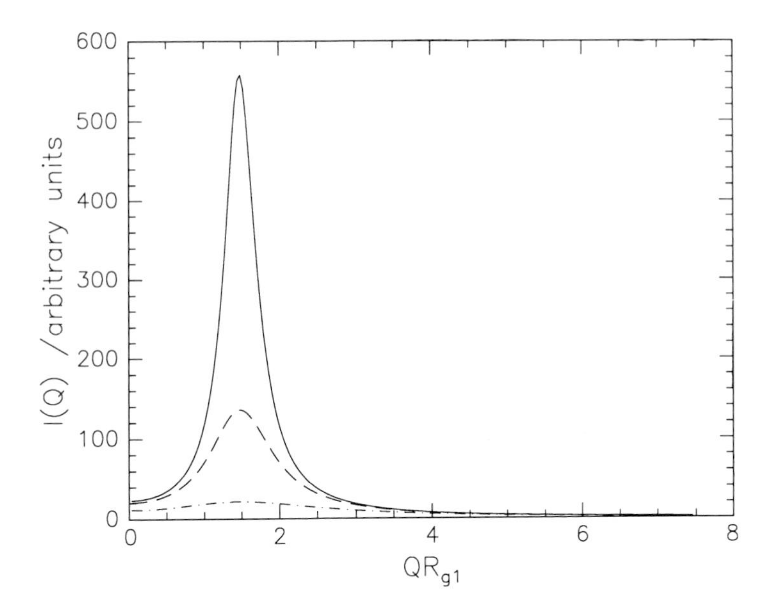

Figure 8.8 Influence of (χN) on the intensity of scattering from a symmetrical block copolymer in the homogeneous phase: ——, $\chi N = 9$; – – –, $\chi N = 8.5$; – · – · –, $\chi N = 5$.

$$I(Q_{max}) \approx [(\chi N)_S - (\chi N)]^{-1} \tag{8.21}$$

since:

$$\chi \alpha T^{-1}$$

then:

$$I(Q_{max}) \approx (T_S^{-1} - T^{-1})^{-1}$$

and thus the temperature where $I(Q_{max})^{-1} = 0$ defines the spinodal temperature. This temperature was obtained by extrapolation of the reciprocal of the experimental maximum scattered intensities as a function of $1/T$. For block copolymers that have chemically distinct blocks, the incompatibility of the two blocks results in the transition temperature to the disordered state being at high temperatures (for copolymers which do not have an unrealistically low molecular weight), and thus thermal degradation may be a problem. To overcome this, Connell and Richards [41] used SANS to investigate the thermal behaviour of concentrated solutions of diblock copolymers of styrene and isoprene. Earlier small-angle X-ray work by Shibayama and Hashimoto [42–45] had shown that the ordered structures characteristic of the solid state are maintained in such concentrated solutions. Connell and Richards [41] looked at two copolymers with different volume fractions of deuterostyrene blocks.

The spinodal temperatures were obtained by extrapolating the maximum intensity as a function of T^{-1}, as outlined above, for each solution investigated. Values for the spinodal temperature of the bulk were obtained by noting the dependence of these spinodal temperatures on the concentration of copolymer. By this means, the spinodal temperature for two diblock copolymers of total number-average molecular weight 82×10^3 were 400 K and 415 K for deutero-styrene weight fractions of 0.19 and 0.38, respectively. The dependence of the interaction parameter on temperature was also evaluated and a partial phase diagram evaluated. It was in reasonable agreement with that predicted by Leibler.

Concentrated solutions of block copolymers have also been used to investigate the kinetics of ordering by using SANS. A Landau–Ginzburg theory for the kinetic process has been suggested by Hashimoto [46], and the expression for the change in scattered intensity with time following a sudden 'quench' from one equilibrium state to a different state is:

$$I(Q,t) = [I(Q,0) - I(Q,\infty)] \exp(2R(Q)t) + I(Q,\infty) \tag{8.22}$$

where $I(Q,0)$ is the intensity in the initial equilibrium state, $I(Q,\infty)$ is the intensity in the new equilibrium state, and $R(Q)$ is an amplification factor which is given by:

$$R(Q) = L_0 Q^2 [- S(Q)^{-1}]$$

where $S(Q)$ is given by the Leibler expression (eqn (8.20)), and L_0 in this equation is an Onsager coefficient characteristic of the diffusion of the copolymer molecules. The form of $R(Q)$ is determined by the value of χN. For $\chi N < (\chi N)_S$, $R(Q)$ is negative and the composition fluctuations decay, but if $R(Q)$ is positive the composition fluctuations grow and the system is unstable and eventually phase-separates. Moreover, as Figure 8.9 shows, $R(Q)$ is Q-dependent, being positive between upper and lower critical Q values. Homopolymer blends have only an upper critical Q value, and the maximum in $R(Q)$ is located at $Q = 0$. By electronically 'gating' the detector, it was possible to follow the change in scattered intensity after a sudden quench from a high temperature to a series of low temperatures. Depending on the value of Q, the quench led to either a decrease or an increase in scattered intensity with time following the quench, and the variation of $R(Q)/Q^2$ displayed a lower critical Q value and a maximum at a finite value of Q but no upper critical value of Q. The absence of the upper critical value was attributed to the contribution of long-range ordering effects, especially for deep quenches. Overall, the evidence from these studies and others indicates that the random-phase approximation is a wholly valid description of the homogeneous state of block copolymers.

8.2.5.3 Liquid-crystal polymers. Polymers with liquid-crystal moieties either as substituents or as part of the main chain have been of interest for a

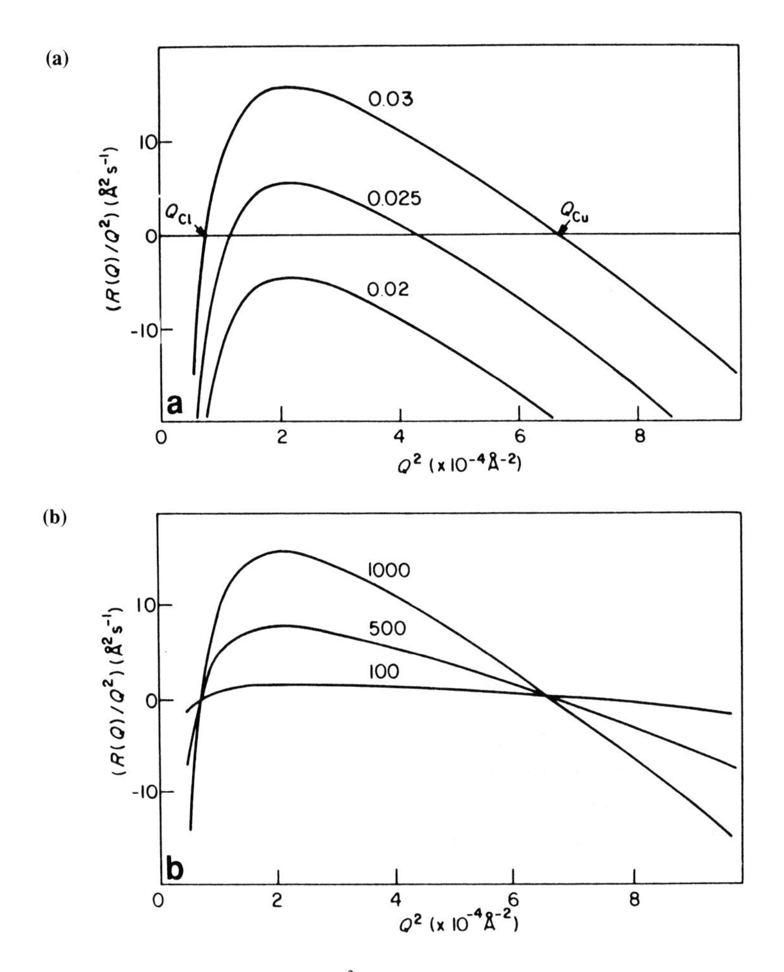

Figure 8.9 (a) Influence of χ on $R(Q)/Q^2$; values of χ marked on each curve. (b) Influence of Onsager coefficient on $R(Q)/Q^2$; values of L_0 marked on each curve.

number of years [47]. The former offer the possibility of novel electro- and mechano-optic effects; the latter are generally materials of high strength in the chain-axis direction. Considerable effort has been devoted to establishing the organisation of these materials but surprisingly little use has been made of small-angle neutron scattering. This may be due to the difficulty of synthesising the deuterium-labelled polymers needed for SANS and, in the little work that has been reported, fully deuterated polymers were not available.

MacDonald *et al.* [48] used SANS to determine the solid-state configuration of a main-chain liquid-crystal polymer that had residues of hydroxybenzoic acid, hydroquinone and isophthalic acid in the main chain. The last-named residue could not be deuterated. This polymer had a crystalline to nematic transition at ~ 270°C and like many such polymers it is a polyester. Polyesters are susceptible to transesterification (that is, the random scission and recombination of the molecules) on heating. This reaction leads to 'scrambling' of the labelled segments over all the molecules if samples containing a portion of labelled chains are heated. The kinetics of this reaction had first been observed by SANS on poly(ethylene terephthalate) [49], and a proper analysis of the kinetic scheme was published later by Benoit *et al.* [50]. This reaction takes place very rapidly in the thermotropic polyester studied by MacDonald *et al.* [51], and leads to an apparent decrease in the molecular weight with time as the average length of the deuterium-labelled segments becomes shorter due to transesterification (Figure 8.10). Analysis of these data gave an activation energy of 174 kJ mol^{-1}, and suggested that the reaction was initiated by an active chain-end mechanism. Evidently, the rapidity of such reactions precludes determination of the configuration of such thermotropic polyesters in the liquid-crystal phase. However, in a subsequent paper, MacDonald *et al.* [48] showed that the polyester molecule in the solid state initially had a random-coil configuration, but, on heating to increasingly higher temperatures below the nematic transition temperature, the molecule became stiffer and more rod-like. Over the temperature range 175–250°C the persistence length of the molecule increased from 9 Å to 44 Å suggesting that there is some pre-ordering before the crystalline to nematic transition temperature.

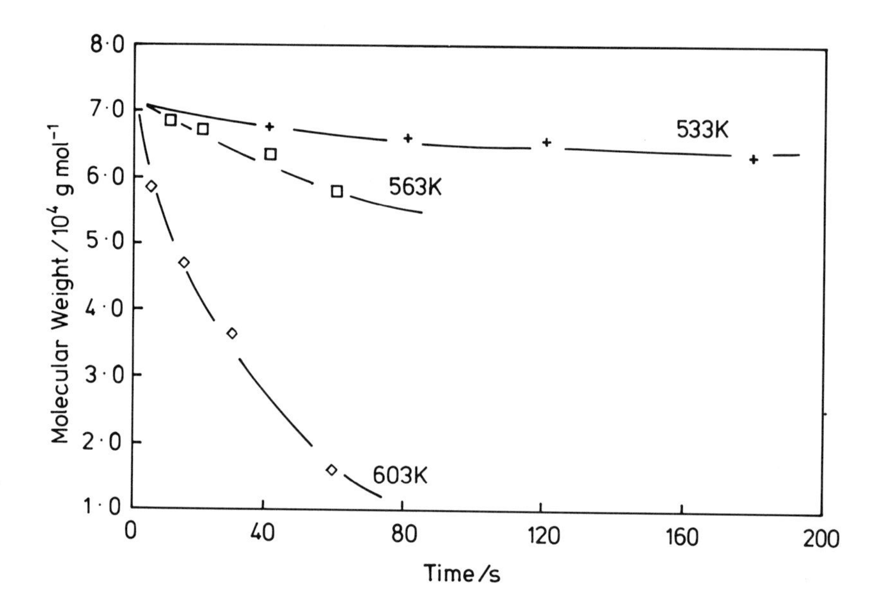

Figure 8.10 Variation in molecular weight of an aromatic polyester as a function of time at each of the temperatures indicated.

SANS from side-chain liquid-crystal polymers has been discussed by Keller *et al.* [52]. Generally these materials are held at elevated temperatures in a magnetic field and the radii of gyration parallel to and perpendicular to the applied magnetic field determined as the polymer passes through the several phases (amorphous glass, smectic, nematic and isotropic). For the methacrylate polymer:

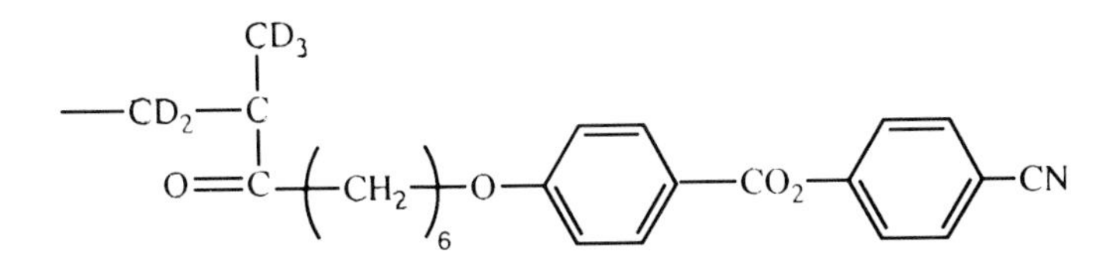

the radius of gyration perpendicular to the magnetic field was greater than that parallel to the field in the glassy, smectic and nematic phases, although this was smaller than that observed when the –CN group is replaced by an $-OC_4H_9$ group. This behaviour was attributed to confinement of the polymer backbone to one smectic layer and there being increased overlap between the mesogenic groups in the labelled –CN polymers. A side-chain liquid-crystal polymer with a siloxane backbone was also reported [52]. For this polymer the radius of gyration parallel to the field was greater than that perpendicular to the field in the smectic and nematic phases. Although this could be explained as being due to the increased flexibility of the Si–O–Si-bond in the main chain, the authors also pointed out that only one methyl group in the side chain was deuterium-labelled and this could lead to artefacts in the SANS data (*cf.* remarks on Rawiso's work [15] earlier).

8.3 Neutron reflectometry

The surface structure and composition of polymers, the nature of the interface between polymers, and the organisation of how polymers spread at air–liquid interfaces and are absorbed at solid–liquid interfaces are relevant to a variety of applications. Amongst these one can list the surface 'fractionation' of polymer chains due to the molecular-weight dependence of surface tension, the extent of the diffuse interface in polymer mixtures, the stabilisation of colloidal dispersions, and the use of polymers as Langmuir–Blodgett films and as model membranes. Although many techniques have been developed to study solid surfaces (SIMS, ESCA and Auger spectroscopy, LEED and attenuated total reflectance infrared spectroscopy), they are limited by either the length scale that they can probe or the resolution in length over which the composition variation is sensed. Polymer surfaces and interfaces present a problem in that the typical length scale is of the order of the radius of gyration but the resolution that may be required may be on a length scale of tens of angstroms. Ellipsometry is an optical-reflection technique that has been available for many years but it cannot provide a length-scale resolution of 10 Å, and moreover sometimes assumptions have to be made about the refractive index of the surface, which may not be valid. Ion-beam analysis methods [53] (forward recoil spectrometry, Rutherford back-scattering, nuclear reaction analysis) are able to penetrate polymer surfaces to great depths and provide a direct description of the variation in composition with depth from the surface. Unfortunately, their resolution is of the order of 100–300 Å, and for many systems this is not sufficient for detailed work. Nonetheless, ion-beam methods can provide vital complementary information and significant insight.

Reflection of X-rays and neutrons from a surface responds to the variation of electron density and scattering-length density, respectively. Penetration depths can be several thousand angstroms and resolution ~ 10Å. Hence, in principle these techniques should be ideal for providing information on the nature, composition and size of polymer surfaces and interfaces. However, a direct description of the surface is not provided; the predictions of a model have to be compared with literature data, and in many cases finding a unique solution may be difficult. Only neutron reflectometry is dealt with here, since this has been applied to a wider range of polymer systems than X-ray reflectometry.

8.3.1 Theory

Figure 8.11 shows a neutron beam incident at a glancing angle θ on a surface. Three processes take place: specular reflection (where the angle of reflection equals the angle of incidence); transmission and refraction of the beam in the bulk of the sample; and scattering from the bulk, which will eventually form part of the background signal. If the specimen consists of several discrete layers, then at each interface there will be additional specular reflection due to the transmitted beam. These specularly reflected beams interfere with each other to produce a reflectivity profile which is determined by the variation in neutron refractive index normal to the surface. In what follows, it is assumed that the surfaces and interfaces are perfectly smooth; a correction for surface roughness is introduced later. For a uniform medium the refractive index is given by:

$$n = 1 - (\lambda^2/2\pi)\rho + i\lambda\Sigma_a/4\pi \tag{8.23}$$

where Σ_a is the macroscopic absorption cross section of the material. Generally, $\Sigma_a \approx 0$ and eqn (8.23) simplifies to

$$n = 1 - (\lambda^2/2\pi)\rho \tag{8.24}$$

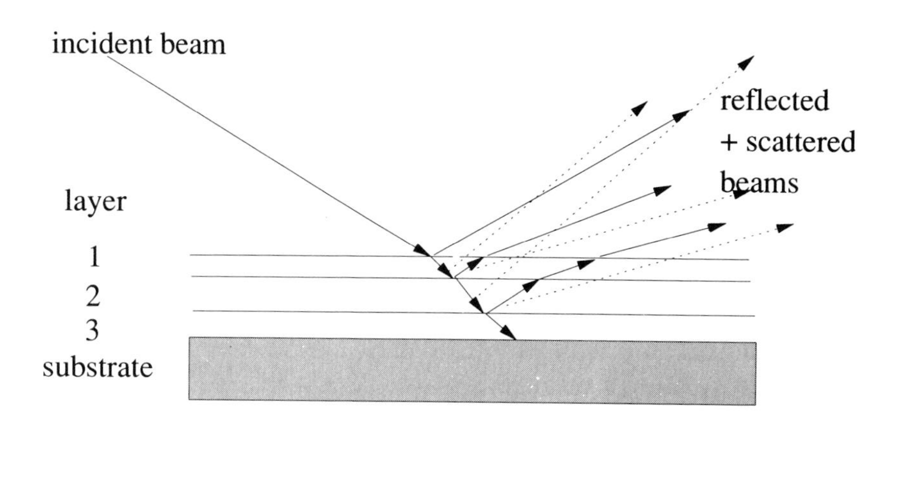

Figure 8.11 Schematic diagram of the reflection of a neutron beam from a thin film of several layers on a substrate.

From eqn (8.24) we note that since $\rho > 1$ for most materials, then $n < 1$ and total reflection will take place when the incident angle satisfies Snell's Law:

$$\cos\theta_C = n$$

i.e.

$$\theta_C = \lambda(\rho/\pi)^{\frac{1}{2}}$$

for small values of θ_C, the critical angle for total reflection. The application of optical relations to neutron reflectometry is not confined to calculation of the critical angle, since the amplitude of the reflected beam can be calculated from the Fresnel coefficients for reflection and transmittance at the interface. For a smooth surface, the Fresnel reflectivity is the square of the amplitude of the reflected beam and is given by:

$$R_F = \left[\frac{Q - (Q^2 - Q_C^2)^{\frac{1}{2}}}{Q + (Q^2 - Q_C^2)^{\frac{1}{2}}}\right]^2 \tag{8.25}$$

where Q is the scattering vector normal to the interface between the two media and Q_C is the critical value below which total reflection takes place, $Q_C = 4\pi^{\frac{1}{2}}(\rho_2 - \rho_1)^{\frac{1}{2}}$.

Replacing this latter expression into eqn (8.23) and for $Q >> Q_C$:

$$R_F = 16\pi^2(\rho_2 - \rho_1)^2/Q^4 \tag{8.26}$$

If the medium 2 is a layer of finite thickness on a thick substrate layer of scattering-length density, then, for $Q >> Q_C$:

$$R_F = 16\pi^2[(\rho_2 - \rho_1)^2 + (\rho_2 - \rho_3)^2]/Q^4 \tag{8.27}$$

i.e. the reflectivity at large scattering vector is proportional to the sum of the squares of the difference between the 'jumps' in the scattering-length density in the specimen. If the variation in refractive index normal to the surface is continuous, this can be approximated by a number of layers parallel to the surface with the same change in scattering-length density on going from layer to layer (hence the layer thickness may change). Each layer can be defined by a characteristic optical matrix $[M_k]$:

$$[M_k] = \left[\frac{\cos\beta_k - (1/k_k)\sin\beta_k}{-ik_k\sin\beta_k\cos\beta_k}\right] \tag{8.28}$$

where $k_k = (2\pi/\lambda)\sin\theta_k$ and $\beta_k = (2\pi/\lambda)n_k d_k \sin\theta_k$ and d_k is the thickness of the kth layer. The reflectivity amplitude matrix is given by the product of the individual optical matrices for each of the layers in the specimen. Denoting the elements of the reflectivity amplitude matrix by M_{ij}, then the amplitude of the reflected beam is:

$$A_R = \frac{(M_{11} + M_{12}k_k)k_1 - (M_{21} + M_{22})k_n}{(M_{11} + M_{12}k_n)k_1 + (M_{21} + M_{22})k_n} \tag{8.29}$$

and $R_F = |A_R|^2$. The formulation of the reflectivity as a product of matrices means that simulations and data analysis are suitable for computer solution. Since the refractive index and the thickness of each layer appear explicitly in eqn (8.26), then in principle the composition (proportional to n via ρ; see eqn (8.24)) and the thickness of each layer are obtainable from the reflectivity as a function of Q.

Surface roughness will give the surface some diffuse character and the reflectivity will be suitably modified to:

$$R = R_F \exp(-4k^2 \sin\theta_0 \sin\theta_1 \langle\sigma\rangle^2) \tag{8.30}$$

where $\langle\sigma\rangle^2$ is the mean-square roughness, i.e. σ is the standard deviation of the interface from the average position of the interface presuming a Gaussian distribution about this average position. Figure 8.12 shows simulated reflectivity profiles and the influence of roughness.

Reflectivity can also be described as the motion of the neutron in one dimension subjected to a pseudopotential whose magnitude is related to the scattering-length density of the medium. Using the Schrödinger equation to describe the perpendicular component of the motion eventually results in the

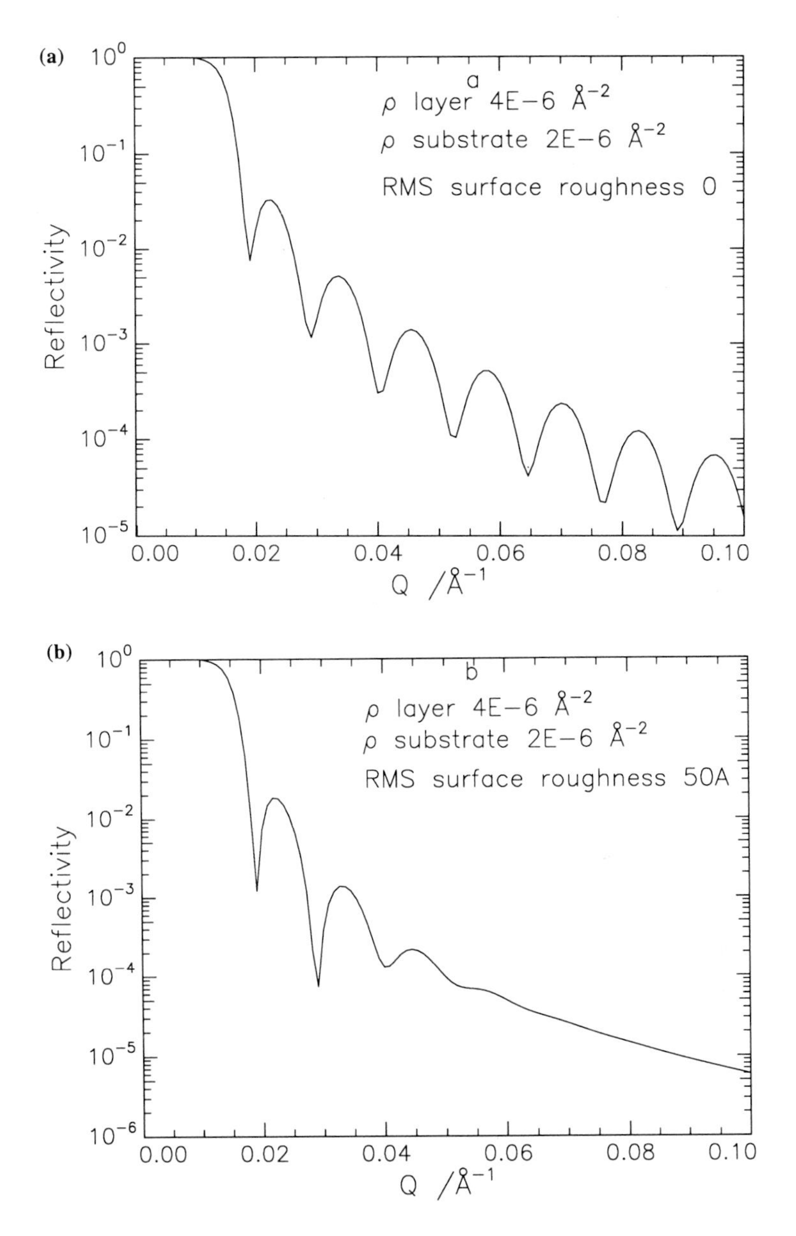

Figure 8.12 Simulated reflectivity profiles for a solid layer 500 Å thick: (a) Smooth surface, (b) rough surface with the Gaussian roughness indicated.

same equations for reflectivity being obtained as set out for the optical matrix description above. A detailed exposition of the theory of reflectivity is available in the monograph by Lekner [54].

8.3.2 Experimental

The fundamental aim of a specular reflection experiment is to measure the reflectivity of the test specimen as a function of the scattering vector, Q, perpendicular to the surface. There are two broad classes of neutron reflectometers: (i) fixed wavelength of neutron beam and variable incident angle, and (ii) fixed angle of incident beam and variable wavelength. Type (i) instruments are found at steady-state reactor sources (Institut Laue–Langevin, Grenoble, France; Brookhaven National Laboratory, New York, USA; and at the National Institute of Standards and Technology, Gaithersburg, Maryland, USA). Type (ii) instruments use time-of-flight methods and are available at pulsed neutron sources (ISIS at the Rutherford–Appleton Laboratory, UK; Los Alamos National Laboratory, New Mexico, USA, and Argonne National Laboratory, Illinois, USA). Figure 8.13 shows the CRISP reflectometer at ISIS in the Rutherford–Appleton Laboratory in the UK. The pulsed neutron beam obtained from the collision of protons with a uranium target is moderated by passage through hydrogen at 20 K, giving an effective wavelength range of 0.5–15 Å. The chopper, C, defines the wavelength band and to some extent suppresses frame overlap; the frame-overlap mirrors, F, reflect out neutrons with $\lambda > 13$ Å from the beam. The beam, inclined at 1.5° to the horizontal, is collimated by the cadmium slits S_1 and S_2 to give beam dimensions of 40 mm width and 0.5–6 mm height. The sample is located at S and the specularly reflected beam is detected at D by either a one-dimensional position-sensitive detector or a single ^{3}He detector. The incident-beam intensity is monitored by the monitor M, located immediately before the sample. The sample position can accommodate a variety of environments (crystals, goniometers, Langmuir troughs, etc.), and thus a very broad range of experiments and materials can

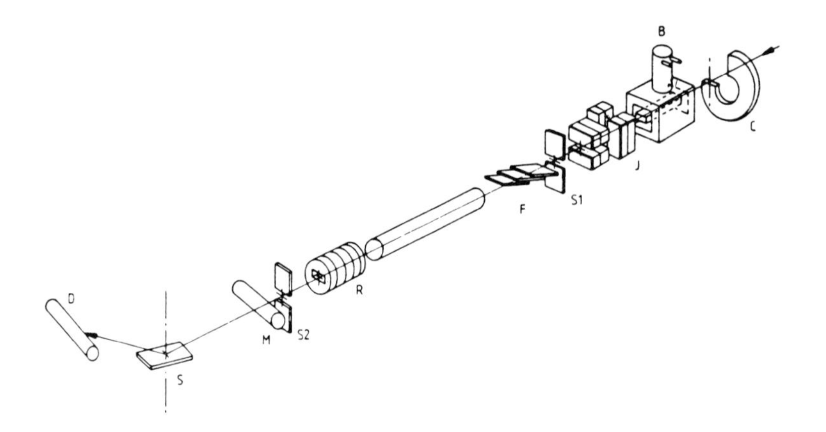

Figure 8.13 The CRISP reflectometer at the Rutherford–Appleton Laboratory. C, chopper; S1, S2, collimating slits; F, frame overlap mirrors; M, beam monitor; S, sample; D, detector. Reproduced with permission from ref. 55.

be investigated. Additionally, the incident neutron beam can be polarised and thus the surface magnetic structure of materials may also be investigated.

8.3.3 Applications of neutron reflectometry [55]

Neutron reflectometry has been applied to three classes of polymer systems: (1) polymers adsorbed from solution to an interface (generally the solid–liquid interface); (2) solid polymer films; and (3) polymers spread as monolayers. Russell [56] has surveyed results up to mid-1990, and so attention here is concentrated on results obtained since that time.

8.3.3.1 Polymer adsorption from solution. An overview of much of the theory and experimental methods relating to this field has been given by Cosgrove [57], and early results concerning the adsorption of a styrene–methyl methacrylate block copolymer from carbon tetrachloride solution on to quartz have been discussed by Satija *et al.* [58]. Cosgrove *et al.* [59] used a fairly simple experimental approach to investigate the adsorption of polystyrene and a block copolymer of polystyrene–poly-2-vinylpyridine on to mica from cyclohexane and toluene solutions. The solutions were simply placed on top of the mica substrate and the neutron beam was brought in through the solution layer. The authors admit to difficulties with controlling the liquid-layer thickness and there will have been considerable incoherent scattering. Nonetheless, the reflectivity of the diblock copolymer adsorbed from toluene could be fitted equally well by a shifted Gaussian density profile or by a Scheutjens–Fleer [60] model for tails extending from the surface. Both gave a root-mean-square layer thickness of 45 Å. For deuteropolystyrene adsorbed from cyclohexane, the authors [59] claim that a Scheutjens–Fleer model for physisorbed polymer gave the best fit to the reflectivity profile, but at other higher concentration Gaussian profiles appear to be the better fit. Layer thickness was about 80–100 Å. It was stated that the data really need to be extended to higher Q ranges to be able to discriminate between the various models proposed. Lee *et al.* [61] investigated the adsorption of poly(ethylene oxide) (PEO) from water on to amorphous quartz. In this case both the incident and the reflected neutron beam travelled through the quartz block. There were problems from the roughness of the quartz surface and the apparent penetration of the quartz by the polymer solution. The model used to fit the reflectivity profiles was a concentration profile of PEO segments with a maximum some 20 Å from the surface. The Scheutjens–Fleer model of polymer adsorption could not be tested, since a much higher-molecular-weight polymer is needed, but it appears that most of the adsorbed polymer is within two statistical segment lengths of the surface, i.e. the polymer is a dense mat on the amorphous quartz surface. Perhaps the fullest study reported to date on polymers adsorbed from solution is that of Field *et al.* [62]. Block copolymers of styrene and ethylene oxide were used, where the ethylene oxide block was small and essentially anchored the copolymer to the surface of the single-crystal quartz block used as the adsorbing solid. Solutions of the copolymer were made in deuterotoluene and, as in the adsorption of PEO discussed earlier, the neutron beam was brought in through the quartz block. The data were analysed on the basis of the mean-field theory of Milner *et al.* [63], which views the adsorbed molecules as a 'brush' on the surface and each molecule as strongly stretched. A parabolic density profile of segments at the surface is predicted by this theory which gives the layer thickness as:

$$L_0 = (12/\pi)^{\frac{1}{3}} (w\sigma)^{\frac{1}{3}} N \tag{8.31}$$

where N is the degree of polymerisation, w is an excluded-volume parameter and σ is related to the mean separation between anchor points. The expression for the volume fraction profile used to fit the data was:

$$\phi(z) = A - BZ^n \tag{8.32}$$

where A, B and n were variables in the fitting process. For the three molecular weights used, $n \approx 2$, the value for a parabola. However, there were large error limits on these values. For the highest molecular weight, the value of n was in no way appropriate for a parabolic density profile. It was pointed out that at low Q values, the authors could not distinguish between a parabolic density profile and one described by an error function. On the basis that the description of the polymer brush requires a correction of the end density which is exponential, it was proposed that the density profile is parabolic with a Gaussian tail at long distances from the quartz surface. The layer thicknesses varied from 450 Å to 1240 Å and scaled with N as $N^{0.6}$, in good agreement with theory. Furthermore, these layer thicknesses were also in good agreement with values from force-balance measurements. A note of caution should be added, as the block copolymers did not all have the same poly(ethylene oxide) content; moreover, in one case the molecular weight of the poly(ethylene oxide) block was ~ 6000.

8.3.3.2 Solid polymer films. Interdiffusion of polymers has been much studied using neutron reflectometry in combination with other techniques such as ion-beam analysis. The experimental set-up usually consists of spin coating a layer of hydrogenous polymer on to a substrate and then laying a film of deuterated polymer on top of this film. The deuterated polymer film is separately spun on a substrate and floated off on water. Subsequently it is picked up on to the hydrogenous polymer film and dried, and then the whole assembly is annealed above the glass transition temperature for increasing times. Depending on the thickness of the layers, the neutron reflectometry profile consists of a series of maxima, which on annealing decrease in amplitude (Figure 8.14). By fitting these profiles to an appropriate model, the increase in the width of the interface between the two layers can be obtained and its time-dependence compared with the predictions of reptation theory. In the reptation model, the polymer molecule is confined to a tube defined by the surrounding molecules. For very short times, a segment is not constrained in its motion by the surrounding chains, and at a time τ_e the displacement of the segment becomes comparable to the tube diameter. Above this time, the molecule can only move along the tube, but for times less than the characteristic Rouse relaxation time, τ_R, segmental motion is not correlated for the whole molecule. For times greater than τ_R but less than the time needed for the molecule to disengage from its original tube, τ_d, the motion is governed by the dynamics of the contour-length fluctuation. After the time τ_d, the diffusion path is governed by the reptation process and has Fickian diffusion characteristics. The time dependence of the root-mean-square displacement of the segments in each of these time regions is [64]:

$$\phi(t)^{\frac{1}{2}} \approx \begin{cases} t^{\frac{1}{4}} & t \leqslant \tau_e \\ t^{\frac{1}{8}} & \tau_e \leqslant t \leqslant \tau_R \\ t^{\frac{1}{4}} & \tau_R \leqslant t \leqslant \tau_d \\ t^{\frac{1}{2}} & \tau_d \leqslant t \end{cases}$$

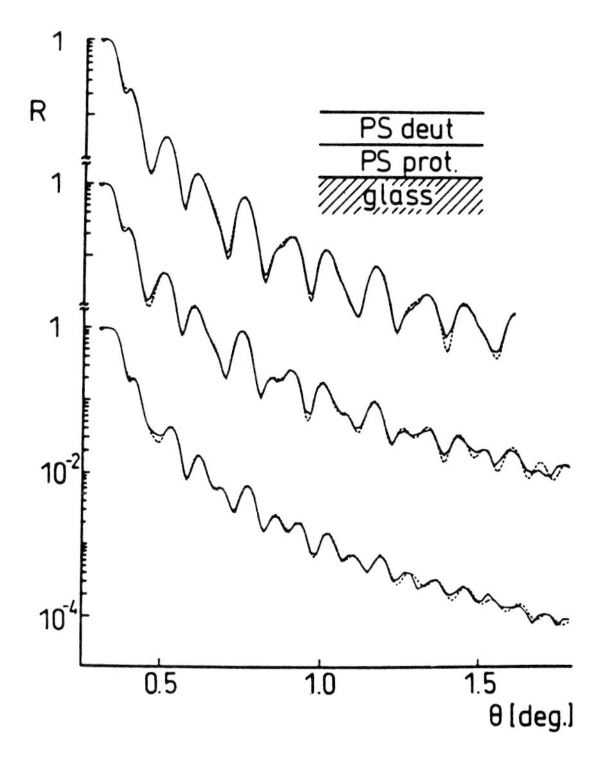

Figure 8.14 Neutron reflectivity profiles for a deuterated–protonated polystyrene bilayer annealed at 120°C for 0 min (top curve); 2 min (middle curve); 3900 min (bottom curve). Curves are displaced for clarity. Broken lines are fits to the data. Reproduced with permission from ref. 67.

Stamm *et al.* [65] observed three time regimes for the interdiffusion of polystyrene, but the annealing was insufficiently long to explore the Fickian diffusion regime. Karim *et al.* [66] observed this regime as well as the smaller time regimes in their experiments on interdiffusion of polystyrene. Reiter and Steiner [67] combined neutron reflectometry and nuclear reaction analysis to examine the interdiffusion of hydrogenous and deuterated polystyrene. The neutron reflectometry profiles were all for times $< \tau_d$ and were fitted by two superimposed error functions, one of which described the Rouse motion and the other the reptation motion. Each of these had its own interfacial width and the model included a parameter which was the fractional contribution of the reptation motion to the interfacial profile. This latter parameter increased as the annealing time increased. Extrapolating this to where the fraction has a value of unity gave 1×10^5 min for τ_d. The exponents for the time-dependence of the Rouse-motion contribution and the reptation contribution were 0.07 and 0.17 respectively; the exponent for the total broadening of the interface in the time regime explored by neutron reflectometry was 0.24. Nuclear reaction analysis explored a time region greater than τ_d and the exponent obtained from these data was 0.53, in reasonable agreement with theoretical prediction.

Interdiffusion of two different polymers has not been widely reported. Fernandez *et al.* [68] discussed the interdiffusion of deuteropoly(methyl methacrylate) into solution-chlorinated polyethylene (SCPE) (a compatible polymer pair). Annealing of the bilayer was done at 120°C above T_g (and also above the phase separation temperature). Swelling of the SCPE at the interface was observed, due to penetration by the poly(methyl methacrylate), resulting in a shift of the interface into the PMMA layer. The shape of the interface evaluated from the neutron reflectometry was that of a step followed by a Gaussian tail of poly(methyl methacrylate) into the SCPE. This profile is very close to that predicted by Brochard and de Gennes [69] for the dissolution of a polymer by low-molecular-weight solvent. Sauer and Walsh [70] used neutron reflectometry to study the interface between polystyrene and poly(vinyl methyl ether) after annealing at 80°C. Interfacial thicknesses obtained using a symmetric density profile were 120 Å and 140 Å for annealing at 2 min and 12 min, respectively. However, there was evidence that the density profile was asymmetric, and a better fit to the reflectivity data was obtained using the profile shown in Figure 8.15. Spectroscopic ellipsometry on the same specimens suggested that the interfacial width was ~ 400 Å for a symmetric interface. Use of the asymmetric profile of Figure 8.15 rationalised this apparent disagreement between the interfacial widths obtained by the two techniques. The dependence of the interfacial width on annealing time was linear and supported Case II diffusion of poly(vinyl methyl ether) into the polystyrene. The authors noted that the polystyrene dissolves rapidly into the poly(vinyl methyl ether), and hence the interface stays relatively sharp for the molecular weights used by them (polystyrene, 104 000; poly(vinyl methyl ether), 99 000).

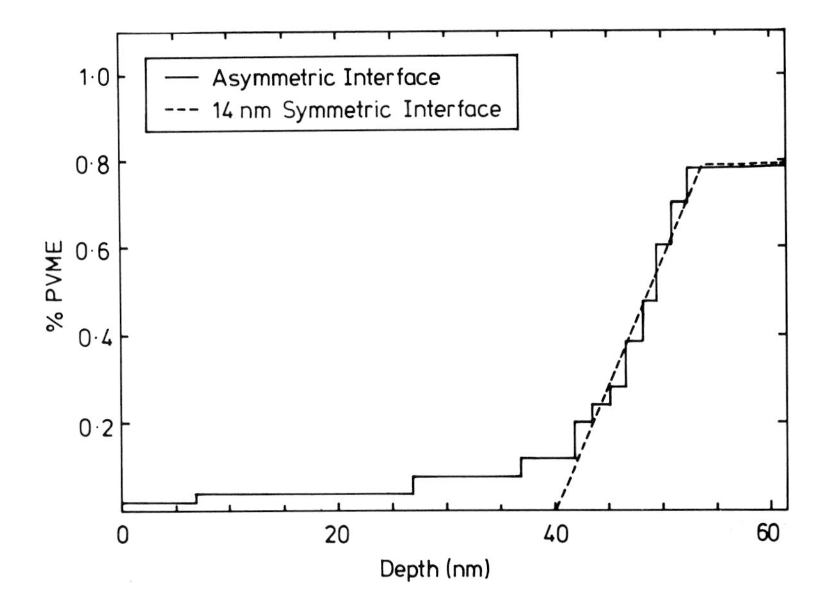

Figure 8.15 Interfacial profile used to fit the neutron-reflectivity data obtained for the interface between polystyrene and poly(vinyl methyl ether). Reproduced with permission from ref. 70.

Jones *et al.* [71, 72] made a comprehensive study of the enrichment by deuteropolystyrene of the surface of a mixture of the deuterated polymer with the hydrogenous polymer after annealing near the boundary of the co-existence curve. The unfavourable interaction between deuterated and hydrogenous polystyrenes (see earlier) and the lower surface tension of the deuterated polymer lead to a surface density profile of deuteropolymer, the shape of which is very similar to that predicted by the mean-field description of Schmidt and Binder [73]. Subsequently, by end-functionalising the polystyrene with trimethylsilane, ion-beam analysis showed that enrichment now took place at the silicon substrate, and the density profile from neutron reflectometry had a maximum at ~ 100 Å away from the silicon surface [74]. A more detailed treatment of the thermodynamics of wetting transitions in polymer mixtures has been published by Carmesin and Noolandi [75]; for systems which have a favourable χ a macroscopically thick wet layer of the lower-surface-tension component should be formed at the surface. Such a layer has been observed by using neutron reflectometry on annealed mixtures of deuterated and hydrogenous poly(methyl methacrylate) [76].

8.3.3.3 Polymer monolayers. Polymeric spread films at the air–water interface are examples of polymers in pseudo-two-dimensional situations. The nature of such films is germane to the preparation of Langmuir–Blodgett films, polymers as membranes, and the stabilisation of emulsions. Relatively little has been published to date on such spread films. Henderson and Richards [77] investigated spread films of the tactic isomers of poly(methyl methacrylate), and showed that the films always contain water and air. The thicknesses of the films were ~ 18 Å for the syndiotactic polymer and ~ 15 Å for the isotactic polymer; moreover, the isotactic polymer film never contained more than 50% polymer whereas the syndiotactic polymer film attained ~ 90% polymer content. Subsequently [78] these data were analysed by using the kinematic approximation and partial-structure factor analysis developed by Thomas [79] and applied extensively to surfactants. In the kinematic approximation the reflectivity is given by:

$$R(Q) = (16\pi^2/Q^2)\,|\,\rho(Q)\,|^2 \tag{8.33}$$

For a polymer film containing water then the variation of scattering-length density normal to the surface is:

$$\rho(z) = b_p n_p(z) + b_w n_w(z) \tag{8.34}$$

where $n_i(z)$ is the number-density distribution of species i with scattering length b_i and p and w refer to polymer and water respectively. Fourier transformation of eqn (8.34) and replacement into eqn (8.33) gives

$$R(Q) = (16\pi^2/Q^2)(b_p^2 h_{pp}(Q) + b_w^2 h_{ww}(Q) + 2 b_p b_w h_{pw}(Q)) \tag{8.35}$$

where h_{ij} are the partial structure factors describing the correlations between species i and j. Hence, if reflectivity experiments are done under three different contrast conditions, the partial structure factors can be obtained by solving three simultaneous equations of the form of eqn (8.33). Furthermore, the form of these partial structure factors is known for a few simple cases. Figure 8.16 shows the fit of a uniform-layer model to the experimental form factor obtained for syndiotactic poly(methyl methacrylate) spread on water at a surface concentration of 1.5 mg m^{-2}. It was also shown that the water was uniformly distributed in the layer. Henderson [80] has also investigated spread layers of poly(ethylene oxide) on water. A single uniform layer is not formed with this polymer; the best fit to the data was for a two-layer model, which was marginally better than an exponential distribution of poly(ethylene oxide) segments at the surface. The upper layer (nearer the air) has a thickness of ~ 20 Å and a polymer volume fraction rising from 0.03 to 0.2 as the surface concentration increased from 0.2 to 0.7 mg m^{-2}. The lower layer also had a constant thickness but with a value of ~ 45 Å, and the polymer volume fraction was constant at ~ 0.02. This description is reminiscent of the adsorbed polymers discussed earlier. Hodge *et al.* [81] have used neutron reflectometry to ascertain the structure of derivatives of copolymers of styrene and maleic anhydride spread at the air–water interface. The copolymers had been treated with long-chain alcohols to give half-esters, and it was found that the alkyl chains were

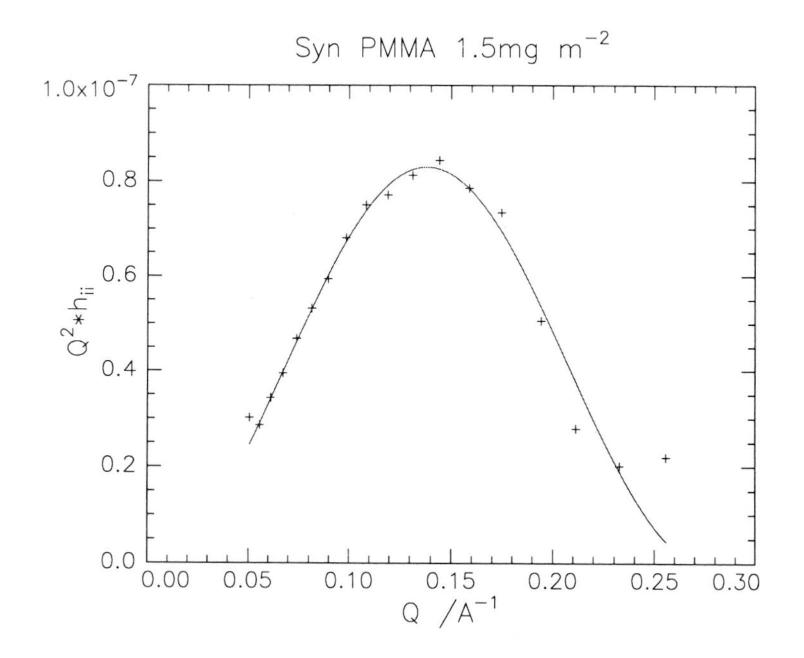

Figure 8.16 The product $Q^2 h_{ii}(Q)$ for a monolayer of syndiotactic poly(methyl methacrylate) spread on water; $h_{ii}(Q)$ is the partial structure factor for the methyl methacrylate units. The solid line is a fit to the data for a uniform layer of thickness 23 Å.

completely out of the water subphase at high pressures, although the thickness of the alkyl layer suggested that they were tilted away from the surface normal. The main-chain backbone and the phenyl groups were immersed in the aqueous subphase and had a layer thickness of ~ 15 Å, which increased as the surface film was compressed.

8.4 Conclusion

This chapter has been deliberately confined to the applications of neutron scattering and neutron reflectometry to polymer systems. Many of the scattering laws and equations are directly transferable to X-ray scattering and reflectometry (with proper correction for the more prevalent absorption effects); however, the power and flexibility available by deuterium labelling is not at hand for X-rays, and for this reason attention has been restricted to neutron scattering. Even with this restriction it has been impossible to discuss properly all the applications of neutron scattering and neutron reflectometry in the available space. It is abundantly clear that more people are using the techniques to provide unique information on the systems that they are studying. Moreover, it is also clear that one does not have to be an expert to extract useful information from the data, although experience can help in obtaining more subtle and deeper insights. Neutron-scattering techniques are not restricted to a few areas of research; they are a powerful means of obtaining structural and dynamic information on liquids and solids of all types. Hopefully the few examples discussed here are sufficiently convincing.

Acknowledgements

I thank all my past and present students who, by their constant questions, have initiated questions in me and from which I (and hopefully they) have gained much experience in the use of neutron scattering in polymers. It would also be remiss not to mention the very great dependence that anyone doing neutron scattering has on the host establishments and particularly the instrument scientists there. From my own experience I would like to single out the following: Dr Adrian Rennie (now at Cambridge University), Manuel Cruz and Dr Peter Timmins, who are (or were) all at the Institut Laue–Langevin, and Dr Jeff Penfold, Dr Stephen King, Dr Richard Heenan and Dr John Webster, all the Rutherford–Appleton Laboratory. The foresight of the funding agencies of the neutron-scattering establishments should also be recognised in providing the basic resource with which the research can be done.

References

1. G. D. Wignall (1987). *Encyclopedia of Polymer Science and Engineering*, Vol. 10, 2nd edn, Wiley, New York.
2. R. W. Richards (1989). Chapter 6 in *Comprehensive Polymer Science*, Vol. 1, G. Allen and J. C. Bevington (eds), Pergamon, Oxford.
3. D. M. Sadler (1989). Chapter 32 in *Comprehensive Polymer Science*, Vol. 1, G. Allen and J. C. Bevington (eds), Pergamon, Oxford.
4. R. W. Richards (1986). Chapter 1 in *Developments in Polymer Characterisation—5*, J. V. Dawkins (ed.), Elsevier Applied Science, London.
5. S. W. Lovesey (1984). *Theory of Neutron Scattering from Condensed Matter*, Oxford University Press, Oxford.
6. O. Glatter (1982). Chapter 4 in *Small Angle X-ray Scattering*, O. Glatter and O. Kratky (eds), Academic Press, London.
7. O. Kratky and G. Porod (1949). Rec. Trav. Chim. Pays-Bas, **68**, 1106.
8. H. Yamakawa (1984). *Ann. Rev. Phys. Chem.*, **35**, 23.
9. P. Sharp and V. A. Bloomfield (1968). *Biopolymers*, **6**, 1201.
10. A. Guinier and G. Fournet (1955). *Small Angle Scattering of X-rays*, Wiley, New York.
11. W. Burchard (1978). Chapter 10 in *Applied Fibre Science*, Vol. 1, F. W. Happey (ed.), Academic Press, London.
12. R. G. Kirste and R. C. Oberthür (1982). Chapter 12 in *Small Angle X-ray Scattering*, O. Glatter and O. Kratky (eds), Academic Press, London.
13. A. Z. Akcasu, G. C. Summerfield, S. N. Jahshan, C. C. Han, C. Y. Kim and H. Yu (1980). *J. Polym. Sci. Polym. Phys. Ed.*, **18**, 863.
14. C. Tangari. G. C. Summerfield, J. S. King, R. Berliner and D. F. R. Mildner (1980). *Macromolecules*, **13**, 1546.
15. M. Rawiso, R. Duplessix and C. Picot (1987). *Macromolecules*, **20**, 630.
16. P. G. de Gennes (1979). *Scaling Concepts in Polymer Physics*, Cornell University Press, London.
17. M. Warner, J. S. Higgins and A. J. Carter (1983). *Macromolecules*, **16**, 1931.

18. P. E. Tomlins and J. S. Higgins, (1988). *Macromolecules*, **21**, 425.
19. I. G. Voigt-Martin, K. H. Leister, R. Rosenau and R. Koningsveld (1986). *J. Polym. Sci. Polym. Phys. Ed.*, **24**, 723.
20. A. D. Buckingham and H. G. E. Hertschel (1980). *J. Polym. Sci. Polym. Phys. Ed.*, **18**, 853.
21. F. S. Bates and G. D. Wignall (1986). *Macromolecules*, **19**, 932.
22. D. Schwahn, K. Hahn, J. Streib and T. Springer (1990). *J. Chem. Phys.*, **93**, 8383.
23. J. W. Cahn (1961). *Acta Metallogr.*, **9**, 795.
24. K. Binder (1983). *J. Chem. Phys.*, **79**, 6387.
25. F. S. Bates, S. B. Dierker and G. D. Wignall (1986). *Macromolecules*, **19**, 1938.
26. S. Sakurai, H. Hasegawa, T. Hashimoto, I. G. Hoargis, S. L. Aggarwal and C. C. Han (1990). *Macromolecules*, **23**, 451.
27. H. Jinnai, H. Hasegawa, T. Hashimoto and C. C. Han (1991). *Macromolecules*, **24**, 282.
28. H. Hasegawa, S. Sakurai, M. Takenaka, T. Hashimoto and C. C. Han (1991). *Macromolecules*, **24**, 1813.
29. S. Sakurai, H. Jinnai, H. Hasegawa and T. Hashimoto (1991). *Macromolecules*, **24**, 4839.
30. G. ten Brinke, F. E. Karasz and W. J. MacKnight (1983). *Macromolecules*, **18**, 2179.
31. A. I. Nakatani, H. Kim, Y. Takahashi, Y. Matsushita, A. Takano, B. J. Bauer and C. C. Han (1990). *J. Chem. Phys.*, **93**, 795.
32. C. C. Han, B. J. Bauer, J. C. Clarke, Y. Muroga, Y. Matsushita, M. Okoda, Q. Trancong, T. Chang and I. C. Sanchez (1988). *Polymer*, **29**, 2002.
33. B. J. Bauer, R. M. Briber and C. C. Han (1989). *Macromolecules*, **22**, 940.
34. M. G. Brereton, E. W. Fischer, C. Herkt-Maetzky and K. Mortensen (1987). *J. Chem. Phys.*, **87**, 6144.
35. E. Helfand and Z. R. Wasserman (1982). Chapter 4 in *Developments in Block Copolymers—1*, I. Goodman (ed.), Applied Science Publishers, London.
36. F. S. Bates and G. H. Fredrickson (1990). *Ann. Rev. Phys. Chem.*
37. L. Leibler (1980). *Macromolecules*, **13**, 1602.
38. K. Mori, H. Tanaka and T. Hashimoto (1987). *Macromolecules*, **20**, 381.
39. K. Mori, H. Tanaka, H. Hasegawa and T. Hashimoto (1989). *Polymer*, **30**, 1389.
40. F. S. Bates and M. A. Hartney (1985). *Macromolecules*, **18**, 2478.
41. J. G. Connell and R. W. Richards (1990). *Macromolecules*, **23**, 1766.
42. M. Shibayama, T. Hashimoto and H. Kawai (1983). *Macromolecules*, **16**, 16.
43. T. Hashimoto, M. Shibayama, H. Kawai, H. Watanabe and T. Kotaka (1983). *Macromolecules*, **16**, 361.
44. M. Shibayama, T. Hashimoto, H. Hasegawa and H. Kawai (1983). *Macromolecules*, **16**, 1427.
45. M. Shibayama. T. Hashimoto and H. Kawai (1983). *Macromolecules*, **16**, 1434.
46. T. Hashimoto (1987). *Macromolecules*, **20**, 485.
47. C. McArdle (1989). *Side Chain Liquid Crystal Polymers*, Blackie, Glasgow; and A. M. Donald and A. Windle (1992). *Liquid Crystal Polymers*, Cambridge University Press, Cambridge.
48. W. A. MacDonald, A. D. W. McLenaghan and R. W. Richards (1992). *Macromolecules*, **25**, 826.
49. J. Kugler, J. W. Gilmer, D. Wiswe, H. G. Zachmann, K. Hahn and E. W. Fischer (1987). *Macromolecules*, **20**, 1116.
50. H. C. Benoit, E. W. Fischer, H. G. Zachmann (1989). *Polymer*, **30**, 379.
51. W. A. MacDonald, A. D. W. McLenaghan, G. McLean, R. W. Richards and S. M. King (1991). *Macromolecules*, **24**, 6164.
52. F. Moussa, J. P. Culton, F. Hardawin, P. Keller, M. Lambert, G. Pepy, M. Mauzac and H. Richard (1987). *J. Phys. (Paris)*, **48**, 1079.
53. R. A. L. Jones (1992). In *Polymer Surfaces and Interfaces II*, W. J. Feast, H. S. Munro and R. W. Richards (eds), Wiley, Chichester.
54. J. Lekner (1987). *Theory of Reflection*, Nijhoff, Dordrecht.
55. J. Penfold and R. K. Thomas (1990). *J. Phys. Condens. Matter*, **2**, 1369.
56. T. P. Russell (1990). *Mater. Sci. Rep.*, **5**, 173.
57. T. Cosgrove (1990). *J. Chem. Soc., Faraday Trans.*, **86**, 1323.
58. S. K. Satija, T. P. Russell, E. B. Sirota, C. T. Hughes and S. K. Sinha (1990). *Macromolecules*, **23**, 3860.
59. T. Cosgrove, T. G. Heath, J. S. Phipps and R. M. Richardson (1991). *Macromolecules*, **24**, 94.
60. J. M. Scheutjens and G. M. Fleer (1979). *J. Phys. Chem.*, **83**, 1619; and (1980). *J. Phys. Chem.*, **84**, 178.

61. E. M. Lee, R. K. Thomas and A. R. Rennie (1990). *Europhys. Lett.*, **13**, 135.
62. J. B. Field, C. Toprackcioglu, R. C. Ball, H. B. Stanley, L. Doi, W. Barford, J. Penfold, G. Smith and W. Hamilton (1992). *Macromolecules*, **25**, 434.
63. S. T. Milner, T. A. Witten, M. E. Cates (1988). *Europhys. Lett.*, **5**, 413.
64. M. Doi and S. F. Edwards (1986). *The Theory of Polymer Dynamics*, Oxford University Press, Oxford.
65. M. Stamm, S. Huttenbach, G. Reiter and T. Springer (1991). *Europhys. Lett.*, **14**, 451.
66. A. Karim, A. Mansour, G. P. Felcher and T. P. Russell (1990). *Phys. Rev. B*. **42**, 6846.
67. G. Reiter and U. Steiner (1991). *J. Phys. II (Paris)*, **1**, 659.
68. M. L. Fernandez, J. S. Higgins, J. Penfold and C. Shackleton (1991). *J. Chem. Soc., Faraday* Trans., **87**, 2055.
69. F. Brochard and P. G. de Gennes (1986). *Europhys. Lett.*, **1**, 221.
70. B. B. Sauer and D. J. Walsh (1991). *Macromolecules*, **24**, 5948.
71. R. A. L. Jones, E. J. Kramer, M. H. Rafoulovich, J. Sokolov and S. A. Schwarz (1989). *Phys. Rev. Lett.*, **62**, 280.
72. R. A. L. Jones, L. J. Norton, E. J. Kramer, R. J. Composto, R. S. Stein, T. P. Russell, A. Mansour, A. Karim, G. P. Felcher, M. H. Rafoulovich, J. Sokolov, X. Zhoer and S. A. Schwarz (1990). *Europhys. Lett.*, **12**, 41.
73. I. Schmidt and K. Binder (1985). *J. Phys. (Paris)*, **46**, 1631.
74. R. A. L. Jones, L. J. Norton, K. R. Shull, E. J. Kramer, G. P. Felcher, A. Karim and L. J. Fetters, (1992). *Macromolecules*, **25**, 2359.
75. I. Carmesin and J. Noolandi (1989). *Macromolecules*, **22**, 1689.
76. I. Hopkinson and R. W. Richards, unpublished work.
77. J. A. Henderson, R. W. Richards, J. Penfold, C. Shackleton and R. K. Thomas (1991). *Polymer*, **32**, 3283.
78. J. A. Henderson, R. W. Richards, J. Penfold and R. K. Thomas (1992). *Macromolecules*, in the press.
79. T. L. Crowley, E. M. Lee, E. A. Simister and R. K. Thomas (1991). *Physica*, **B174**, 143.
80. J. A. Henderson (1992). Ph.D. Thesis, University of Durham.
81. P. Hodge, C. R. Towns, R. K. Thomas and C. Shackleton (1992). *Langmuir*, **8**, 585.

9 Mechanical and rheological testing

M. FLANAGAN

9.1 Introduction

Mechanical and rheological testing of polymers or polymer compounds needs to be carried out for a number of reasons. These include:

Monitoring the quality and consistency of raw materials and products made from these.
Establishing the consistency (and thus control of) a process.
Determining the merits of new materials and processes.
Establishing the suitability of a particular design to the application.
Determining some estimate of service life under typical conditions.

It should always be remembered that no matter how far-sighted the test designer, test results can never predict precisely how a polymer compound may process or meet service requirements. Despite these limitations, some form of testing will be required in virtually all polymer applications.

Polymer materials behave in many respects in a different manner from more conventional materials like metals and ceramics. Thus, although certain tests, like tensile-strength, hardness and impact, can be similar to those employed with metals, many of the tests used in the polymer industry are specific to polymers.

The types of test carried out can be thought of in the following manner:

Basic testing: measuring a distinctive property that may be important in the potential application, e.g. tensile, tear or impact properties.
Laboratory testing of specific articles: often relating more directly to end-use, but testing designed to produce a result on a reasonable timescale, e.g. burst testing of a hose under conditions and pressures never reached during service.
Service tests: intended to simulate the treatment that an article will receive in service.
In-service tests: monitoring of an article in working service over a period of time.

This chapter concentrates on the basic or more common tests employed widely in the polymer industry.

9.2 Standards

The results obtained from tests can vary considerably. This may be due to many factors, such as differences in the shape or size of the test piece or perhaps its preparation. The storage of the material, test method or conditions can all make substantial differences to test results. Polymer behaviour is particularly sensitive to the preconditioning of the test piece and the actual test conditions. Comparable test results can never be obtained unless care is taken to ensure uniform temperature, humidity levels and testing rates. To ensure reproducible results that can be directly compared to results from other laboratories, standardised equipment design and test procedures must be used. Standards also help in indicating suitable test conditions and techniques as well as revealing possible sources of test error.

Organisations responsible for producing standards include:

International Standards Organisation (ISO).
National standards organisations (BSI, ASTM, DIN etc.).
Trade organisations and companies.

Great care must be taken in obtaining the latest edition of standards, reading them carefully and paying attention to detail. Poor testing technique invariably leads to poor test results being obtained. It is always best to test to an existing published standard test method if possible, preferably one from ISO.

9.3 Conditioning

Specimens always need to be conditioned prior to testing. This involves keeping the test material under the specified test conditions for a specified time before testing can take place. This is to enable the specimen to reach equilibrium with its testing environment. A maximum and minimum delay period from processing or shaping the polymer material will also be specified in standards (often 16 h). Items used in dynamic applications must be mechanically preconditioned prior to testing; this is to overcome the initial strain softening or 'Mullins effect' experienced in the early part of cyclical or dynamic tests.

9.4 Standard test conditions

BS903 and BS2782 specify standard conditions as either (a) 23°C and 50% relative humidity or (b) 27°C and 65% relative humidity. For elevated or reduced test temperatures, ISO specifies the following preferred temperatures: −70, −55, −40, −25, −10, 0, 40, 55, 70, 85, 100, 125, 150, 175, 200, 225 and 250°C

9.5 Types of testing

The testing of plastics and rubber can be divided into several areas.

9.5.1 Rheological testing

This is used to obtain information relating to flow behaviour of the polymer when in a semi-liquid state. Test conditions may relate to those experienced during processing or, for quality-control purposes, may be of an arbitrary nature. Typically, for plastics, tests include melt stiffness and viscosity. Rubbers and thermoset plastics require additional cure-related testing. Recent improvements in software design have enabled splitting of the viscous and elastic response of polymers to the specific shear deformations experienced during testing.

9.5.2 Mechanical properties

These include both destructive and non-destructive tests. The former may include tensile, compression or shear testing, and other product-related tests such as tear, abrasion, flexural and impact testing. Non-destructive tests may vary from simple visual examination and weight, density or hardness tests to ultrasonics, X-ray analysis, thermography or other sophisticated techniques.

9.5.3 Thermal testing

These tests relate to the temperature-dependence of the properties of a polymeric material. Common tests include softening behaviour (particularly of thermoplastics) and low-temperature flexibility, for example Vicat softening point, heat-deflection temperature and Gehmann, Clash and Berg apparatus testing.

9.5.4 Electrical testing

A material's behaviour in the proximity of electrical fields or discharges and the static build-up and dicharge behaviour of a material are tested. Tests include volume and surface resistivity, permittivity, breakdown strength, insulation resistance, power (or loss) factor and tracking resistance.

9.5.5 Environmental testing

Establishment of a material's resistance to the typical operational working environment may involve testing singly or in combination with heat, oxygen, ozone, light (usually UV), weathering, chemical resistance, radiation and biodegradation.

9.5.6 *Barrier/transmission properties*

These tests are related to the permeability of the material to air, moisture and other liquids or gases. Staining behaviour and transmission or spread of stain may also be of interest.

9.5.7 *Optical properties*

These are often of importance in plastic products and include colour, surface gloss, light transmittance and haze as well as refractive index.

9.5.8 *Flammability behaviour*

Testing to establish the burning behaviour of polymer materials is of increasing importance. Typical tests include limiting oxygen index (LOI), spread of flame and many specialist burn tests. Other tests establish smoke and toxic-gas evolution levels on burning. Ablation (burn through) testing may also be included. Burning behaviour is particularly difficult to simulate in laboratory tests as it is dependent on conditions, material form and quantity and type of ignition.

9.6 Rheological testing

Information relating to the rheological or flow properties of polymers is required by resin manufacturers, compounders, processors and machinery manufacturers. The simplest checks used to indicate the uniformity of material may involve melt flow index (MFI) in the case of thermoplastics or Mooney viscosity in the case of rubbers. Complete characterisation of a polymer melt will be needed for material research or mathematical modelling of polymer processing. The use of a capillary rheometer will be required for this to give plots of shear stress as a function of shear rate, over a series of temperature and pressure profiles. Some indication of the viscoelastic nature of the material will also be required to help predict processing problems like die swell. Recent developments in oscillating and rotational rheometers for use with rubber materials have enabled the splitting of viscous and elastic response components. This has been of great use in identifying faulty batches and in prediction of processing behaviour. The more common testing techniques are described below.

9.6.1 *Thermoset elastomers*

Traditionally testing of unvulcanised rubber compounds before processing involves evaluation of plasticity (or, more correctly, viscosity-related properties) and scorch/curing behaviour.

The Wallace plastimeter is used to indicate material 'plasticity' but has more recently lost favour to more comprehensive test equipment such as the

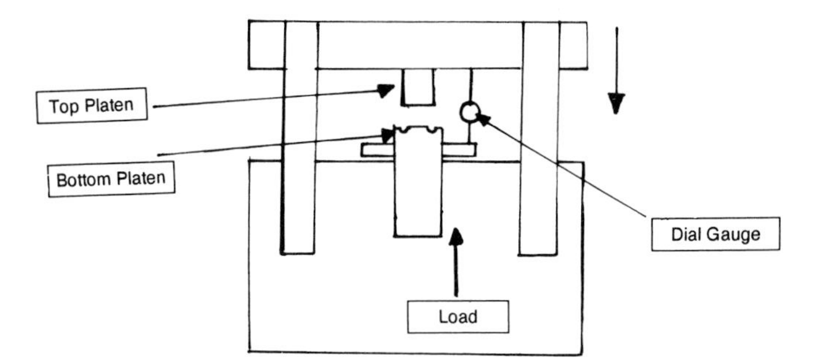

Figure 9.1 Wallace rapid plastimeter.

oscillating disc rheometer (ODR) and moving die rheometer (MDR). The Wallace rapid plastimeter is a compression-type instrument thet uses a small cylindrical testpiece (Figure 9.1). The sample is precompressed to 1 mm and then further compressed by a load of 10 kg. The final thickness is measured in hundredths of a millimetre and given as a Wallace Plasticity Number, a figure that indicates the stiffness of the compound at the test temperature of 100°C rather than true plasticity. The test is of limited use, operating in compression and at low shear rate (1 s^{-1}), but it still offers a quick indication of compound viscosity useful for quality-control purposes. It still finds use as such with rubber compounds and is often used to assess the mastication of rubber.

The Mooney plastimeter is a rotational shearing-disc viscometer (Figure 9.2). The temperature of the machine cavity dies can be preset to typical

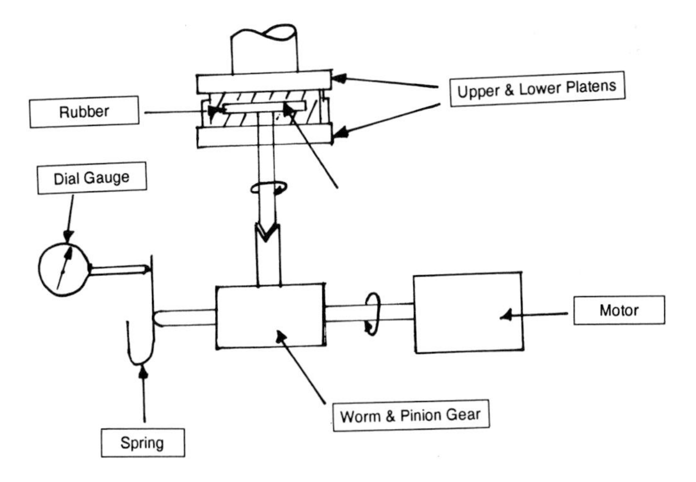

Figure 9.2 Mooney plastimeter.

processing temperatures usually 100, 120, 130, 135°C, etc. The resistance offered by the material to the rotating disc is indicated as Mooney viscosity units. The test was, and still is to some extent, popular in the industry. It is still used as a check on incoming raw rubber (often at 100°C) and for indicating the viscosity and scorch safety of rubber compounds. The latter test is carried out at the higher temperatures and is indicated by recording the Mooney viscosity versus time of test. As the rubber begins to crosslink (or 'scorch') the Mooney viscosity rises. The time in minutes is then noted when the Mooney viscosity rises by an arbitrary number of units above the minimum value recorded, often 5, 8 or 10 points. The 'viscosity' test usually consists of a 1-min warm-up and 4-min run, much longer than the Wallace plasticity test. The 'scorch' test (see Figure 9.3) can take a considerable time. The disc rotation of 2 rpm again gives a low shear rate (1–2 s^{-1}). These tests give some guide to the processing characteristics of the rubber compound but do not indicate the curing behaviour. Curemeters can be used to establish this. These continuously and automatically monitor and record the progress of vulcanisation with time at the required set temperature. Their major use is for quality control or evaluation purposes and to establish the curing characteristics of new compounds.

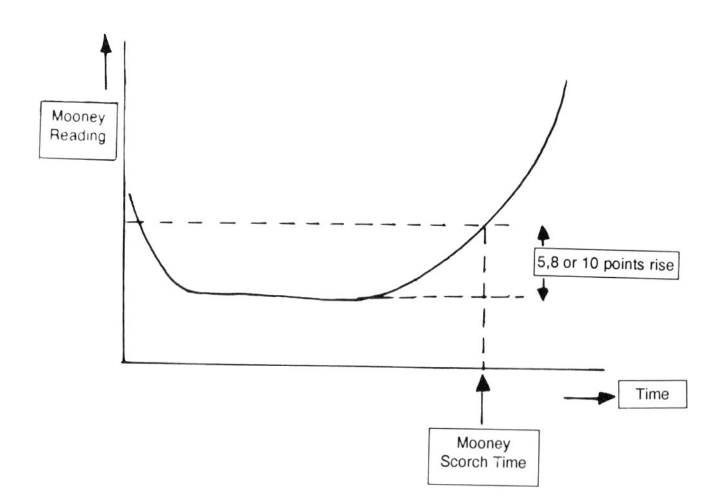

Figure 9.3 Mooney Scorch curve for a rubber compound at a set temperature.

9.6.2 Curemeters for rubber compounds

Current industrial practice in the rubber industry is for these instruments to measure a property, usually stiffness, changing progressively with time at a set temperature. The temperature may be set so as to equate to typical cure conditions, or may be of an arbitrary nature if used for quality-control testing. The sample of unvulcanised rubber used will have a hot stiffness or 'viscosity' before any crosslinking occurs. This will increase from a minimum value to a maximum as the curing reaction proceeds. Early machines such as the Wallace Shawbury Curometer were only able to establish guide cure times, due to lack of calibration of the stiffness readings. Most modern machines will produce a fully calibrated S-shaped cure curve. Cure time at a given temperature is established by the time to reach (say) 90% of the stiffness change. Test instruments provide information on minimum stiffness ('viscosity' or 'modulus'), scorch time, cure rate, maximum stiffness, degradation, and 'reversion' or 'Marching modulus' behaviour. The latter information is a guide to the stability of the 'cured' properties with length of 'curing' time. Machines of this type include the Monsanto oscillating disc rheometer (ODR), the JSR Curelastomer and similar ODR-type machines. Strangely, the Wallace Curometer until recently still had the advantage of an almost 'isothermal' result, due to the small sample size and its rapid heat-up time, so this machine still continues to be sold and used as an inexpensive guide to degree of cure versus cure time at a range of temperatures. By contrast, the ODR machines tend to overestimate the cure time, especially at high temperatures, due to their larger sample size. Nevertheless, it is their use over the last twenty years, particularly in quality control, that has led to much greater moulded-part accuracy and reductions in reject levels [1].

9.6.3 Oscillating disc rheometers

The ODR, with its ability to show complete cure curves in a very short period of time, has led many rubber manufacturers to use this type of instrument as a batch-control test [2,3]. A standard test is documented in ASTM D-2084 and BS1763, Part 10, 1977. The three phases in transforming rubber compound from the uncured to the cured state can be characterised by the cure curve produced (Figure 9.4).

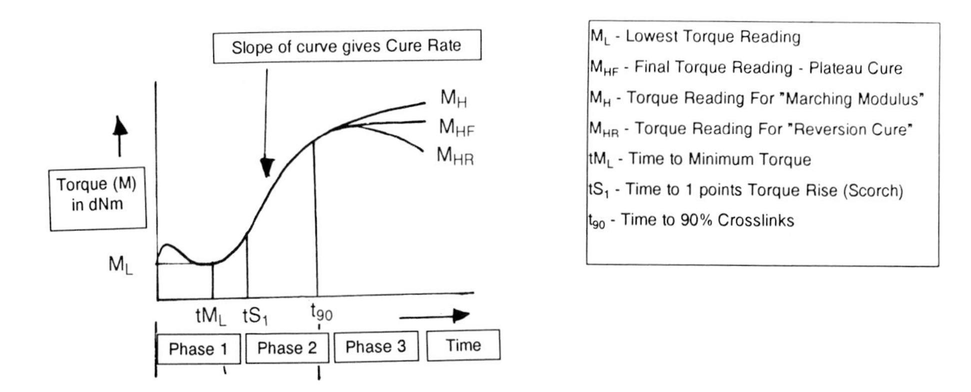

Figure 9.4 Torque versus time curve for an oscillating disc rheometer.

Phase one relates to the 'processability' of the compound into a suitable form by (e.g.) extrusion, calendering or mouldability. The ODR measures the 'viscosity' of the compound as a guide. It should be noted that this 'viscosity' is that established by means of the ODR and does not relate to other viscosity measurements.

Phase two is the curing or vulcanisation phase, as the stiffness of the rubber compound increases with time. The speed at which this occurs determines the length of time the rubber must be left to reach the desired cured state.

Phase three is the final attainment of maximum cure and can be correlated to the physical properties of the cured compound, primarily modulus and hardness.

ODR machines establish the change in compound stiffness by monitoring the change in torque when shearing a rubber sample between a biconical disc and an accurately temperature-controlled cavity (Figure 9.5). The concept of a rotational test was a carry-over from the Mooney-type machine, and a full cure curve can be obtained due to oscillatory movement of the disc. Various compound stiffnesses can be accommodated by the use of alternative angular deflections of ± 1, 3 or 5°. Recent improvements in accuracy have resulted from the use of a microprocessor to collect multiple torque values at incremental strains to determine torque by using Fourier transform analysis. The ODR has the advantage of a rapid view of the three phases in one test and, by increasing test temperatures for quality-control purposes, results can be obtained in as little as 3 min. However, the ODR is not as sensitive to viscosity variations as the Mooney viscometer, and so many manufacturers use both instruments. Both the ODR and the Mooney operate at low shear rates ($1.5\ s^{-1}$) whereas shear rates typical of mixing ($10–100\ s^{-1}$), extrusion ($10–1000\ s^{-1}$) or injection moulding ($100–10000\ s^{-1}$) are much higher. This is significant due to the typical pseudoplastic flow behaviour of most polymers and can lead to misleading predictions of 'processability' of rubber compounds. Until recently neither test machine measured the elasticity of the

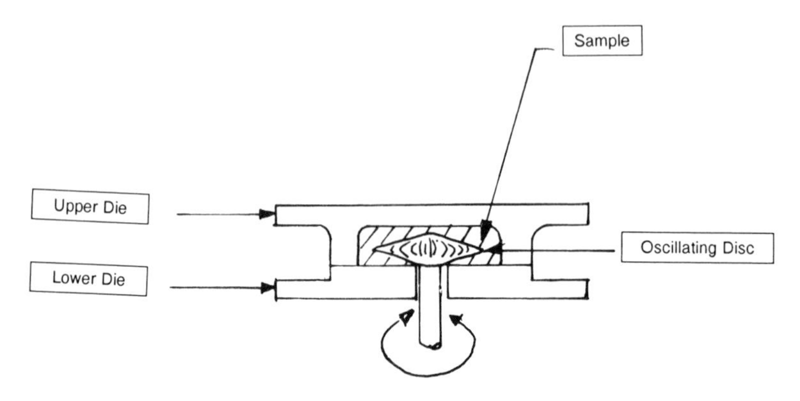

Figure 9.5 Oscillating disc rheometer (Monsanto). The force required to shear the rubber is monitored leading to a complete 'cure curve' of the compound.

rubber compound. Now ODR machines can either be bought or be retrofitted with software to establish the viscous/elastic response of the rubber during the cure cycle. This latter development was as a respose to the newer, more informative, cure traces produced by the 'rotorless' curemeters. Typical ODR machines are currently produced by Monsanto, Zwick, and Barco, whilst the Gibtre Rheo Check and Negretti MV/ODC systems can also be integrated to function as a Mooney shearing disc viscometer as well as an oscillating disc curemeter. This flexibility is possible by the use of microprocessor control and a stepped motor drive with a novel design of gearbox offering variable speed, continuous rotation or oscillating motion.

9.6.4 *Rotorless curemeters*

The higher curing temperatures required by processors, especially those using injection-moulding and continuous-vulcanisation systems, has placed greater demands on the curemeter. Faster quality-control tests are required and more fundamental cure-rate information is also necessary for process computer modelling (e.g. 'Fillcalc' and 'Mouldflow' CAD systems for injection moulding). The newer rotorless instruments have faster temperature recovery, due to smaller samples with thinner sections, and as a result show faster cure traces. Frictional losses associated with rotors has been eliminated and more stable and 'isothermal' boundary conditions results in more reliable information. These machines also allow measurement of loss modulus and angle throughout the test, thus providing further valuable processability information. Several machines are currently on the market complying with ISO/6502 for rotorless curemeters. All machines use directly heated dies and microprocessor temperature control. The lower die oscillates at 1.7 Hz and the reaction torque is measured at the upper die. The sealed, rotorless moving die system improves the capability to detect compound differences and minimises sample slippage at high test strains. As with most equipment, the use of microprocessors has increased both the quantity and the accessibility of information available from these instruments. The reaction torque systems employed provide the typical ODR-type cure trace (torque or elastic modulus versus time, S') and can also produce the viscous modulus (S'') and loss angle (tan δ) versus time curves (Figure 9.6). This provides valuable data of relevence to process behaviour, such as extrusion and calender die swell, rubber 'nerve' and processability, and also to cured hysteresis behaviour.

The two machines most widely used in the UK are the Wallace precision cure analyser (PCA) and the Monsanto moving die Rheometer (MDR) (Figure 9.7). The PCA has a 'top-hat' shaped sample of 1.4 cm^3 volume. Heat-up to within 0.3°C of the set temperature is achieved within 10 s of the platen's closing. The microprocessor system continuously monitors torque, displacement frequency, temperature, test time, press force, test number and date. Data output can be direct to a printer or via a microcomputer. Operating range is

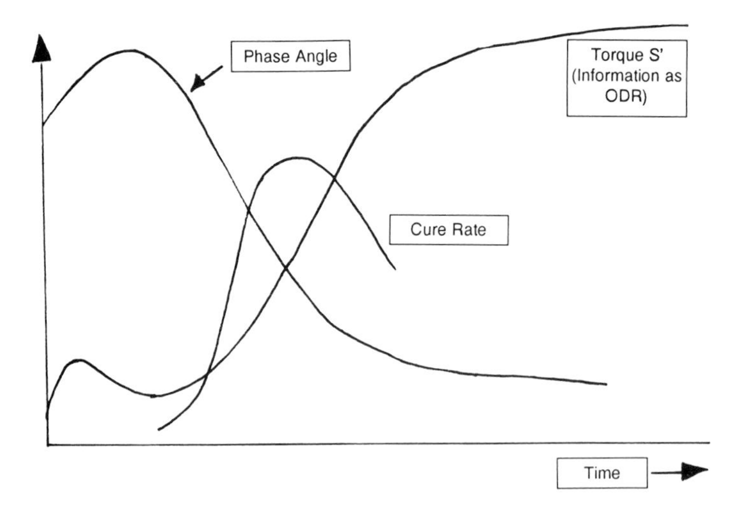

Figure 9.6 Typical information from a modern rubber curemeter.

up to 300°C. The MDR has a graded disc sample with heat-up times of less than 15 s, oscillatory arcs of ± 1, 3 or 5° similar outputs and a temperature range of 100–300°C. As with ODR machines, quality-control Gating and SPC software can be obtained for MDR machines.

The Gottfert Elastograph uses a 4.2 cm^3 radially grooved round sample, an oscillation frequency of 50 cpm, a range of angular amplitudes, and temperature control to 230°C.

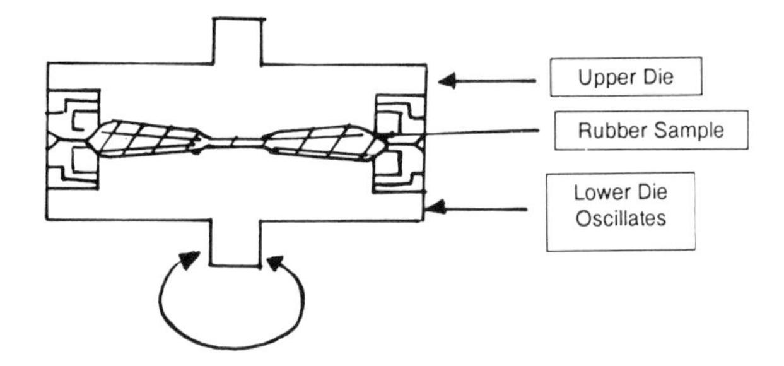

Figure 9.7 Moving die rheometer (Monsanto). A pressurised rotorless curemeter for rubber.

9.6.5 Variable-torque rheometers

Rotational-type rheometers have an advantage over extrusion types when a longer time-scale study of polymer shearing is required. This is used to advantage in the study of viscosity change by crosslinking, gelation or degradation of polymers. They find use in establishing the 'processability' of thermoplastics, thermosets, elastomers and many other plastifiable materials. The instruments consist of a horizontally mounted heavy-duty motor drive fitted with a torque sensor. Most machines are fitted with a temperature-controlled mixing chamber allowing a trace of torque versus time (a 'plastogram') to be produced. The mixing chamber is also fitted with a thermocouple to monitor the material's temperature as the cycle proceeds. Various blades can be used to simulate industrial mixer designs (cam, roller, 'Banbury', etc.). The mixer set-up with, typically, a 50 cm^3 chamber is popular in monitoring PVC compound quality and can also be used for producing viscosity/scorch data for rubber and thermosets. Its usefulness for the latter may be due to its wide acceptance of different forms of materials and its suitability for use as a miniature mixer. It can also be used to study the thermal/shear breakdown behaviour of polymers, again notably of importance in PVC and rubber processing. As a rotational viscometer, it is criticised as generating neither uniform nor controllable flow, non-constant strain history, non-isothermal conditions due to thick layers of polymer, no account of elasticity and inseparable viscous heating effects. These may prove problematic in establishing fundamental rheological information but are typical of process situations, and so the machine has wide usage in quality-control and research and development. A major advantage of the machine is the modular design, which allows the attachment of a variety of mixing and other heads. The latter include a range of measuring extruders of both single- and twin-screw designs. Many die types can be fitted to these, including ribbon, round, tubing, wirecoating, and filmblowing. A Garvey die (used to identify extrusion or injection mouldability of rubbers) and a capillary viscometer (allowing fundamental apparent viscosity versus shear rate data to be produced) can also be fitted. Slit capillary dies [4] with several temperature and pressure sensors fitted along the die length can monitor pressure-drop values. These can be converted to corresponding shear-stress values. Rod dies can be used as an alternative, allowing high shear rates typical of processing to be studied. By the use of rod dies of the same diameter, but with different length-to-diameter ratios, die entrance effects can be eliminated. Pressure drop is calculated from pressure before and after the die, and shear rate is determined automatically from the rate of melt throughput.

Swelltest die heads also allow contactless estimation of die-swell to be obtained. A "Planetarimixing" head also allows the study of powdery materials (again useful in PVC processing). Temperatures up to 300°C can be studied. Current equipment of this type is produced by Brabender ('Plasticorder'), Haake ('Rheocord') and Hampden Instruments (variable torque rheometer).

9.6.6 Melt flow index (ASTM D 1238, BS2782 Method 720A, ISO R113)

Melt flow index (MFI) or rate (MFR) remains one of the most popular quality-control tests for incoming raw thermoplastic materials [5].

The test consists of a weight-loaded piston falling under the action of gravity to force a heat-softened polymer through a rod die of specified dimensions (Figure 9.8).

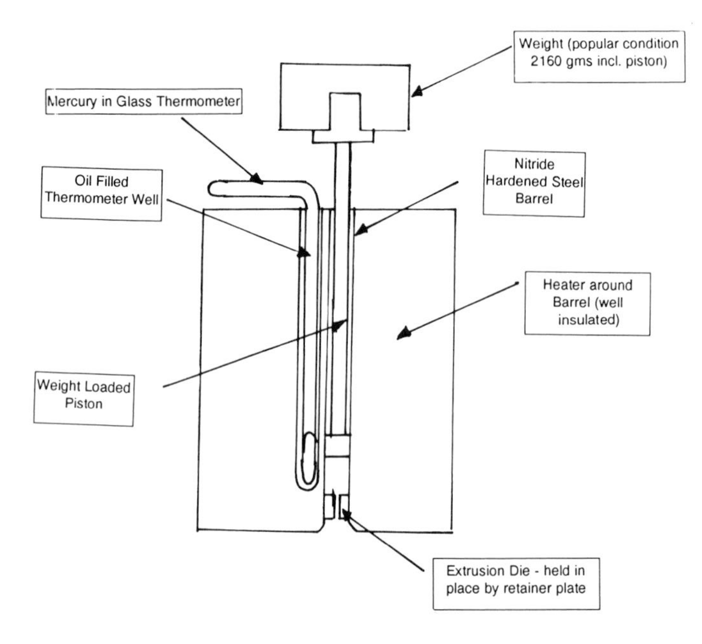

Figure 9.8 Melt flow index/rate apparatus (BS 2782).

The shear rate during the test depends on the polymer and the conditions used, polymers of lower molecular weight tending to flow through the die faster than higher-molecular-mass material and hence experiencing a higher shear rate. Shear rates never appear to exceed 10 s^{-1} and so no prediction of processability can be assumed, due to the pseudoplastic shear thinning of polymer melts at the higher shear rates typical of processing. Conditions for LDPE generally consist of 190°C with a 2.16 kg weight. Three samples are cut at constant time intervals from the extruding melt. Reference marks are used on the piston to improve reproducibility. The pore-free samples are weighed and averaged, and then the MFR is reported as the output in 10 min. Results are reported as 'MFR (190. 21.2) = 2.3', where 190°C is the test temperature, 21.2 is the load on the piston in newtons and 2.3 is the resultant MFR.

The reasons for the continuing use of melt flow indexers are low cost (about 10% the price of a capillary rheometer), simplicity and speed of use. Improved PID heating controls and more precise linear barrel-bore machined from a nitride-hardened steel or specially engineered ceramic has led, with recent models, to more reproducible results. Melt indexers can be obtained in a range from basic utility models with manual cut-offs and operation to more sophisticated (and expensive) semi-automatic machines. These may include motor-driven piston, measurement of piston displacement, automatic cut-off of samples, automatic calculation of MFR from piston displacement, and variable melt density and temperature control from 100 to 400°C. Furthur options offered include automatic weight shifters that can enable the MFI to obtain data previously obtained only by capillary rheometry: flow rates through a restricted-bore die are measured at different applied loads.

9.6.7 Capillary rheometers

Molten polymers are viscoelastic materials, and so study of their behaviour can be complex. Polymers are also non-ideal in behaviour, i.e. they do not follow the Newtonian liquid relationship of simple liquids like water, where shear-stress is proportional to shear strain rate. Unlike Newtonian liquids, polymers show viscosity changes with shear rate, mainly in a pseudoplastic manner. As shear rate increases there is a reduction in melt viscosity. This is true of both heat-softened plastics and rubbers. Other time-dependent effects will also arise with polymer compounds to complicate the rheological process behaviour. These may be viscosity reductions due to molecular-mass breakdown or physical effects due to thixotropic behaviour, or viscosity increases due to crosslinking/branching reactions or degradation. Generally these effects will be studied in rotational-type rheometers and the extrusion-type capillary rheometer.

Since common processing techniques for polymers involve shear rates of about 100–100 000 s^{-1}, there is no substitute for the comprehensive study of shear-stress versus shear rate over the typical processing windows of shear rate and temperature. Clearly, one-point tests such as melt flow index cannot be used as a guide to processability since the shapes of pseudoplastic curves are not identical (Figure 9.9).

Capillary rheometry consists of forcing a polymer material from a heated barrel through a capillary die of known size and shape. The rheometer can be an attachment to a vertical tensometer machine, or more likely a dedicated machine using a ram or screw to shear the material at a range of rates. Screw extruders fitted with slit or rod dies can be used, but more commonly plunger/ram-type machines are the norm.

Ram-extrusion types may be driven by screw or servo hydraulic systems. The material [6] is packed into the heated barrel and held at the chosen temperature to preheat fully. After about 30 min, the material is extruded at low ram speed and the pressure at die entry is measured. Once a consistent value has been obtained, the ram speed is increased and pressure is re-measured. This procedure is continued in a series of steps, either manually or automatically. Thus, a series of pressure readings against increasing speeds is obtained (say 0.2–200 mm min^{-1}) for a single loading of the barrel. Samples

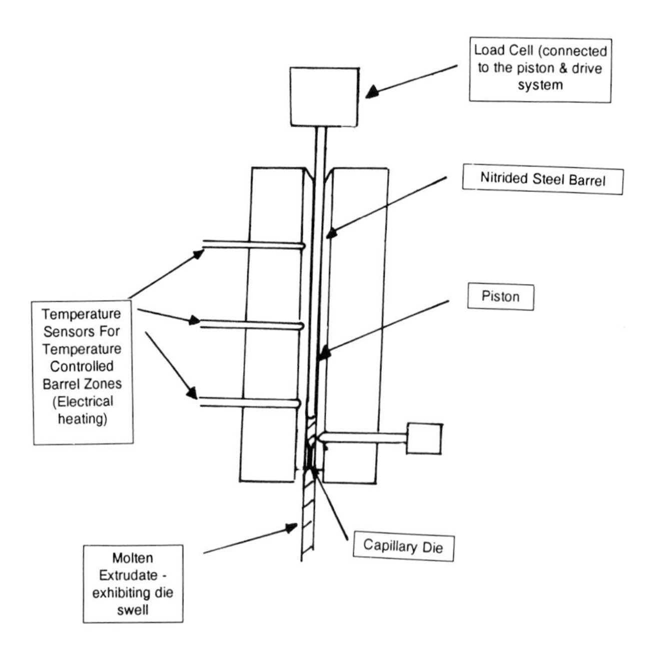

Figure 9.9 Schematic diagram of a capillary rheometer.

of the extrudate can be collected at each ram speed for die-swell measurement and examined for melt fracture. The results obtained are displayed as a shear stress/shear rate curve, where shear stress is obtained from the die entry pressure and shear rate usually from the plunger speed:

$$\text{Shear stress} = PR/2L$$

where P = die entry pressure ($\mathrm{N\,m^{-2}}$), R = die radius (mm), and L = die length (mm).

$$\text{Shear rate} = 4Q/\pi R^2$$

where Q = volume flow rate ($\mathrm{mm^3\,s^{-1}}$).

More accurate measurements can be made by taking 'end effects' into account by using a long and short die. This can be done by carrying out separate runs on the material on a single-bore single-barrel machine, or by the use of a twin-bore single-barrel machine. The latter enables long and short die tests to be carried out simultaneously or gives twice the throughput of samples. Twin-bore double-barrel machines, with each barrel equipped with a separate drive, allow testing of one barrel whilst material is cleaned out or preheated in the second, allowing improvements in operator productivity.

Automatic control and sequencing on advanced machines can be preselected via a keyboard to give much faster testing, use less material, need less operator skill, give greater accuracy and reproducibility, allow computation of derived data (including Bagley and Rabinowitsch corrections), provide presentation of reports and complete data storage and retrieval. Newer machines allow ramped or preselected speed inputs, constant shear-stress extrusion (previously only constant shear rate), set shear rates (previously only piston speeds), high-speed data capture (impossible by eye or recorder), and automatic relaxation experiments during the run.

Automatic running die-swell measurements can be made by using laser scanning. This can be carried out in a temperature-controlled chamber at the die exit for isothermal die-swell. Together with stress-relaxation experiments (by stopping the piston descent), this can provide information on the visco-elastic nature of the material. High die-swell and long relaxation times would indicate a more elastic material. Many rheometers also include automatic MFR calculations on thermoplastics.

Manufacturers of capillary rheometers also make available haul-off and melt-tensioning measurement in a temperature-controlled environmental chamber to study the melt strength and behaviour of the polymer. This is of great relevence to many film- and fibre-forming processes as well as in extrusion, blow-moulding and other forming methods. Other ancillaries available include non-contacting extrudate measuring to detect frictional heating effects due to shearing through the extruder die.

Another current development is the use of in-line capillary rheometers. Two main types available are those that can be used by polymer or compound producers and those for use by processors such as in extrusion of polymers (Figure 9.10). Granular or powder-fed types can automatically sample materials separated from the main production stream. Response time can be very similar to that of an extruder-mounted, melt-fed rheometer, and unlike an

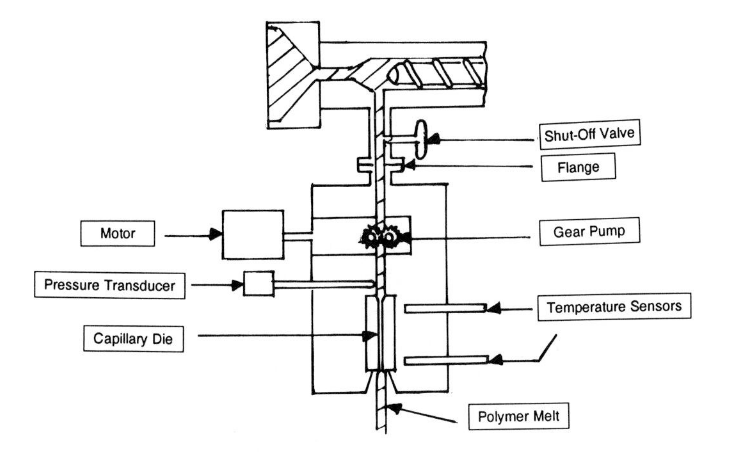

Figure 9.10 Bypass capillary rheometer, automatically monitoring polymer rheology during processing (Gottfert).

extruder system the remelting system used causes only small changes in polymer properties. (Screw extruder types can lead to further mixing and degradation effects before rheological measurements are made.) The on-line rheometer offers frequent measurement of viscosity, MFI or viscoelasticity. This is of great importance for feedback control of computerised material-production plant. The melt-fed on-line rheometer again can be included in feedback control of (say) extrusion equipment. Under processing conditions polymers can undergo rapid changes in molecular structure. On-line rheometers offer MFI determinations between 0.1 and 200, and emulation of standard ASTM loads at process or standard temperatures. The viscosity measurement range is 10^2–10^7 poise. Calculations are automatic and the results can be fed back to the processing equipment by a closed loop.

9.6.8 More recent developments in rheology and cure determination

The Strathclyde rheometer was originally developed for monitoring the cure behaviour of polymer systems, but has since found application as a rheometer in its own right. It has now been used to test materials from polyethylene and polyester to raspberry jam. The instrument uses a probe connected to a spring of known force constant, which is driven at a constant amplitude and frequency by a linear motor. As it reciprocates in the sample its position is continuously monitored by a transducer, which detects changes in amplitude resulting from changes in viscosity. Viscoelasticity is detected by phase differences between the motion of the motor and that of the probe. True rheological data is produced within the temperature range of 30–300°C. The instrument uses a disposable probe and sample container and thus is particularly suited to sticky, liquid-based systems. It can also follow the curing process to completion as well as tracking changes in real and imaginary viscosities of materials as they pass from the liquid to the solid phase by chemical reaction or by heat. All in all, the true versatility and extent of use of this relatively simple and inexpensive piece of equipment has yet to be established.

RAPRA also supply a vibrating needle curemeter (VNC) for use with liquid resin systems such as unsaturated polyester, epoxy and polyurethane systems [7].

The TMS rheometer from Negretti is effectively a modification of the Mooney viscometer principle for use with elastomeric materials. The elastomer is introduced into a cavity containing a shearing disc by injection, and measurements of torque under fixed or variable conditions of temperature and rotor speed can be made. Effective simulation of a variety of process conditions can be made. The instrument has found particular use in the investigation of rubber-to-metal adhesion behaviour, a problem of importance to mould fouling and extrusion processing.

Another technique currently used to monitor the progress in cure of liquid or dough-based polymer compounds (especially polyester and epoxy systems) is dielectric measurement. This has been used successfully for some years to monitor cure of DMC (dough moulding compound) by subjecting a compression-moulding plaque to a periodic electric field during the moulding and cure cycle. This technique has been refined to give the DEA (dielectric analyser), an instrument to put alongside DTA, DSC, TGA and TMA in providing information on polymer chemistry, rheology and chemical mobility of a material. In this case, only a few milligrams of sample in solid, paste or liquid form are required. The dielectric sensors contact the sample and are disposable. A sinusoidal voltage is applied via the electrodes/sensors, and sample response is measured as a function of time, temperature and frequency. Different sensor types are available for different sample forms. The DEA is compatible with other thermal analysis techniques [8].

9.7 Mechanical properties

Mechanical or physical testing of polymer materials is carried out to obtain numerical values for mechanical properties, in order to identify or classify materials. These tests are also often used to assess the ageing or chemical resistance of materials, but are of limited use in predicting product performance.

Any testing carried out should give consideration to the following points:

Polymers are viscoelastic materials and properties will depend on temperature, humidity, speed or timescale of test and history of samples.

Information obtained by testing is limited and no accurate prediction of performance can be made without extensive product testing.

Variability will be found in the test results due to variation in the sample material, sample preparation, test procedure and test machine accuracy.

Test procedures should always be designed to minimise these sources of error.

Change in the tensile properties of a material are also useful indicators of the degradation of a material, and this is used to monitor ageing and chemical resistance of polymer materials. Tensile properties are also often obtained as indicators for quality-control purposes [9].

Testing usually consists of securing a standard test sample between two sets of grips. One set of grips is fixed and the other is attached to a moving crosshead and load-cell arrangement. The sample used is usually of a dumb-bell shape so that breaking occurs in the central area away from the grips region. Crosshead speed depends on the nature of the polymer material. Extensible materials such as elastomers are tested at 500 mm min^{-1} crosshead speed, while rigid materials such as fibre-reinforced composites may use speeds as low as 1 mm min^{-1}. The general aim is to use a speed that allows all polymers to be tested in the same timescale. Standards allow a choice of speeds to achieve this [10]. It should be noted that test speed will influence

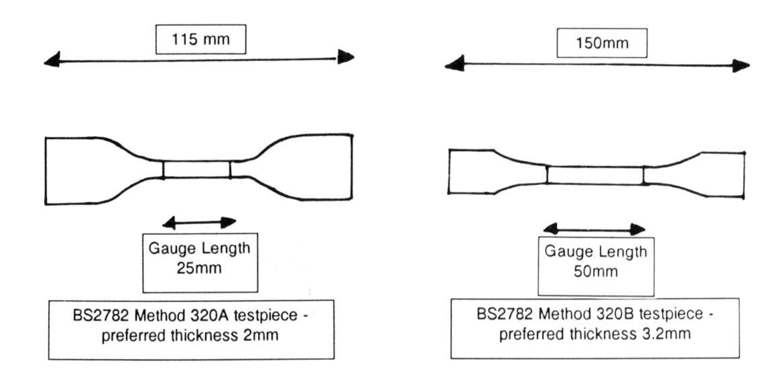

Figure 9.11 Typical tensile test pieces.

tensile behaviour dramatically, slow speeds on extensible materials being nearer to drawing and fast speeds on rigid materials nearer to impact.

Machines measure the force necessary to elongate and break the specimen and also usually determine the accompanying elongation. The latter can be achieved by measurement of crosshead movement in the case of a straight sample, but, due to excessive stress in the grip areas and resultant premature failure, dumb-bells are preferred (Figure 9.11). With dumb-bell samples, elongation is determined either by manually monitoring the separation of two gauge marks (20, 25 or 50 mm apart) or by the use of clip-on or non-contacting extensometers. The latter automatically gives a signal for display via recorder or computer VDU or printer. Force values are converted to stress values by dividing by sample cross-section area, whilst elongation is given by the extension in gauge length divided by the original gauge length. Elongation is given as a percentage figure, whilst strain is shown as a fraction. In so doing, the results in theory are independent of sample size, but in practice consistent standard sample size is necessary for reproducible results.

$$\text{Tensile stress (MPa)} = F/A$$

where F = force (Newtons) required to stretch the test piece, and A = cross-section area of test piece (mm^2).

$$\text{Tensile strain} = \text{Change in length/original length} = (l_1 - l_0)/l_0$$

where l_1 = length between gauge marks (mm), and l_0 = original gauge length (mm).

$$\text{\% Strain or elongation} = (l_1 - l_0)/l_0 \times 100\%$$

$$\text{Modulus} = \text{Stress/strain}$$

$$\text{Young's modulus} = \text{Stress/strain in linear portion of stress–strain curve}$$

The elastic modulus ('modulus of elasticity', 'Young's modulus' or 'tensile modulus') is the ratio of the applied stress to the strain it produces in the region where strain is proportional to stress, i.e. in the initial straight-line portion of the stress–strain curve. Modulus is primarily a measure of stiffness, and plastic parts should be designed such that service behaviour normally falls in this linear region.

In applications where rubbery elasticity is necessary, high ultimate elongations may be desirable. At these higher elongations stress–strain behaviour is non-linear, yet some estimation of material stiffness is required. To accommodate this rubber engineers have traditionally used 'rubber modulus' as a guide. 'Rubber modulus' is not a true modulus value, but is measured as a stress value at a given elongation; e.g. M100 represents that tensile stress (N mm^{-2}) obtained at 100% elongation in the sample. M100, M200, M300 are popular figures to quote (Figure 9.12). Rubber articles are rarely subject to extensions of above 100% in service, nor are they subjected to high tensile stress, yet despite this, high tensile strength and elongations at break are often specified. Some relaxation of these often unnecessary requirements has been required to allow the use of more modern materials such as grades of thermoplastic elastomers. Other, more product-related, tests are often more suitable.

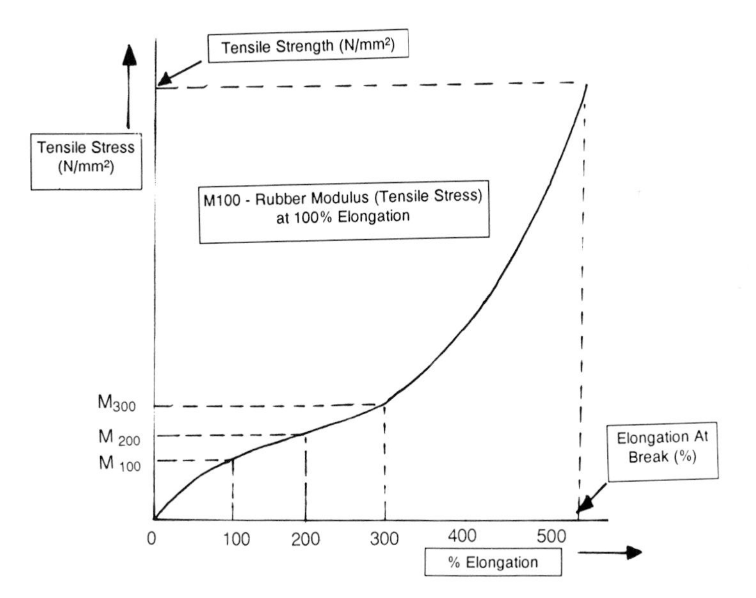

Figure 9.12 Typical stress/elongation curve for a rubber compound in tension.

Another consideration with highly extensible materials is that calculated tensile stress is based on the original cross-section area measurement, and true stress values will be considerably higher due to thinning of the sample. This must be considered when making comparisons against more rigid materials. In plastic materials moderate elongation is of use to permit absorption of rapid impact and shock. The area under the complete stress–strain curve will be a guide to overall toughness of the material. High tensile strength and low elongation at break tend to show a material that would be brittle in service.

Test pieces used are usually of a flat dumb-bell shape, often produced from specially moulded sheet or cut or machined from products. Dumb-bell shape varies dependent on material type and test standards used, so reference must be made to the relevent standards (e.g. ISO R527, BS 2782 Method 320 (1976), ASTM D638 (1977)) (Figure 9.11). Sample preparation can lead to significant errors in tensile properties, so care must be taken to use sharp sample cutters and standard sample thicknesses (using samples preferably 2–4 mm thick in most cases). Buffed or machined samples can give low tensile values and damage should be avoided in ejecting samples from cutters. Orientation and anisotropy can be significant features of polymer products, so testing may be carried out in two mutually perpendicular directions to assess these effects.

9.7.1 Tensile test equipment

Most tensile test machines are vertical machines using electronic load cells to measure force values. These have replaced the earlier pendulum-type machines, thus eliminating the frictional and inertia errors associated with these. They operate at a constant rate of traverse (CRT), i.e. constant cross-head speed, and due to their versatility are often referred to as 'universal' testing machines (Figure 9.13). An ideal machine would consist of an infinitely rigid load frame and load cell and a massless test fixture. This is clearly impossible, but equipment should be chosen so as to minimise any of the resultant errors by non-compliance. Measurement and control at low cross-head speeds is always more difficult than at higher speeds. Suitable grips need to be selected that are as light as possible yet grip without allowing the sample to slip. Grips for tensile applications include wedge, eccentric roller, Gavin and pneumatic grips. Wedge-type grips are often the only type capable of preventing slippage with fibre-reinforced composites. The friction between the test specimen and the gripping faces of the wedge can be obtained by having different surface patterns on the wedge faces or by the use of emery paper to reduce slippage. Lighter, self-tightening roller and Gavin-type grips are preferred with elastomers. Pneumatic grips give a positive reproducible gripping pressure from a few kilopascals to tens of megapascals, and reduce operator-induced errors. They are also popular with elastomers and flexible thermoplastics. Many other types of grips are available for film, textiles, laminates, etc., and are described by Brown [5] and Whelan [14]. Whichever grip system is used, the grips must ensure alignment in the direction of strain and not allow any bending, shearing or twisting of the sample.

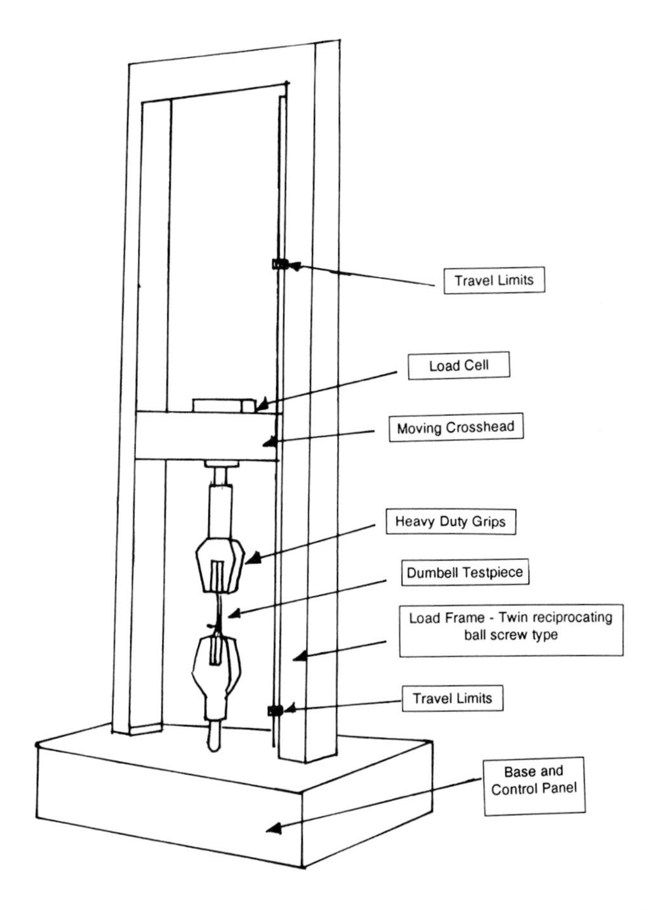

Figure 9.13 Universal tester load frame, set up for tensile testing of rigid plastics.

Load frames for tensile testing are available in a range from less than 1 kN to over 300 kN. Environmental chambers that fit around the sample can be used to study low- or high-temperature behaviour, and are typically available to control temperatures in the range – 70–250°C.

9.7.2 Extensometers for use in tensile testing

In most cases simple determination of tensile strength is not sufficient. Elongation at break and often full stress–strain behaviour analysis is required. Elongation of the sample must be measured. Many problems are associated with the accurate measurement of extension of polymers [11]. There is a wide range of extensions to be measured, from less than 1% for stiff materials like GP polystyrene to over 1000% for certain elastomer systems. There is a significant reduction in cross-section area in extending the polymer beyond the initial few percent, and specimens used do not have a constant cross-section area. As stated earlier, the use of dumb-bell samples means that any measurement of elongation or strain must be carried out on the parallel-waisted section of the dumb-bell. This usually involves marking two parallel gauge lines perpendicular to the sample length in this area. The separation of these marks is then used to calculate elongation (the original length being taken as the initial distance between the gauge marks).

9.7.3 Extensometry techniques

The simplest method of estimating elongation is by the use of a hand-held scale [11]. The scale is held alongside the gauge lines marked on the test specimen and elongations are noted as extension of the sample proceeds. This method is particularly inaccurate at the higher test speeds used for rubbers and flexible thermoplastics (500 mm min^{-1}). As the sample extends both gauge marks move, and it is only possible to view one mark at any time, so errors are inevitable.

Cross-head movement or grip separation is sometimes taken as a measure of sample extension. This is clearly only true of a parallel strip sample having the same cross-section throughout its length. Such samples inevitably break in the grips area due to the pressure imposed by the grips [11]. For some time these were the only methods available to most testers and they did enable comparisons of materials to be made, if in a somewhat coarser manner than possible today.

Clip-on (contacting) extensometers consist of two grips that are clamped to the narrow section of the dumb-bell. They are initially set at a standard gauge length apart (typically 20, 25 or 50 mm). As the extensometer grips separate, a generated electrical signal corresponds to the extension of the dumb-bell. Lightweight, counterbalanced extensometers are required and these must give minimum distortion and minimum additional stress to the test piece. The extensometer grips must always accurately follow the test piece with no slippage, yet not exert too large a gripping load (breakage must occur between the extensometer grips). Due to the precision required at low elongations with rigid plastics and the high extensions necessary with elastomers, more than one extensometer will be required to cover the full range of polymer materials. To this end most manufacturers supply a low-elongation extensometer capable of readings of 0–80% with ± 1% accuracy (usual gauge length 50 mm) and a high-elongation extensometer to give readings of 0–2000% based on gauge lengths of 20 and 25 mm to an accuracy of better than ± 1%. Clip-on extensometers are also available that measure transverse dimensional changes (i.e. sample width or thickness) as well as combined units that measure both axial and transverse strains. These are useful in determination of Poisson's ratio and also in the shear testing of composite and engineering plastics.

9.7.4 Non-contacting extensometers

Non-contacting optical-follower extensometers are widely available from a variety of manufacturers, and free the user from the slippage/gripping problems of contacting types. The two optical heads are locked on to two, usually white, painted marks or stickers placed on the dumb-bell at the gauge length. As the sample extends the gauge marks separate and an electrical signal is generated. Extensometers are available which use visible light, pulsed infrared (Figure 9.14) or lasers as the detecting beams. Whichever system is used, there must be good contact between the gauge marks and the specimen. Any paint or adhesive used must not have any deleterious effect on the sample. The optical heads must be able to follow the marks in normal illumination and in an environmental cabinet if at all possible. The radiation incident on the sample must not cause any deterioration or raise its temperature [11].

Visible-light systems were amongst the first systems used in non-contacting extensometers and were initially considered to be the 'Rolls Royce' system. However, the infrared system is less prone to problems caused by incident light on the sample, and by using very short infrared pulses (less than 5 μs) negligible heating occurs in the sample. Initial problems with environmental cabinets have been overcome, by detection through a glass door and the use of internal finishes that are non-reflective to infrared. Laser systems typically use a beam that scans the sample some 300 times per second. The beam can scan through two layers of glass and so can be used for both cryogenic and elevated-temperature testing [12].

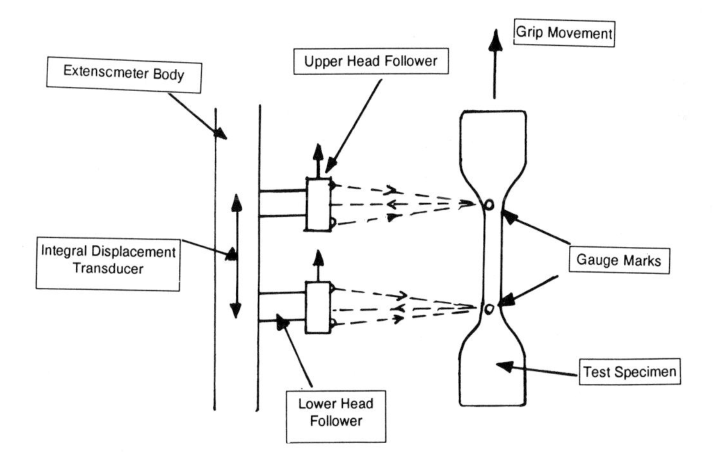

Figure 9.14 Schematic view of an infrared non-contacting extensometer.

Most non contacting type extensometers are stand-alone units and can be used with other manufacturers' tensometer equipment if specified at purchase.

Often high-extension and low-extension models are offered to give the degree of accuracy required. Horizontal versions are also available.

9.7.5 Definition of terms used in tensile testing [5,13]

Tensile stress (nominal)	The tensile force per unit area of the original cross-section within the gauge length carried by the test piece at any given moment. Units are MPa = MN m^{-2} = N mm^{-2}.
Tensile strength (nominal)	The maximum tensile stress (nominal) sustained by a test piece during a tension test.
Tensile stress at break	The tensile stress that occurs at break of the test piece.
Yield stress	The tensile stress at which occurs the first marked inflection of the stress–strain curve, where an increase in strain occurs without any increase in stress.
Gauge length	The original length between two marks on the test piece over which the change in length is determined.
Strain	The change in length per unit original length of the measured gauge length of the test specimen. It is expressed as a dimensionless ratio.
Percentage elongation	The strain produced in the test piece by a tensile stress, expressed as a percentage of the gauge length.
Percentage elongation at break, or at maximum load	The elongation at break, or at maximum load, produced in the gauge length of the test piece, expressed as a percentage of the gauge length.
Proportional limit	The greatest stress which a material is capable of supporting without any deviation from the proportionality of stress to strain (Hooke's Law).
Elastic modulus in tension (Young's modulus)	The ratio of the tensile stress to the corresponding strain below the proportional limit. The stress–strain relationship of many plastics does not conform to Hooke's Law throughout the elastic range but deviates therefrom even at stresses well below the yield stress. For such materials the slope of the tangent to the stress–strain curve at a low strain is usually taken as the elastic modulus.
Secant modulus	In general, the ratio of stress to strain at any given point on the stress–strain curve.
Rubber modulus	The tensile stress at a given tensile strain, e.g. Modulus 100%, 200%, 300%.

9.7.6 Modern tensile systems and software

Many systems now available can be driven by a computer keyboard and have full microprocessor data-acquisition systems. The most popular systems are IBM PC-compatible. (IBM is a registered trade mark of International Business Machines.) Standard features include keyboard control of machine functions, load cells for use in both tension or compression, displacement by extensometer, cross-head or time, overload protection, automatic zeroing of force and displacement, and much more. The use of microcomputers in control and data handling has permitted the use of real-time graphics, automatic force ranging and curve scaling, automatic calculation of results, a variety of display modes (e.g. force/displacement, force/time, displacement/time), expansion of result curves, and the use of saved operating or data-handling programs.

Results can be printed or saved to disk or other medium, with the form and information produced limited only by the software in use. Typical information obtained can include: Young's modulus, proof stress, yield point, area under curve, average load, load at a given deflection, deflection at a given load, secant modulus, median of peak heights, peak load after initial peak, flexure modulus, n value, k factor, true stress and true strain. Printing of standard test reports and statistical analysis are also possible.

Cross-head control can often be programmed on more advanced machines to cycle against load or displacement. Stress or strain rate may also be controlled, and longer-term creep testing may be carried out at set stress levels.

As with other areas of testing, it has been the introduction of inexpensive microcomputer control and data-acquisition/display systems that has improved both the precision and the range of information available to the polymer engineer.

Vertical tensometers are available in a variety of frame sizes, with available width between screws ranging from around 150 mm on 1 kN machines to over 500 mm on larger machines (twin-screw types). Before a machine is chosen, consideration must be given to attachments and extras required. Maximum traverse distances of around 1 m allow sufficient travel for all but the most extensible materials. Suppliers will manufacture tailor-made equipment if required.

Modular systems are available that are particularly suited to quality-assurance applications, but in many respects, with the availability of inexpensive 'universal' systems and a range of load cells, dedicated software can be used to control quality-assurance testing without impairing the versatility of a machine. Load cells used in the UK should comply with BS1610 (1985) to grade 1.

9.8 Flexural tests

The stiffness of plastics has generally been low, and this has been a limitation to their use. This situation has changed considerably recently, with the expansion in use of advanced fibre-reinforced composite materials. The stiffness or rigidity of plastics when subject to bending or flex is of great and increasing importance with their use in engineering and structural applications. Consideration must also be given to the variation in stiffness of polymers with temperature. Excessive stiffening may occur at low temperatures, whilst creep or softening may occur at elevated temperatures.

Flexural properties are measured most often in plastics by using a three-point bending jig attached to a 'universal'-type tensometer. The test consists of measuring the stress developed at the surface of a prescribed test piece supported near each end and loaded at the centre (Figure 9.15). The test piece is deflected at a set rate (standard cross-head separation speed), dependent, as in tensile testing, on the type of plastic to be tested.

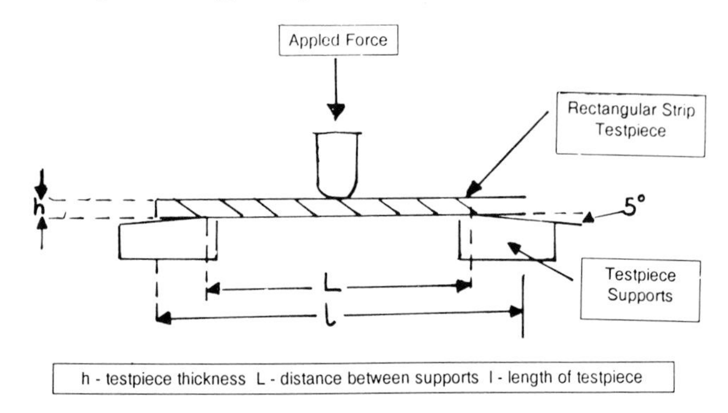

Figure 9.15 Three-point bending (flexural) test set-up (BS 2782).

Samples are usually rectangular bars of a length at least twenty times the thickness. Preferred dimensions for BS 2782 [15] samples are a width of 10 mm and a depth of 4 mm. Span distance is set to 15–17 times the thickness.

A load versus deflection curve can be produced by a recorder, and from this, or directly from machine load/movement readings, measurement of flexural strength (crossbreaking strength), breaking strain, flexural stress at a conventional deflection (often 1.5 times the thickness) and elastic modulus in flexure can be carried out. The nature of the material's failure will also indicate if its behaviour is brittle (low-strain failure of less than 0.05) or tough (sample bends appreciably or yields). Other relevant standards include ISO 178 (1975), ASTM D 790 (1971) and DIN 53452 (1977).

For a rectangular cross-section, flexural stress (σ_f) is calculated as:

$$\sigma_f = \frac{3}{2} FL/bh^2$$

and modulus of elasticity in flexure (E_b) as:

$$E_b = \frac{L^3}{4bh^3} \times \frac{\Delta F}{\Delta d}$$

where L is the distance between supports, b is the width of the test piece, F is the force applied, h is the thickness of the test piece, ΔF is the change in force in the linear portion of the load–deflection curve, and Δd is the change in deflection corresponding to ΔF.

9.9 Impact testing

Rate of deformation is of vital importance in determining the behavioural nature of a polymer. It has already been noted that high rates of deformation lead to a more brittle behaviour whilst low rates of deformation often allow ductile yielding. In many applications, plastics may be subject to sudden shock loadings, and some estimation of their effect must be made. Thus, a variety of impact tests have been derived, each of which studies material behaviour at one point on the general curve of strength properties as a function of speed of testing [5]. Test pieces can vary from a small 50 mm × 6 mm × 4 mm rectangular bar (BS 2782 Part 3 Method 359 (1984), ISO 179) to the front of a lorry cab. Methods standardised for materials testing usually have the advantage of ready measurement of the energy required to break an arbitrary-shaped test piece, information not easily calculated from tensile and flexural tests. This situation does not necessarily hold true in the light of recent developments, but the widespread acceptance of impact tests such as Charpy and Izod pendulums should ensure their use in material comparisons.

Impact tests fall into two groups:

(a) Instruments where a pendulum of known energy strikes a specimen of known size and shape, e.g. Charpy and Izod tests.
(b) Tests where darts, weights or other impactors are allowed to fall freely through a known height on to the sample. In this case, impact strength can be calculated from a combination of the minimum height and weight required to cause fracture of the sample.

9.9.1 Pendulum impact tests

Type (a) tests include the Izod (BS2782 Method 306A, ISO 180) [16, 22] and the Charpy (BS2782 Method 359, ISO 179) [17]. The Izod test uses a notched specimen clamped in a vice and struck by a swinging pendulum (Figure 9.16). The energy lost by the pendulum in breaking the specimen is registered by a moving pointer on a dial or electronically. Results of the test are often reported as energy per notch width ($J\,m^{-1}$) or energy per area of sample fractured ($J\,m^{-2}$). Despite the popularity of this test, specimens are totally unrepresentative of plastics parts (i.e. are very thick) and contain notches. Notch size and shape may differ with standards (as is also the case with Charpy) and the

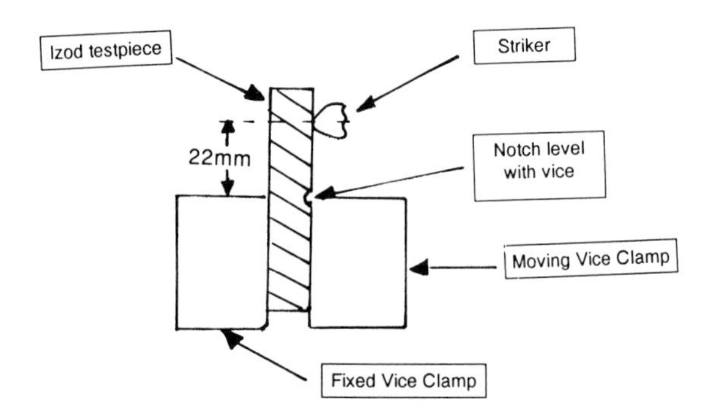

Figure 9.16 Izod impact test: arrangement of specimen. View from the front of the apparatus (BS 2782).

results are very notch-dependent. Specimen production and freedom from voids can also prove a problem.

The Charpy test involves resting a sample on two anvils and then striking the sample with a pendulum (Figure 9.17). Notched or un-notched specimens can be used, the notch acting as a stress concentrator. Unlike the Izod, the Charpy notch is placed on the opposite side of the sample to the impact. Izod is an example of a cantilever-beam action while Charpy is a supported beam type. A range of sample sizes are specified for the Charpy (BS 2782 specifies lengths of 50, 80, 120 and 125 mm) and the smaller sizes are easier to prepare than the Izod. Flexible materials will need to be notched to avoid bending and no-break situations, and with rigid materials Charpy has the added advantage of some study of the notch-sensitivity of materials (i.e. test both notched and un-notched).

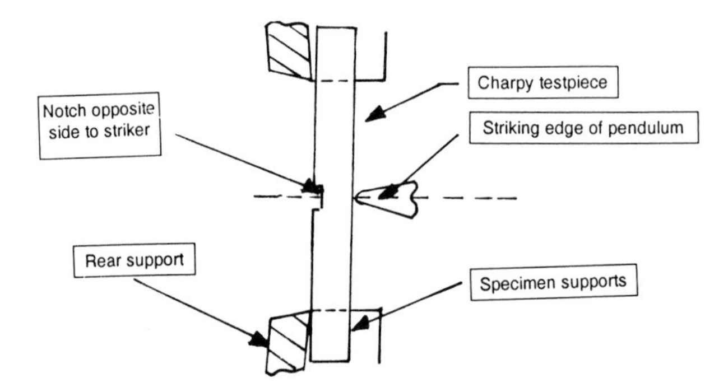

Figure 9.17 Charpy (supported-beam) pendulum-impact test. View from above the specimen.

Test specimens can be prepared by compression, injection moulding or other methods, but, as with other testing, the preparation method will influence test results.

For an un-notched test piece:

$$W_s = W/A$$

where W = measured impact energy (kJ), A = fracture cross section (m^2), and W_s = impact strength ($kJ\ m^{-2}$).

For a ñotched test piece:

$$W_k = W/A_k$$

where W_k = notched impact strength ($kJ\ m^{-2}$), and A_k = residual cross-section (m^2).

9.9.2 *Falling-weight impact tests*

Falling-weight tests allow a loaded indentor to fall a specified height on to a flat disc sample. In its simplest non-standardised form, a ballbearing can be allowed to fall on to a product surface and some comparative estimation of inpact resistance can be made. The falling dart method is a popular test with plastic film (BS2782 Method 352D) [18]. A dart with a hemispherical end is dropped from a known height on to the centre of a circular piece of film clamped by a vacuum. The mass of the dart can be adjusted by means of detachable weights which fit over the shaft of the dart. The weight which causes the fracture of 50% of the specimens is determined. When this is multiplied by the height from which the dart is dropped, the impact strength is obtained. A similar test for plastic disc samples (57–64 mm × 1.52 mm) is detailed in BS2782 Method 306B & C [19].

Falling weight impact strength tests (FWIS) are possibly the most useful type of impact test as they can often be carried out on products and require little specimen preparation. End-use service performance can be more easily simulated, especially with the more advanced instrumented models [20]. This form of testing is popular for film, sheet and pipe, besides general plastic products.

9.9.3 *Instrumented impact testing*

The most traditional methods to measure impact strength of polymers are Charpy and Izod. These are used for homogeneous materials and primarily as quality-control or material-selection indicators. Results, especially with composites, are difficult to relate to residual strength [21] and conflict in some cases with service experience [22]. Methods nearer to actual service experience have been devised, using FWIS testers where energy and mass ranges can be altered accordingly. This approach still fails to provide all the information needed to characterise the material. In addition, the process is tedious and labour-intensive, requiring repetitive testing to obtain a single result.

There is a great need, for specification and research and development purposes, to be able to assess material performance quickly. The main feature of instrumented impact testers is the ability to follow the entire history of contact between the indentor (whether pendulum or falling mass) and the specimen, thus showing all the variations in the mode of failure during the test. All testers of this type use a force transducer fitted to the impactor and produce in the simplest form a force versus contact time-curve (Figure 9.18). Instrumented testers are available that can carry out Charpy, Izod, falling-weight or dart-drop tests, in addition to quality-control/development tests on finished products such as extruded pipes and sections, film, sheet and moulded articles. Instruments may be of a modular design, such that by use of alternative anvils all forms of impact test can be carried out using the same test frame.

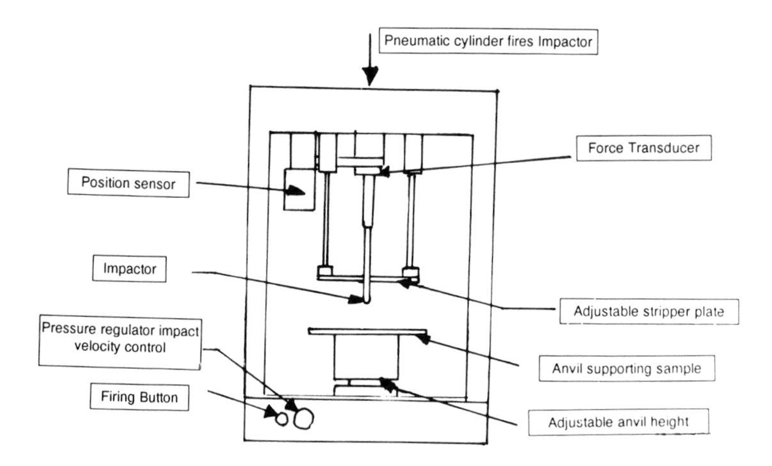

Figure 9.18 Instrumented dart-type impact tester. There are various types of impactors, anvils and force transducers.

The most popular form of instrumented impact tester is the falling-weight type, but the information gained is similar to that for the pendulum type. FWIS types use a striker that is sufficiently heavy to break the specimen easily and without decrease in speed [20]. A force transducer is mounted behind the 'nose' of the striker, and information obtained from this can be used to obtain not only the energy to break the specimen but also the the energy to yield the load and deformation at yield and at final failure. A full F–t curve is obtained that can indicate crack initiation and propagation, produce sample stiffness and indicate the mode of failure. The speed and/or displacement of the striker may be measured photo-electrically. The test can be carried out quickly with only a few specimens to study behaviour over a range of temperatures if necessary. If the speed of impact of the dart V_0 remains virtually unchanged during the test, Energy of material at impact (E_m) is:

$$E_m = V_0 \int F \, dt$$

where F is force [20,23].

If the dart speed changes then a correction is required to E_m:

$$E = E_m(1 - E_m/4E_0)$$

where E_0 is the kinetic energy of the dart at the moment of impact.

The deflection (W) of the specimen at the impact point can be taken as the length travelled by the striker and can be calculated via a double integration of the F–t curve. So a whole series of curves including W–t, F–W, velocity, time, etc., can be obtained, all allowing further study of material failure mechanisms. As a result the equipment can be of great use in fundamental material research in fracture analysis.

The use of microcomputers has allowed rapid computation of results from the signals generated, and these are processed into the desired form for display on VDU, printed or stored to disk. A detailed analysis of the curves produced can also provide interesting information about the material's dynamic behaviour [23]. From the example F–T curve, F_i can be assumed to be the force necessary to initiate the fracture, under dynamic conditions; up to this point the force acts only in an elastic fashion, with no visible damage on the specimen. The nature of the fracture initiating at the value F_i depends on the sample's material and geometry as well as the boundary conditions. From the point F_i onwards the curve represents the 'toughness' of the material and thus its ability to sustain further application of force under dynamic conditions. Brittle or tough behaviour will also be apparent. Further parameters can be calculated to give the initiation energy and the energy to propagate the fracture. The ratio of these two values can indicate the toughness number or ratio of the material. Clearly by variation in material composition/structure and/or test conditions a fuller characterisation of fracture analysis can be made.

Test equipment may consist of a free-falling striker (as with Ceast, Yarsley and Daventest equipment) or a driven impactor (as with the ICI high-speed pneumatic system).

9.10 Creep, stress relaxation and permanent set

The response of a polymer to an applied stress or strain depends on the rate or time period of loading. The behaviour is not wholly elastic but is time-dependent and consequently polymers are often referred to as viscoelastic materials. So at low temperatures and high strain rates they behave more like elastic solids, while at low strain rates and higher temperatures they behave more like viscous liquids. Thus, when a component is under continuous stress or strain in an application, longer-term testing than simple tensile or flexural tests will be required.

9.10.1 *Stress relaxation*

Stress relaxation refers to the decay in stress with time when a polymer sample or component is held at constant strain. Over a period of time the chains will move by a coiling or uncoiling mechanism to take up a more favourable arrangement. The rate of this stress decay or relaxation is of particular importance when designing products such as rubber sealing gaskets and O-rings. Testing can be carried out on tensile-type equipment fitted with a load-cell force measuring system (Figure 9.19).

Stress relaxation at time t, R_t, is given by:

$$R_t = (F_0 - F_t) \times 100/F_0$$

where F_0 is the initial force to cause the strain, and F_t is the force at time t to maintain the same strain.

Stress relaxation is rarely measured in plastics, creep testing being preferred. There are a number of standardised test methods for rubber as detailed by Brown [24].

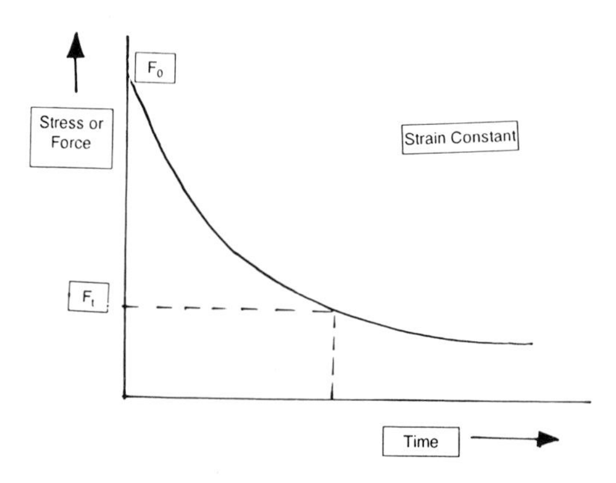

Figure 9.19 Typical stress relaxation curve in tension.

9.10.2 *Permanent set*

When a rubber or elastomeric component is under stress or strain for a period of time the entangled network of polymer chains can begin to undergo 'viscous' non-returnable flow. This 'permanent' flow or deformation is known as 'set'. The resistance to permanent set is important to many 'sealing' applications of rubber materials.

Permanent set tests can be carried out in tension by applying a tensile load for a time period and measuring the permanent change in gauge length after the load is removed. Care must be taken when setting up such tests, as temperatures, sample shape, time periods of loading and relaxation will clearly all affect results.

More commonly, compression-set tests are used on rubber-type materials. Rubber-seal type materials are used in compression because the test apparatus is simple and inexpensive to construct. Despite this popularity, stress-relaxation testing gives a more useful insight into the suitability of a material for seals [5].

Constant-stress tests use a heavy metal spring applied to a disc-shaped sample. The spring has a precalibrated stress–strain behaviour, and so, by means of a screw, various stress levels can be applied to the sample and maintained over a long period of time.

Constant-strain tests are more popular and simpler to use. In this case, sample discs of rubber are compressed between two polished steel plates to a predetermined strain by using spacers and bolts. The permanent set is usually quoted as the unrecovered percentage of the applied deformation, and is not related to the original sample height. The BS 903 rubber test specifies a cylindrical test piece 13 mm in diameter and 6.3 mm high. The sample dimensions are especially important in compression due to the 'shape factor' effect. Recovery time before measurement of final set value is usually 30 min., and must be specified as this creep recovery is time-dependent.

9.10.3 Creep

The term creep refers to the total time-dependent deformation of a material under stress [27]. Material manufacturers and product designers need to have a knowledge of the deformation that products will undergo in long-term service conditions. Even relatively small loads applied over a long period of time can lead to significant deformation of plastics. A storage tank may continue to distort long after it is filled with a liquid, and bottle crates no longer 'stack' after long-term use. Increasing engineering use of polymers means that data on the creep behaviour of a material are required, as has been the case for some time for applications like bridge bearings, gears, pipes and suspension systems. Creep data are important to both the plastics and the rubber engineer.

Creep can be largely recoverable with polymers, unlike metals. This will be particularly significant if the product is used intermittently, where the relative rates of creep and recovery will influence material suitability.

Creep tests are carried out by applying a weight or load to a polymer sample in a temperature-controlled environment. Most testing is carried out in tension, but compression, shear and flexure may be used if more applicable to service conditions. Creep testing can be carried out on some tensile testers, but, since data may need to be collected over periods of up to a year, dedicated equipment is normally used. Creep data will often be required over a wide range of conditions and test times. Five to ten different stress levels will be required to construct a 'family' of creep curves, and elevated-temperature testing may also be required. Log–log plots of strain versus time are created and extrapolated to give curves to the time period required [25].

Testing is usually carried out on a dumb-bell specimen which has a very long parallel-waisted section of approximately 80 mm. Accurate measurement of the small change in specimen strain is required over periods of months or years.

A weight or load is applied to the dumb-bell by way of a lever system [26]. The small extensions to be measured will require a very sensitive extensometer to be used. Extensometers based on the optical lever principle have been widely used but those based on Moiré fringes or on displacement transducers are more suitable for data aquisition by microcomputer. Over this long period of testing, stress must be maintained to an accuracy of better than 1% and no electrical or thermal drift must occur in the extensometer [26].

Temperature and humidity levels will be critical to test results and sample conditioning prior to testing should be carefully carried out (e.g. moisture pick-up and crystallisation in nylon on storage). Any small differences in test pieces (e.g. crystallinity, degree of crosslinking, moulded-in stress, orientation, etc.) can alter the creep behaviour of the material significantly [5].

Extension measurements are initially taken at short time intervals, but as the creep deformation decreases longer time intervals are chosen. Quality-control tests may only require one stress level, but more fundamental design data will mean testing at a series of creep loads.

BS4618 defines the following:

Creep strain The total strain, which is time-dependent, resulting from an applied stress or system of stresses.

Creep modulus The ratio of applied stress to creep strain.

Creep lateral contraction ratio The ratio of lateral strain to longitudinal strain measured simultaneously in a creep experiment (also known as Poisson's ratio).

Note that this definition does not subtract the instantaneous or elastic strain to obtain creep. This is due to the difficulty in separating the two components of the time-dependent strain required by the Kelvin or Voigt model of viscoelasticity.

Of further note is the fact that constant true stress levels are rarely achieved in creep tests as the cross-section of the sample decreases as the sample extends, thus increasing stress levels (Figure 9.20). Information from a family of creep curves can be rearranged to give an isometric stress–time curve. A line drawn at constant strain across the curves will intercept them at a number of stress–time combinations, and these are used to plot the curve. This information can be used to determine the maximum stress that can be accepted for a specified time.

Similarly, an isochronous stress–strain curve can be obtained by taking points from the family of curves that are intercepted by a line taken at a constant time, say 100 s [26]. This information can be used to show the effect of crystallinity or water absorption on creep properties of a material.

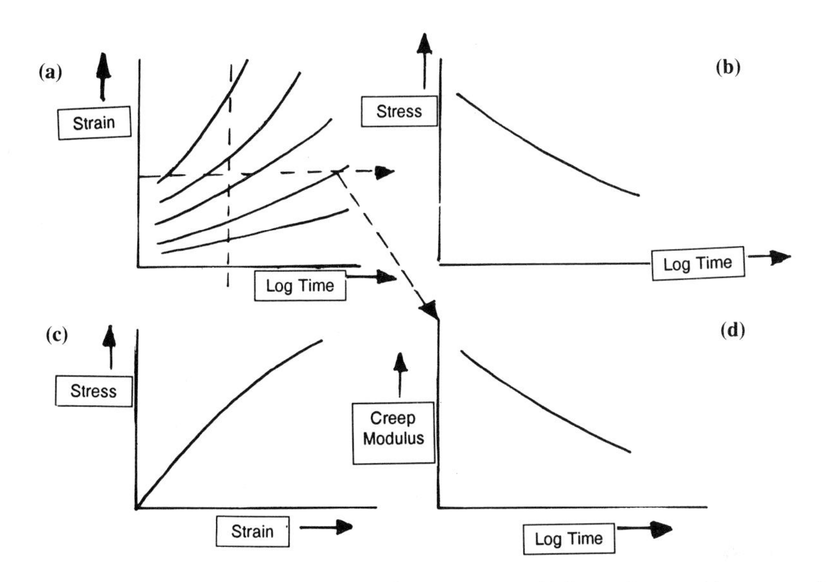

Figure 9.20 Creep test results. (a) Series of creep curves. (b) Isometric stress/time curve. (c) Isochronous stress/strain curve. (d) Creep modulus/time curve. After Whelan and Craft [26].

A third curve can be derived from the basic family of creep curves: creep modulus–time. This is a modified version of the isometric stress–log time curve, substituting creep modulus values at the intercept points for stress values.

References

1. D. Hands and R. H. Norman (1985). A new curemeter, paper presented at the International Rubber Conference, Stuttgart, Germany, 25–27 June.
2. J. A. Sezna (1989). The use of processability tests for quality assurance, *Rubber World*, Jan., 21–29.
3. H. A. Pawlowski & A. C. Perry (1984). A new automatic curemeter, paper presented at Rubberex, Birmingham, England, 12–16 Nov.
4. Anon. (1991). Rheometry testing polymer quality, *European Plastics News*, Oct., 53–54.
5. R. P. Brown (1981). *Handbook of Plastics Test Methods* (2nd edn), George Godwin.
6. A. Whelan and J. Craft (1982). Flow testing—-High shear rate rheometry, *British Plastics and Rubber*, Sept., 57.
7. Anon. (1982). Liquid characterics are fundamental to solid plastics, *British Plastics and Rubber*, Sept.
8. DuPont Co. 2970 Dielectric Analyser Sales/Technical Brochure.
9. A. Whelan and J. Craft (1982). Tensile strength, *British Plastics and Rubber*, Feb., 40.
10. BS2782 (1976). Part 3, Methods 320A–320F, Tensile testing.
11. A. Whelan (1982). *The Physical Testing of Plastics & Rubber*, SPS Technical Ltd, Fareham.
12. Anon. (1990). There is still originality in tensile testing, *British Plastics and Rubber*, July/Aug., 12–13.
13. ISO/DIS 527 Plastics—Determination of tensile properties.
14. A. Whelan and J. Craft (1982). Rigidity, *British Plastics and Rubber*, March, 50.

15. BS 2782 (1978). Method 335A and Method 1005.
16. BS2782 (1970). Method 306A, Impact strength (pendulum method).
17. BS2782 (1969). Method 359/ISO179, Determination of Charpy impact strength of rigid materials.
18. BS2782 (1979). Method 352B, Falling weight impact strength of plastic film.
19. BS2782 (1970). Method 306B, Impact strength (falling weight method with sheet specimen).
20. A. Whelan and J. Craft (1982). Impact—Falling weight, *British Plastics and Rubber*, June, 39–40.
21. G. Darcy (1980). Relationship between impact resistance and fracture toughness in advanced composite materials, paper presented at the AGARD Meeting on the Effects of Service Environment in Composites, 14–16 April.
22. D. F. Adams (1977). Impact response of polymer matrix composites materials, *Composite Materials Testing & Design* (*Proc. Fourth Conference*), ASTM STP 617.
23. I. V. Visconti (1983). The study and behaviour of composite materials using the Ceast instrumented impact tester, paper presented at the Symposium on Fracture Mechanics, Bradford University, 20 April.
24. R. P. Brown (1979). *Physical Testing of Rubbers*, Applied Science Publishers, Barking.
25. N. J. Mills (1986). *Microstructure, Properties and Applications*, Edward Arnold, London.
26. A. Whelan and J. Craft (1982). Creep, *British Plastics and Rubber*, July/Aug., 41.
27. C. Geddes (1984). *Time Dependent Effects of Deformation*, Kelvin Publishing.

10 Polymer microscopy

A. S. VAUGHAN

10.1 Introduction

The term microscopy may be defined as the study and use of the various types of microscope. As such, this field encompasses instrumentation that ranges from the simple stereomicroscope, used to examine the surface appearance of objects at magnifications of the order of 10 ×, to the most advanced high-resolution transmission electron microscope (TEM), the current generation of which boasts line resolutions better than 0.14 nm. Microscopy therefore covers a wide range of dimensional levels. Microscopy also differs from most other experimental techniques in providing data in a spatially resolved form, that is, it leads to images. While this can be considered a great advantage, when it comes to interpretation, and particularly quantification, analysis of images requires considerable experience on the part of the experimenter if unbiased results are to be derived. However, modern microscopes are considerably more than merely magnifying devices. In the polarising optical microscope, conoscopic interference figures may be formed to provide data simultaneously concerning the properties of optically anisotropic crystals along different directions. When electrons interact with matter, many processes occur which may involve the emission of low-energy electrons, X-rays, optical photons, etc. With suitable detectors, appropriate emissions may be collected and used not only for imaging but also, for example, to probe the chemical composition of the specimen. Also, where a beam of electrons passes through a thin crystalline foil, some electrons will undergo elastic scattering, and the diffracted beams may be used to provide crystallographic data. Thus, microscopy includes elements of spectroscopy and crystallography as well as the more traditional concept of producing a magnified image of the object. Indeed, the borders of the realm of microscopy are becoming increasingly blurred as instrument manufacturers develop and commercialise new imaging techniques, such as the scanning secondary-ion mass spectrometer, the scanning acoustic microscope, the Raman microprobe and the scanning tunnelling microscope [1, 2]. To cover the theory and application of the complete range of available imaging and analysis instrumentation relating to microscopy, in the broadest sense of this term, is clearly beyond the scope of a single chapter. Thus, emphasis here will be concentrated upon three particular areas, namely the application of optical microscopy, scanning electron microscopy (SEM) and TEM to the study of polymeric materials.

10.2 Optical microscopy

Optical microscopy has made a major contribution to the development of understanding in many different areas of science. At its simplest level, an optical microscope provides a magnified image of an object, enabling structural features to be observed that would otherwise be beyond the resolution limit of the human eye. However, the optical microscope can be much more than merely a sophisticated magnifying glass. Potentially it may provide structural information, with a resolution of the order of 1 μm, molecular information from the birefringence observed when an optically anisotropic sample is viewed between crossed polars, and even some limited chemical information, for example, through the observation of the colour changes that occur during degradation. In addition to these factors, the optical microscope is also attractive on grounds of cost and the relative ease with which suitable samples may be produced. Taking all these elements together, optical microscopy is a powerful technique, particularly for the study of materials such as polymers, which transmit a reasonable proportion of the light that is incident upon them. In considering the optical microscopy of polymers it is convenient to consider the available techniques under two headings: transmitted light microscopy and reflected light microscopy.

10.2.1 Transmitted light microscopy

When a beam of light travels through a sample it is modified as a result of the optical properties of the material and their spatial variation within the specimen. In samples where the absorption coefficient varies from place to place this naturally leads to contrast in the final image. However, of more interest, as far as many polymers are concerned, are techniques that enable differences in other optical properties to be imaged.

10.2.1.1 Bright- and dark-field illumination. The most straightforward mode of imaging in the optical microscope is by using bright-field illumination. In this, the illuminating beam passes through the sample, and spatial variations in optical density and colour within the specimen result in amplitude contrast in the image. In many situations, notably in biological systems, the inherent differences in optical absorption are enhanced by the use of staining reagents, which are taken up to different degrees by different parts of the morphology. In polymers, bright-field illumination may be used to reveal the presence of foreign particles or absorbing defect structures [3–5]. Figure 10.1 shows a

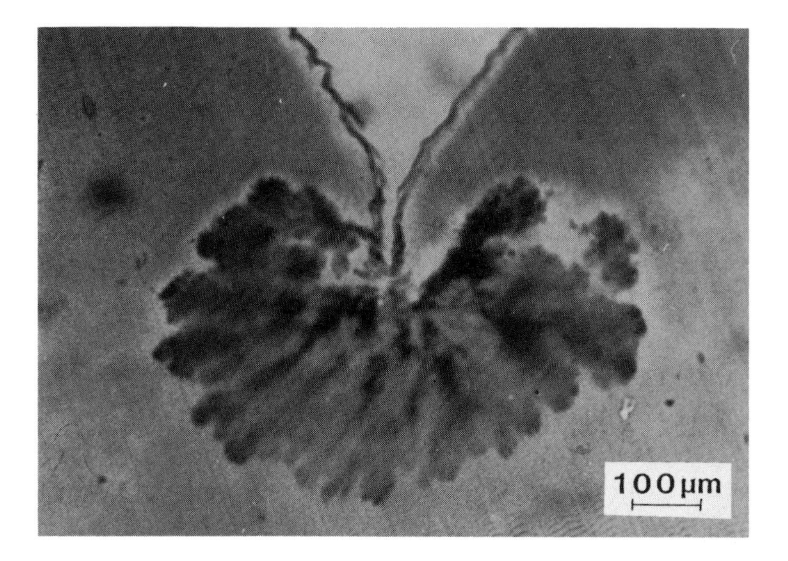

Figure 10.1 Bright-field transmission optical micrograph showing a methylene blue-stained water tree grown from a reamed hole in a sample of medium-voltage polyethylene cable insulation. From Olley *et al.* (1992) [7].

defect structure (water tree) in a cable dielectric. To enhance contrast in bright field, this water tree has been stained with methylene blue [6].

However, most pure polymer systems are transparent and, as a consequence, simple bright-field illumination rarely produces good image contrast. In bright field, the illuminating beam is incident normally on to the sample. However, the incident light may be tilted with respect to the optical axis of the microscope, to give so-called oblique illumination [8]. By using the simple expedient of tilting the incident beam, additional contrast may be generated at discontinuities or boundaries within the specimen. Although it seems unlikely that improved images will result from such a crude approach as deliberately displacing the condenser aperture off the optic axis, or even through the insertion of an object into the illuminating beam in such a way that light only falls on to the sample from one side of the condenser aperture, these techniques do have their uses, as can be seen in Figure 10.2. This micrograph shows a thin film of polystyrene crystallising within a hot stage. Because the lamellae were predominantly oriented within the plane of the film, polarising techniques could not be used for imaging, and neither could phase contrast, because of the geometrical limitations imposed by the design of the hot stage. Although the precise origins of the contrast seen using this approach depend upon many factors, this simple experimental expedient can be remarkably effective, as in this case.

In both bright-field and oblique illumination, the undeviated incident beam, after passing through the sample, falls within the limits of the objective lens.

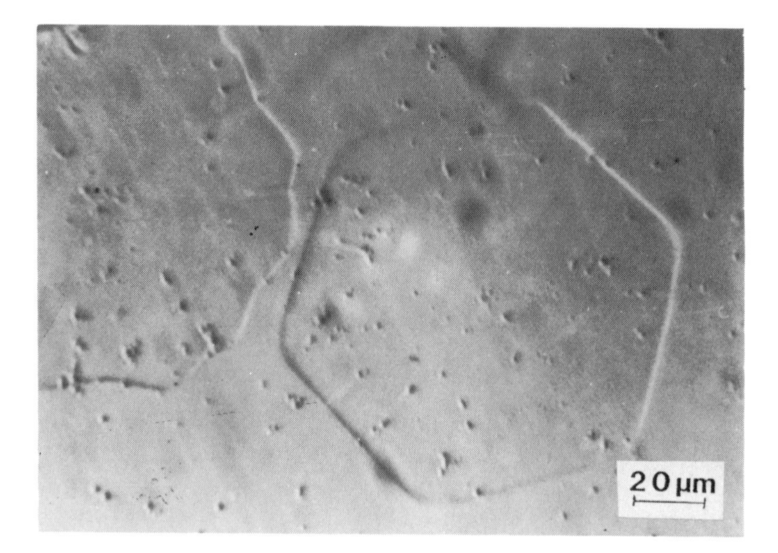

Figure 10.2 Lamellar aggregate growing from the melt in a thin film of isotactic polystyrene at 215°C. Transmission optical micrograph taken using oblique illumination in a hot stage (compare with Figure 10.6).

However, the obliquity of the illumination can be increased to such an extent that undeviated rays pass outside the objective lens. In practice, dark-field

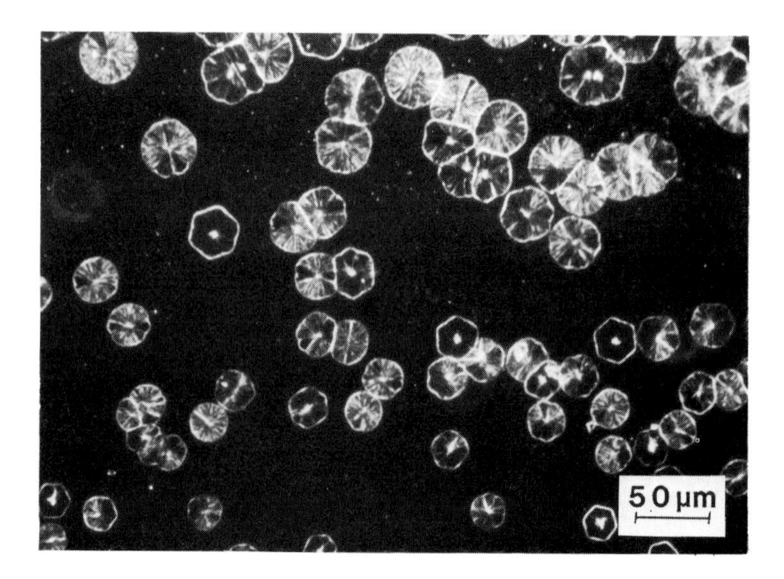

Figure 10.3 Dark-field transmission optical micrograph showing a thin film of isotactic polystyrene partially crystallised at 170°C and then quenched.

illumination conditions are achieved not by a single oblique beam but rather by the use of an annular condenser aperture which produces an illuminating cone of wider angle than that defined by the numerical aperture of the objective lens. Under these conditions the only rays entering the objective lens are those diffracted or scattered by features within the specimen. An example of a dark-field transmitted light image is given in Figure 10.3. As can be seen in this micrograph, which again shows a sample of polystyrene, dark-field illumination is particularly useful for highlighting boundaries.

10.2.1.2 Polarised light microscopy. Polarised light is used in the study of materials in which the refractive index is an anisotropic quantity. The linear nature of polymer molecules implies that such materials will inherently tend to possess anisotropic optical properties at the molecular level. Whether or not this is translated into macroscopic birefringence depends upon the way in which the molecules are packed together. Therefore, well-ordered crystalline regions should be birefringent, whereas the disordered molecular conformations present when a polymer is in solution, in the melt or in a glassy state should confer no optical anisotropy on the system as a whole. However, all of these situations may become birefrigent under circumstances where the sample is subjected to some form of stress field (see Figure 10.4). Thus, polarised light techniques may be used to study liquid crystals [9–11], flowing or sheared melts [12–13],

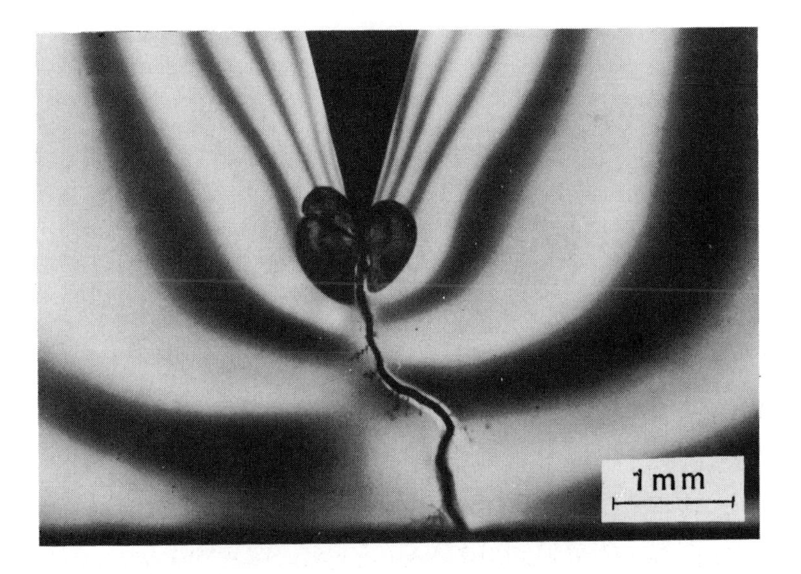

Figure 10.4 Interference effects resulting from the stress field around the tip of a needle embedded in a block of epoxy resin. The electrical breakdown path between the needle and the planar earth electrode can be seen, together with fracture features that are also associated with the failure process. Plane-polarised illumination.

polymers in solution [14–16], and deformed glasses [17–19]. For an account of the theory of polarising microscopy, ref. 20 is recommended.

Perhaps the most widely used application of polarised light microscopy in the study of polymers is in connection with crystalline morphologies [21–30]. Figure 10.5 shows typical spherulites of the α crystal form of poly(vinylidene fluoride), as imaged in polarised light. The extinction pattern seen in Figure 10.5(a) is made up of a black Maltese cross, a feature that is typical of a

(a)

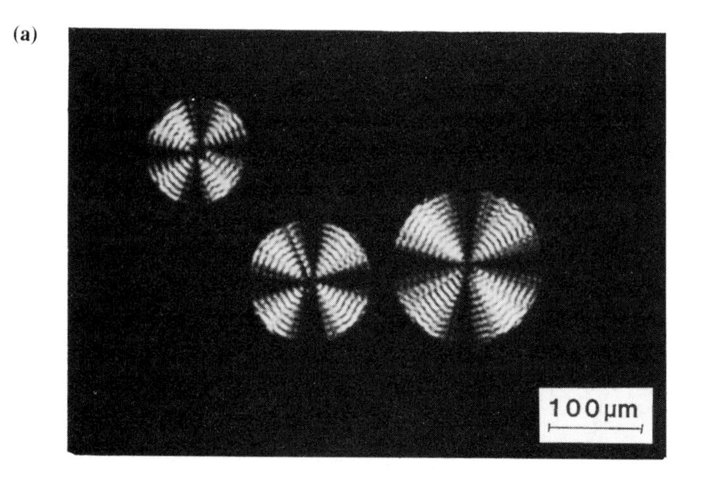

(b)

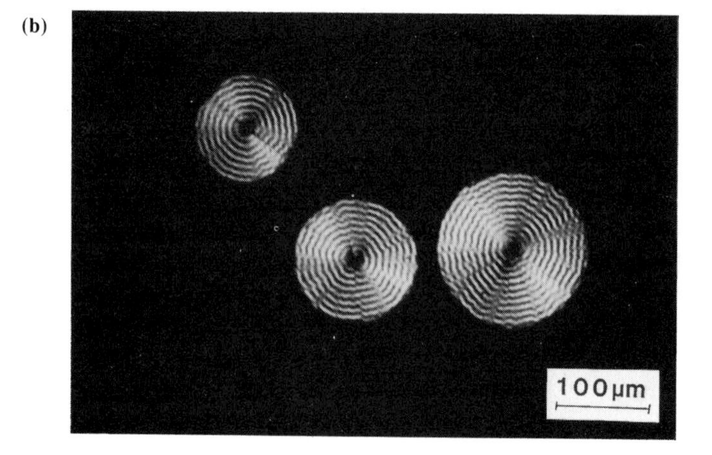

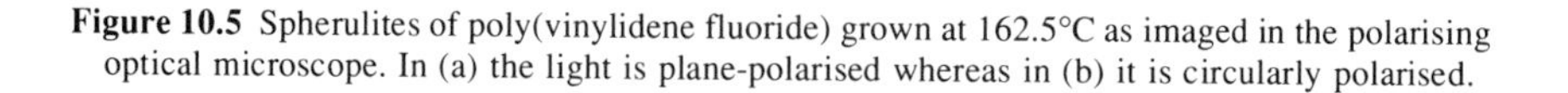

Figure 10.5 Spherulites of poly(vinylidene fluoride) grown at 162.5°C as imaged in the polarising optical microscope. In (a) the light is plane-polarised whereas in (b) it is circularly polarised.

spherulitic structure, together with a number of concentric bands which indicate that the constituent crystalline units are twisting about the spherulite radius [31–33]. These two optical features respectively correspond to regions of zero amplitude and zero birefringence extinction [34]. Thus, within the Maltese cross, the polymer molecules lie parallel to the analyser or the polariser, whereas within the concentric bands the molecular axis is parallel to the direction of illumination. These two forms of birefringence may be distinguished from one another by the simple expedient of inserting crossed quarter-wave plates on either side of the specimen (see Figure 10.5(b)), so that the sample is illuminated in circularly, rather than plane, polarised light. Under these conditions the blackness in the image corresponding to zero birefringence extinction remains, whereas zero-amplitude extinction can no longer occur.

In specimens where there is no overall orientation, the molecular axis cannot always be uniquely defined simply on the basis of the extinction pattern. Where such information is required a λ-plate (alternatively referred to as a first-order red or sensitive-tint plate) may be employed to give additional information. The λ-plate is a fixed compensator that, on insertion at 45° between polariser and analyser, introduces a phase difference of about 550 nm between the two orthogonal transmitted rays. On recombination, this phase difference results in the interference colour first-order red in the absence of any additional birefringence within the system. However, on addition or subtraction of additional small phase changes due to birefringence within the sample, the interference colour changes towards blue or yellow respectively [35]. In practise, this results in a blue colour being observed in regions of the sample where the direction of greatest refractive index is within 45° of the slow direction of the λ-plate, red in regions of zero birefringence and yellow elsewhere. Where the relative refractive indices parallel and perpendicular to the molecular chains are known, for example following the examination of oriented samples, the spatial variation in chain direction may be deduced directly from the colours that are observed.

Using the above techniques, the local molecular chain axis may be deduced from the observed extinction patterns and interference colours. However, for the absolute measurement of birefringence, a variable compensator is required. Many different types of compensator are available, depending upon the optical-path difference to be determined. However, very particular sample morphologies are required if useful data is to be derived in this way (e.g. macroscopic birefringent single crystals). Although absolute birefringence measurements may be useful in certain polymeric systems, such as fibres or oriented films, in most cases qualitative data is sufficient. Therefore, compensators will not be considered further here.

10.2.1.3 Phase-contrast microscopy. In bright-field illumination, variations in optical absorption within the specimen give rise to an image which consists of variations in brightness and which, as a result, is readily visible. Since such a sample affects the amplitude of the light rays passing through it, it is known as an amplitude object. However, some samples are transparent, and the microstructure present manifests itself in the form of variations in refractive index rather than absorption coefficient. In bright-field illumination, such a specimen produces an image of uniform brightness, in which only the phase of the transmitted light varies from place to place. Such an object is termed a phase object. Although such specimens are not directly visible to the naked eye, variations in phase can be made to appear as changes in amplitude (brightness) by using the techniques of phase-contrast microscopy [36] (see Figure 10.6). Phase contrast is particularly well suited to the study of specimens that are thin and highly transparent and, as such, has been used extensively in the study of polymer lamellae and thin films [37–42].

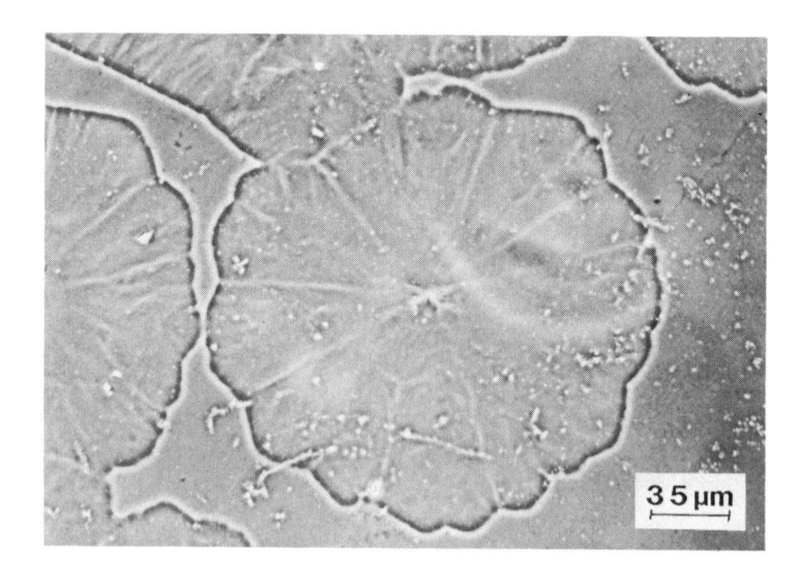

Figure 10.6 Phase-contrast transmission optical micrograph showing a sample of isotactic polystyrene crystallised at 215°C in a thin film (see also Figures 10.2 and 10.19).

10.2.2 Reflected light microscopy

The examination of specimens in transmission necessarily requires that they be thin. Thus, the techniques of transmission optical microscopy are inappropriate for the examination of bulk specimens, or materials such as metals or conducting polymers which absorb strongly at visible wavelengths. Nevertheless, examination of such samples in reflection may provide useful structural information, through the observation of topographical features.

In general, images may be formed in reflection in similar ways to those used in transmission. Thus bright-field [8], dark-field, polarised light [20] and phase techniques [36] have all been developed for use with incident light. In general, the theory and practice of these techniques are similar to those for transmitted light and, therefore, they will not be considered further here. However, there is one class of optical technique that has not been considered thus far, primarily because its main usage is in connection with reflected light. This is interference microscopy [43]. In this, the illumination is split into two or more beams which, on recombination, produce an interference pattern that is related to the sample structure or, in reflection, its surface topography. A particularly useful technique that makes use of this latter approach is differential interference contrast (DIC) or Nomarski microscopy [43, 44].

In DIC, the illumination (white light) is split into two beams that are displaced from one another at the specimen plane by a very small distance. On recombination, small optical-path differences manifest themselves as interference colours. When DIC is used in reflection the surface topography of

Figure 10.7 An etched surface within a sample of medium-voltage polyethylene cable insulation, as revealed in reflection using DIC optics. The dendritic structure of a water tree is evident, together with rectangular regions which correspond to beam damage as a result of prior examination in the SEM. From Olley *et al.* (1992) [7].

the sample is highlighted and a pseudo-three-dimensional coloured image is produced. For best results, the sample surface should be highly reflective and should be mounted precisely perpendicular to the axis of the microscope. Since DIC is a technique for enhancing small surface features (see Figure 10.7), very rough samples often produce better images by using more direct, bright- or dark-field, techniques.

The use of DIC in the study of polymers is particularly well suited to situations such as etched surfaces (see sections 10.3.2.2 and 10.4.2.3), which are macroscopically flat but which nevertheless contain fine topographic details [45]. However, the direct examination of polymer surfaces in reflection is not always entirely straightforward because of the low reflectivity exhibited by many systems. It is therefore often desirable to evaporate or sputter a reflective metallic coating on to the specimen surface prior to examination. While this improves the reflectivity of the sample it also has another less obvious benefit. In low-absorbance systems, the illumination may be reflected back into the objective lens not only from the surface of the specimen, but also from other boundaries within the sample. Particularly at low magnifications, where the depth of field may be considerable in comparison with the size of the surface features, such subsurface effects may give a false impression of the sample topography.

10.2.3 Other optical techniques

Many of the more specialised techniques for the optical examination of specimens have not been covered above, for reasons of space. These include conoscopy [35], infrared microscopy [46], fluorescence microscopy [47], and, of particular interest in connection with polymers, Raman microscopy [2].

10.3 Scanning electron microscopy (SEM)

The optical design of the SEM endows it with a particularly useful combination of virtues [48–50]. Not only is it capable of generating high-magnification images, with a resolution better than 5 nm, but also, at the other extreme of the magnification range, excellent low-magnification (~ 1000 ×) images may be produced even from rough samples, as a result of the instrument's inherently large depth of field. This characteristic makes the SEM particularly useful for examining large objects in, for example, forensic work. In addition to their conventional back-scattered and secondary electron imaging capabilities, most modern SEM systems are also configured such that images may be formed from alternative signal sources. When one or more X-ray detectors are available [49,50], the signal characteristic of a particular element can be displayed on the imaging screen in synchronisation with the scanning electron beam. In this way element maps may be obtained. Similarly, the current generated in the sample (electron beam induced current, EBIC) or the voltage induced by the beam (electron beam induced voltage, EBIV) may be displayed, so as to form an image that is related to spatial variations in the electrical properties of the specimen. While such techniques find many uses in the study of metals, ceramics and semiconductors, the elemental composition and electrical properties of most polymers mean that such approaches are only of use in connection with particular problems (e.g. composite systems and conducting polymers [51]). Discussion here will therefore be limited to the more conventional imaging techniques, that is, those using secondary and back-scattered electrons.

10.3.1 Sample imaging in the SEM

In the SEM a fine beam of electrons is scanned across the specimen surface and an appropriate detector collects the electrons emitted from each point. The amplified current from the detector is then displayed on a cathode-ray tube, which is scanned synchronously with the electron probe. In this way the image is built up, line by line.

10.3.2 Sample preparation

Primarily, the SEM is an instrument for the examination of surfaces (crystallographic information may be obtained from macroscopic crystals, in the form of electron channelling patterns) [1] and, as such, appropriate samples may be examined directly with little or no prior preparation (see Figure 10.8). However, polymers do present some specific problems. Firstly, most polymers are poor conductors of electricity and, as a result, charge rapidly builds up on the surface of the sample as the electron beam is scanned across it. The resulting field then interacts with the incident electron beam, resulting in image distortion. The second problem concerns the molecular changes that are induced in the sample by the impinging electrons [52, 53]. Examples of radiation damage can be seen in Figure 10.7. Such effects serve to restrict the operating conditions under which particularly sensitive specimens may be examined. Generally, low accelerating voltages and low beam currents are required to avoid radiation damage. Whilst these requirements are not a problem in themselves, they may compromise the performance obtainable from the instrument, resulting in lower resolution and reduced signal strength.

10.3.2.1 Conductive coatings. For stable images to be formed it is essential that the charge deposited on the sample surface by the electron beam is able to leak away to earth. Thus, for insulating materials such as polymers, it is usually desirable to coat the specimen with a conducting film prior to examination. The choice of material depends very much upon the information that is required. In many circumstances the coating material chosen is gold, a film

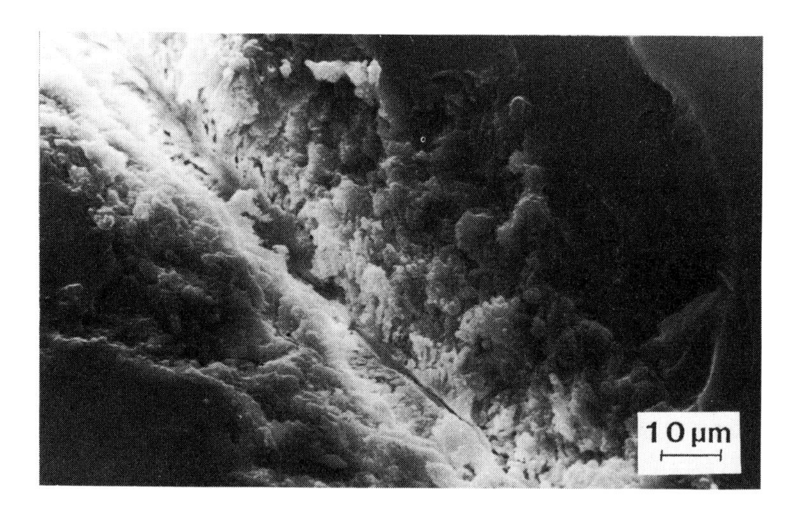

Figure 10.8 SEM micrograph showing the internal appearance of an electrical tree channel (see Figure 10.4). The tree was grown in an epoxy resin block which was then fractured open to reveal the internal structure.

of which is applied either by evaporation or, more conveniently, by sputtering. Typical film thicknesses are of the order of 20 nm [54]. Such a film provides many advantages: it is easy to apply, gives a good leakage path to earth, and gives a high secondary electron yield. However, gold is not ideal in all respects, and for particular problems other materials may be preferable (e.g. gold–palladium alloy or carbon) [54].

While a conductive coating is highly desirable, there are circumstances where this approach is simply not practical, for example where the surface structure is to be monitored at a number of different stages during a complex multi-stage process. When it is essential that uncoated insulating specimens are examined directly, the operating conditions of the microscope must be chosen very carefully indeed. Basically, insulating specimens can only be examined by using low accelerating potentials and the minimum beam current consistent with a reasonable image. In addition, the scan rate can also be an important consideration in minimising sample charging. At television scan rates, for example, the residence time of the beam at any point on the sample is very low, so that charge does not build up sufficiently to interact locally with the incoming electron beam. At the lower scan rates used for photography, this situation may change. One way around this problem is through digital image acquisition and frame averaging within a suitable frame store.

Where external surfaces or fracture surfaces are of interest [55–62] the application of a suitable conductive coating is all that is required in the way of sample preparation for SEM. However, other sample preparation procedures may be used to enhance the surface topography that is present.

10.3.2.2 General chemical etching techniques. Etching involves the preferential removal of one or more structural elements from a sample surface, so that relief develops which is related to the underlying microstructure. Thus, since etching enhances surface topography, it is natural that it should be used in conjunction with a surface imaging instrument such as the SEM. Chemical etching techniques have been categorised under three headings: solvent etching, degradation techniques, and true etching [63]. In principle all of these different approaches may be applied equally well to both SEM and TEM studies of microstructure. However, in practice the suitability of any particular procedure for TEM applications is often limited by the need to replicate the resulting topography adequately. Thus, in practice, many techniques are unsuitable for TEM work, because they result in surfaces that are either too friable or too rough to give good replicas. In such circumstances, the more direct nature of SEM provides an ideal way forward. Discussion of etching techniques will therefore be divided between this section, where those techniques better suited to SEM will be considered, and section 10.4.2.3, where permanganic etching and replication techniques will be described in more detail.

Solvent etching relies upon the preferential dissolution of one element from a multicomponent system. Although the use of the term 'etching' to describe such a process is, perhaps, rather loose, such an approach does, nevertheless, fulfil the basic requirement of enhancing surface topography in a way that is related to the underlying morphology. For the study of systems that do not contain radically different components, i.e. a single polymer rather than a multicomponent blend, careful selection of both solvent and extraction conditions is essential if the formation of artefacts is to be avoided. Nevertheless, solvent etching has been used to study morphology in, for example, polystyrene [64], nylons [65], polyethylene [66] and polypropylene [67]. However, since the actual etching process probably involves partial swelling and molecular reorganisation in addition to selective dissolution, the resulting surfaces may not contain representative fine-scale detail. Such etching process are therefore better suited to the study of, for example, spherulite size distribution within injection mouldings, than of the fine-scale organisation of lamellae within the spherulites themselves, although solvent extraction has been applied successfully to the study of individual melt-crystallised lamellae of polyethylene [68, 69].

The problems of solvent etching described above are primarily associated with the selectivity of the solvent. In a one-component semi-crystalline polymer, for example, extraction of amorphous regions will generally be accompanied by some disruption of the remaining crystal population. One area where selectivity may not be a problem is that of polymer blends; one component may be readily soluble in a given solvent while the other(s) remain completely undisturbed. Many examples exist where this approach has been applied with success. These include blends of polyethylene, polypropylene and an EPDM rubber [70], blends of PVF_2 and PMMA [71] and a PMMA–liquid crystal composite [72]. A similar extraction technique has been used to study network formation during curing in a polyester system [73]. In this case,

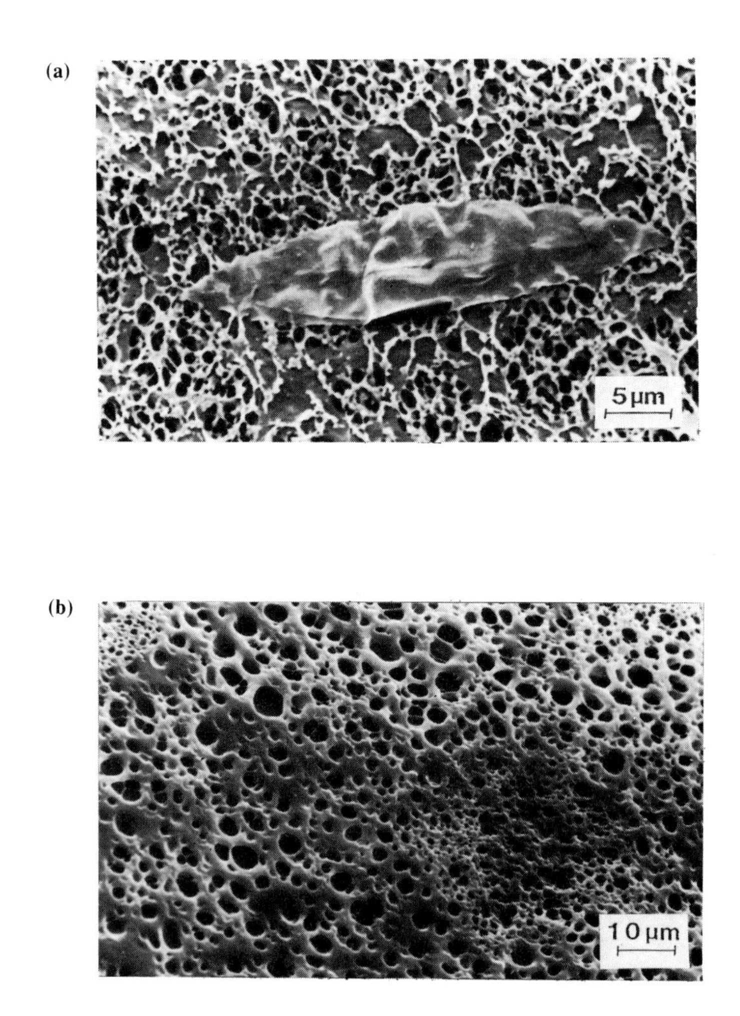

Figure 10.9 Examples of solvent-etched samples, as seen in the SEM; (a) shows an elliptic crystal of linear polyethylene after dissolution of the surrounding matrix in decalin; the crystal is shown sitting on a piece of filter paper; (b) shows the gel/pore structure of a partially cured polyester resin; the unreacted component was extracted by using dichloromethane.

the cross-linking process was terminated at various times after initiation, and the unreacted sol phase was extracted with dichloromethane. Micrographs obtained using solvent-extraction techniques such as those described above are shown in Figure 10.9.

Although referred to as an etching technique by many authors, degradation of polymers by nitric acid [74] is, again, not true etching, in the metallurgical sense. Whereas in metals etching reactions are confined to the surface of the sample, in polymers nitric acid molecules are able to penetrate the interlamellar amorphous regions to considerable distances before reacting. On reaction, nitric acid digests the amorphous molecular conformations [75, 76] leaving a porous aggregate of crystalline material which is mechanically weak and extremely friable [77]. Indeed, subsequent disruption by ultrasonics has been used to give small lamellar fragments which can then be examined in isolation [78]. Related techniques have been developed to remove the non-crystalline material from other semi-crystalline polymers; hydrolytic reagents may be used in appropriate circumstances (e.g. polyesters and polyamides) in place of oxidising species [63].

As far as true liquid chemical etchants are concerned, by far the most successful reagent has been the so-called permanganic etchant, which is particularly well suited to the study of lamellae in semi-crystalline materials. Although, in principle, lamellar morphologies may be studied by SEM it has been demonstrated that for reasons of resolution, replication and examination in the TEM is the more complex but more profitable route. This is not to say that SEM examination of surfaces etched using the permanganic reagent is not of use in many circumstances [79, 80]. Indeed, in some situations it is the only way forward, particularly in multi-component systems where subsequent replication may be a problem (for example, in the study of blends, films or fibres).

While highly selective etchants are not always desirable, there are circumstances where this can be turned to advantage, such as in the study of water trees in cable insulation. Although these objects had been examined for many years by optical techniques, (see Figure 10.1), until recently the details of their microstructure were unknown. By diffusing a low-molecular-mass acrylic monomer into the tree and polymerising it *in situ*, the tree structure can be fixed in place prior to being cut open with a microtome. The acrylic polymer is then etched away to leave the tree visible in the sample surface. In this case the fact that the permanganic etchant removes the acrylic polymer very much more rapidly than the surrounding polyethylene matrix is highly beneficial, since it enables the tree structure to be revealed without complications caused by additional etching of the matrix (see Figure 10.10) [7].

Another area where highly selective etching has been used in conjunction with SEM is the study of conducting polymer films. Here it is the geometry of the samples that presents the major problem. Electrochemically polymerised conducting polymers generally grow in the form of thin films, typically some

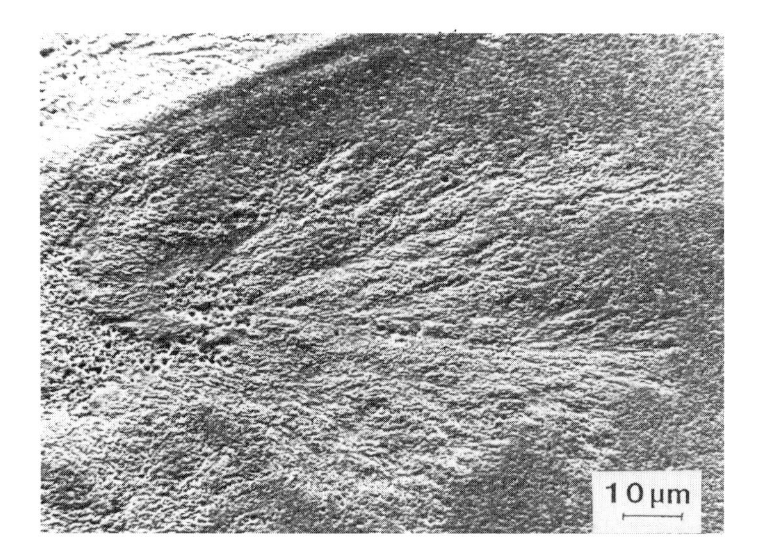

Figure 10.10 SEM micrograph showing an etched filled water tree (compare with Figure 10.7.) From Olley *et al.* (1992) [7].

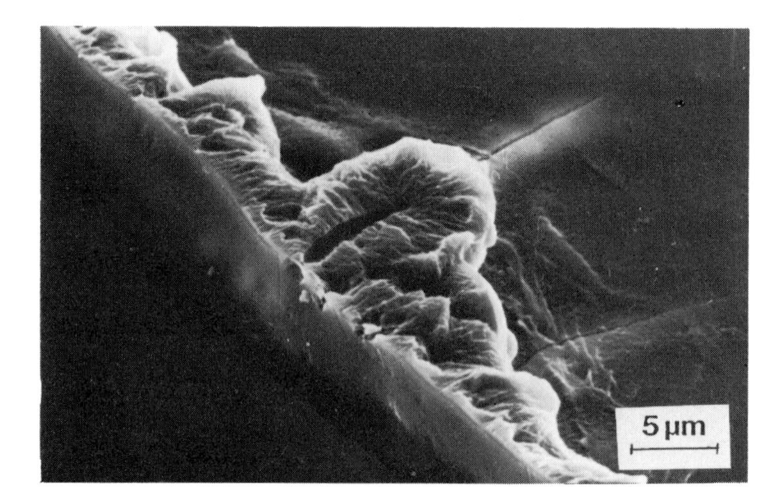

Figure 10.11 Cross-section through a film of polypyrrole. The film was first embedded in an epoxy resin, then microtomed and finally subjected to a two-stage etching process to generate the required relief. The film can be seen standing proud of the embedding material; in addition, internal texture is also apparent within the polypyrrole itself.

tens of micrometres in thickness [81]. The morphologies obtained range from highly fibrillar structures [60] to dense modular films [62]. The internal cross-sectional appearance of such samples may be examined by embedding small specimens within a suitable matrix and then fracturing or microtoming the composite system, so that a cross-section through the conducting polymer is exposed. By varying the etching conditions, either of the components may be etched preferentially, so that surface topographic features are enhanced (see Figure 10.11); clearly, replication would not be possible in this case. This general approach of embedding, fracturing or microtoming and finally etching is useful in many circumstances where the sample geometry or mechanical properties prevent a more direct approach to the study of internal morphologies.

10.3.2.3 Plasma etching. Although used in other areas of materials science, plasma etching has not been greatly developed as a sample preparation technique for use in conjuction with the microscopy of polymers [82–84]. Although general methodologies are well developed (see the excellent review by Egitto *et al.* [85]), the existing techniques have been devised with technological applications in mind, for example, in connection with lithography and polymeric resists in electronics applications. However, it is clear from the selective nature of the process that this approach does have applications, particularly for the study of chemically resistant materials and, perhaps more importantly, composite systems.

10.4 Transmission electron microscopy (TEM)

TEM constitutes a very powerful instrument for the structural study of suitable samples. The potential it offers is not limited only to the formation of high-resolution images, but the ability to form very small-diameter probes means that spatial variations in chemical composition or crystal structure may also be investigated, by spectroscopic and diffraction techniques, respectively. However, to transmit a beam of electrons through a sample necessarily requires the sample to be thin. While having any emergent intensity may be sufficient for some requirements, for optimum imaging, inelastic scattering and multiple scattering should be minimised and, for this, the sample needs to be very thin indeed. While the requirement for a thin sample is shared by all transmission electron microscopists irrespective of their discipline, TEM of polymers does present some particular problems. These may be divided into two groups.

Even when the sample naturally occurs in a form that is thin enough for direct examination in the TEM (e.g. solution crystals [86–89] or samples prepared directly in the form of a very thin film [90–94]), problems are encountered as a direct consequence of the way in which the sample and the electron beam interact. In comparison with inorganic materials, polymers are highly beam-sensitive. Electron irradiation results in chemical changes [52, 53], destruction of crystallinity [95] and mass transport [96]. In addition, the electron beam may also cause heating, which increases the beam-sensitivity of the material still further [97]. Nevertheless, the inherent radiation-sensitivity of polymeric samples may be overcome by the use of appropriate techniques (see section 10.4.1.1). Of potentially greater significance for the study of many systems is the low contrast that is observed when polymers are examined directly by TEM. Ignoring diffraction and dynamical scattering effects [98], bright-field contrast in the TEM arises from two sources, namely, variations in sample thickness from place to place (thickness contrast), and spatial variations in atomic number, that is, electron density ('Z' contrast) [1]. Most polymers, being made up predominantly of light elements, give rise to relatively little scattering and, of more importance, vary little in electron density from place to place. As a result, the image often contains little contrast. However, in particular circumstances the direct examination of sections or thin films has been used to good effect [99, 100], for example, in the study of organometallic copolymers [101].

However, contrary to the initial premise stated above, most polymeric samples, and particularly those of technological or commercial importance, do not naturally tend to occur in the form of a 50 nm-thick film. For bulk samples to be made amenable to examination by TEM they must therefore be subjected to some initial sample-preparation procedure. Indeed, the problems of sample geometry, beam sensitivity and intrinsically low contrast described above are such that they are often best overcome by the indirect examination of the sample. TEM techniques may therefore be divided into those in which the sample is examined directly and those in which a less direct approach is adopted.

10.4.1 Direct examination

10.4.1.1 Sample preparation. Most direct studies of polymers in TEM have involved inherently thin samples. In the case of thin films, sample preparation merely involves picking up a suitably sized specimen on a TEM grid [102, 103], while for isolated crystals, small particles, etc., samples may be prepared for examination simply by dispersion on to a suitable support film [104–108]. In these circumstances, the major experimental problems are associated with beam damage within the sample and, for this reason, experimental conditions must be selected very carefully. Firstly, high accelerating voltages are required, to minimise the inelastic scattering cross-section, and then the sample should only be irradiated for the minimum time and at the minimum intensity consistent with being able to see and/or record the image. (Sometimes it is necessary to record images 'blind', that is, without first seeing them; considerable expertise is required to record images in this way.) Another factor that is known to be important is the sample temperature. Specimen cooling holders may therefore be a useful aid in maximising sample lifetimes [109].

Despite the beam-sensitive nature of polymeric materials, most conventional TEM diffraction and imaging techniques can be employed [34, 110–112]. Figure 10.12 shows a diffraction pattern obtained from a polyethylene sample. While bright-field images of polymer crystals often reveal diffraction contrast (see Figure 10.13), most detailed studies of such structures have relied primarily upon dark-field imaging. Ideally, dark-field images should be produced by tilting the incident beam, so that the required diffracted beam travels through the objective aperture and down the optic axis of the instrument. However, in practice, rapid operation of the instrument is often a greater priority than optimum performance. For this reason, it may be preferable to generate dark-field images by the simple expedient of displacing the objective aperture away from the straight-through beam, so that a dark-field image is formed from an off-axis diffracted beam. The increased aberrations that result from this non-ideal configuration are not significant at low magnifications. An example of a dark-field image can be seen in Figure 10.14, in which pronounced Moiré fringes are evident. In this micrograph the fringe pattern at the arrowed sector boundary is indicative of lattice distortions caused by non-random chain-folding.

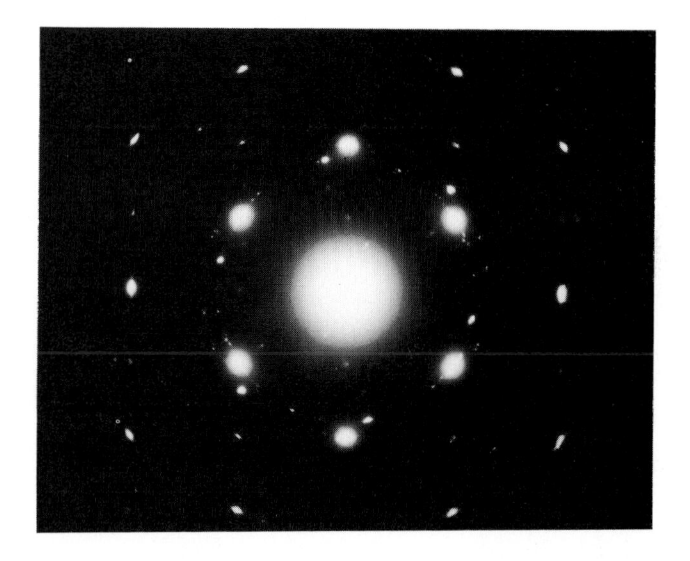

Figure 10.12 Electron diffraction pattern from a solution-grown crystal of polyethylene.

With recent advances in instrument control and digital image acquisition and processing, new possibilities are beginning to appear for the TEM of beam-sensitive samples. One fundamental disadvantage of conventional

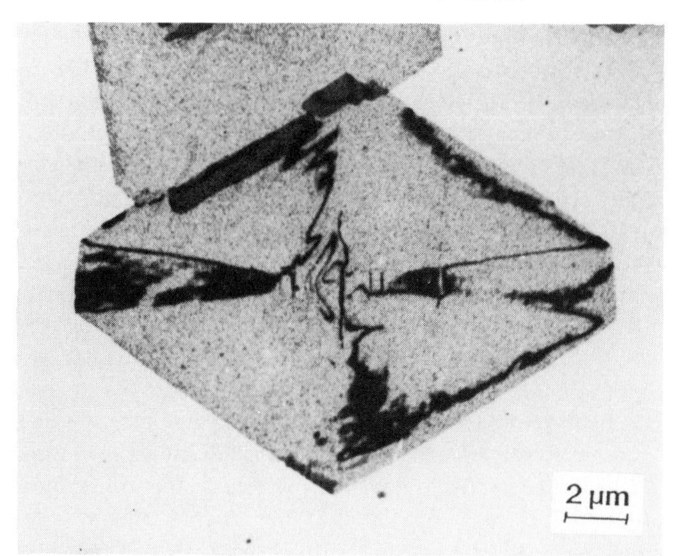

Figure 10.13 Bright-field electron micrograph showing a truncated polyethylene lozenge. The crystal can be seen to be divided into six sectors. From Bassett *et al.* (1959) [113].

bright- or dark-field imaging is that, by only allowing a single beam to pass through the objective aperture, most of the available information is discarded at the objective aperture plane. If multiple bright- and dark-field images are required from a single beam-sensitive sample, this process, which involves discarding information and recording images sequentially, is clearly ineffi-

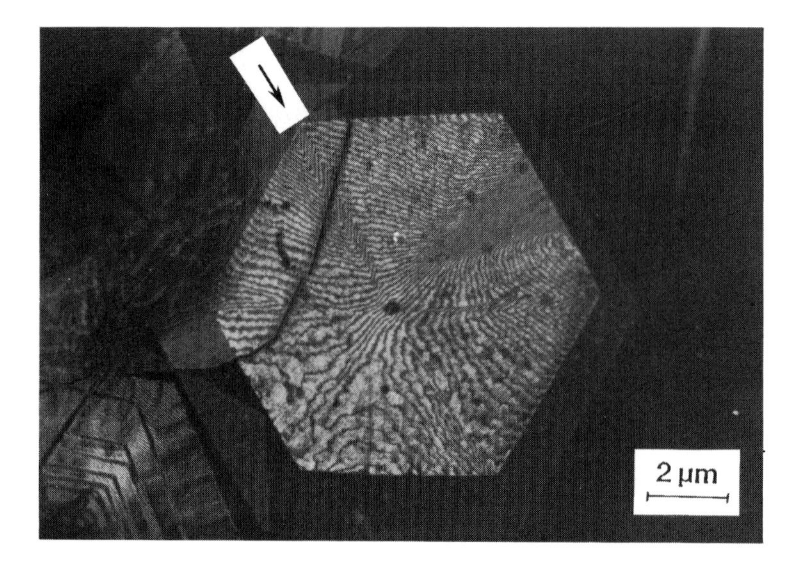

Figure 10.14 Overlapping lamellae of polyoxymethylene. The lattices in the two crystals are offset by a slight rotation so giving rise to the observed Moiré fringes. From Bassett (1964) [114].

cient, resulting in increased irradiation times. However, multiple images have recently been acquired, effectively simultaneously, by using a scanning transmission electron microscope (STEM) probe in conjunction with a multi-element detector [115].

Another type of multi-beam technique that has been employed extensively to study less beam-sensitive materials is so-called lattice imaging. To produce an image where periodicities of less than 1 nm can be resolved necessarily requires high magnifications and an intense, concentrated beam, i.e. conditions not conducive to long lifetimes in beam-sensitive materials. Nevertheless, despite the difficulty of the approach, lattice imaging has been attempted in a number of polymer systems [116–119].

Undoubtedly, in years to come, developments both in the control software and in the instrumentation itself will provide more opportunities for TEM of polymers. One area which would seem to be of particular interest to the polymer microscopist is the formation of energy-filtered images [120]. This technique provides spatially resolved chemical information and, like electron energy loss spectroscopy (EELS), is particularly suited to the investigation of light elements. Energy filtering is therefore well suited to the study of polymers and may be used both to improve image quality by rejecting inelastically scattered electrons and to provide compositional information, for example in polymer blends [121].

10.4.2 Indirect examination

Many of the difficulties mentioned in preceding sections may be obviated by the adoption of less direct examination techniques, particularly where bulk samples are concerned. Many early TEM studies of microstructure simply involved the formation of a replica of a suitable surface, usually either an external surface or one revealed by fracturing the sample to reveal its internal appearance [122–125]. Although this approach is relatively straightforward (see section 10.4.2.3 concerning replication), both external surfaces and internal fracture surfaces may give an overall appearance that is unrepresentative of the bulk microstructure [126]. For this reason the staining and etching–replication techniques described in the following sections were developed.

Replication and shadowing techniques have been used directly in conjunction with other procedures (extraction replication [127]) or specially prepared samples (epitaxially grown crystals [128]), where specific topics are being addressed [129, 130]. Of particular note in this regard is the work of Wittmann and Lotz, who developed a decoration technique for the study of fold surface conformations in lamellar crystals [131].

10.4.2.1 To stain or to etch? Staining techniques are derived from similar biological sample preparation procedures and involve two distinct steps. Generally, some suitable staining reagent is first allowed to diffuse into the sample. After an appropriate period, the sample is removed from the stain and a thin section (typically 10–100 nm thick) is cut and examined in the TEM. Although in some techniques the order of these two stages is reversed, the former approach is generally preferable, since the microtomy of polymers can be problematical [132]. The image that results contains mass contrast that is derived from the distribution of staining material within the sample. Thus, regions of the specimen that react readily with the stain or through which the staining reagent diffuses easily appear dark, whilst other regions appear light. An example of a stained sample of low density polyethylene is shown in Figure 10.15. In this case, the image contains bright lines (crystalline material) separated by distinct dark regions (disordered material); the more disordered a region, the more readily it will take up the stain and the darker it will appear. On the basis of this, much detailed analysis of chlorosulphonated samples of polyethylene has been performed, notably by Voigt-Martin [133, 134]. However, while the relationship between image contrast and sample structure is very direct in this approach, staining techniques are not without major limitations. Examination of Figure 10.15 shows that, in addition to some regions where lamellar detail is good, there are also others where the structure appears blurred, and others that appear devoid of structure altogether. These effects are related to the inclination of the lamellae with respect to the electron beam. Where the electron beam lies within the plane of the lamellae good image contrast results. However, as lamellae become more and more inclined, so

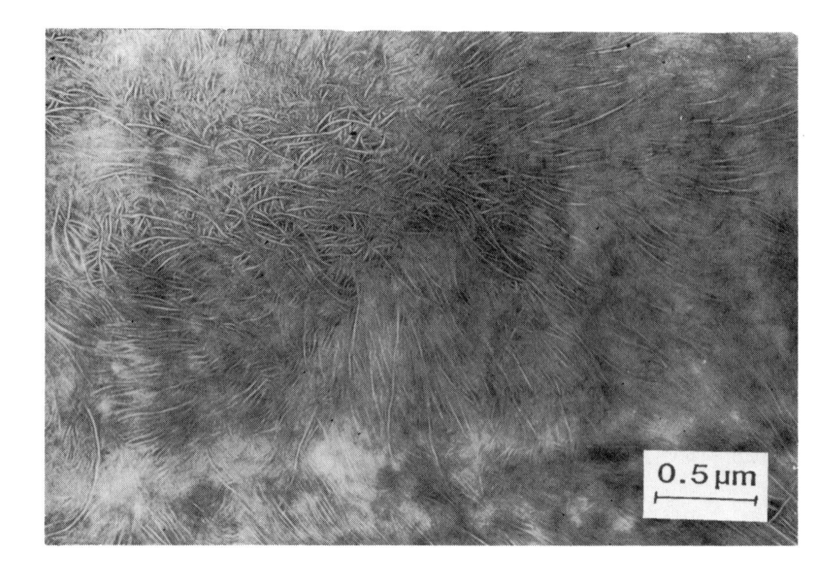

Figure 10.15 Banded spherulite in low-density polyethylene after staining with chlorosulphonic acid. Although the lamellar structure is well resolved at the centre of the object, in other regions morphological detail is less distinct.

amorphous and crystalline regions become increasingly less well differentiated, and the image becomes progressively less distinct. Thus, staining techniques provide good information when the features of interest can be tilted so that they lie edge-on to the beam, but are unsuitable for imaging materials where the structural units are inclined with respect to the beam. Staining has, therefore, been used extensively to study, for example, lamellar morphologies in melt-crystallised systems and domain structures in block copolymers [135–139].

If the above staining methodology can be thought of as having been derived from techniques developed for the examination of biological systems, then polymer etching has its conceptual origins in similar metallographic techniques [140]. In metallurgy, chemical etching renders structural features visible by preferential attack at, for example, grain boundaries and dislocation cores. In this way, relief gradually develops on the surface of an initially polished sample. Thus, etching is a surface technique and the first step is therefore the preparation of a suitable surface. Although this may be an external surface, a non-specific internal surface is generally chosen in preference. In the case of metals, internal surfaces are prepared by cutting and polishing, whereas in polymers fracturing and microtoming are more commonly used techniques. The search for a polymeric etching procedure can be traced back many years [141, 142]. However, it is really only during the last decade or so that significant advances have been made [143]. All etchants share the same objective, namely that of developing surface relief that is related to the underlying microstructure. As such, etching at first appears more directly relevant as a sample-preparation technique for use in conjunction with SEM. However, it is in conjunction with surface replication and subsequent TEM study that etching has most profitably been exploited, particularly in the area of crystalline morphologies.

For TEM study, the surface of interest is first exposed to the etchant and then the resulting surface topography is replicated. As such, in etching techniques it is a copy of the surface topography that is examined, not the specimen material itself. While this avoids problems associated with beam stability and ultramicrotomy, in doing so it raises further questions concerning the validity of the chemical treatment (the possible formation of etching artefacts) and the accuracy of the replica. These potential problems are particularly important where quantitative data are sought. Another apparent limitation of procedures involving surface replication concerns the data that is available; although supermolecular features may be studied, diffraction techniques are not generally applicable and therefore crystallographic or orientational information cannot be obtained.

It is clear from the preceding discussion that direct examination, staining and etching–replication techniques all have their strengths and weaknesses. They are therefore complementary approaches to the study of polymer microstructure and the choice of technique depends very much upon the precise problem being addressed [144].

10.4.2.2 Staining techniques. Many different staining reagents have been applied to polymers as a means of enhancing image contrast in the TEM (see Figure 10.16). All of these rely upon incorporating heavy, electron-dense elements into the structure at particular sites, either through specific chemical reactions or simply through physical absorption. Clearly, of these two processes, chemical reactions are preferred since absorbed reagents, such as iodine, may diffuse out of the sample in the vacuum of a TEM. Staining techniques are generally described as being positive or negative. In positive staining the stain is incorporated within the sample itself, whereas in negative staining the contrast is generated through interactions that are not directly associated with the specimen itself; for example, particles may be revealed by staining the surrounding medium. Indeed, positive and negative staining techniques have even been used in combination [145] to delineate particles and simultaneously study their internal morphology.

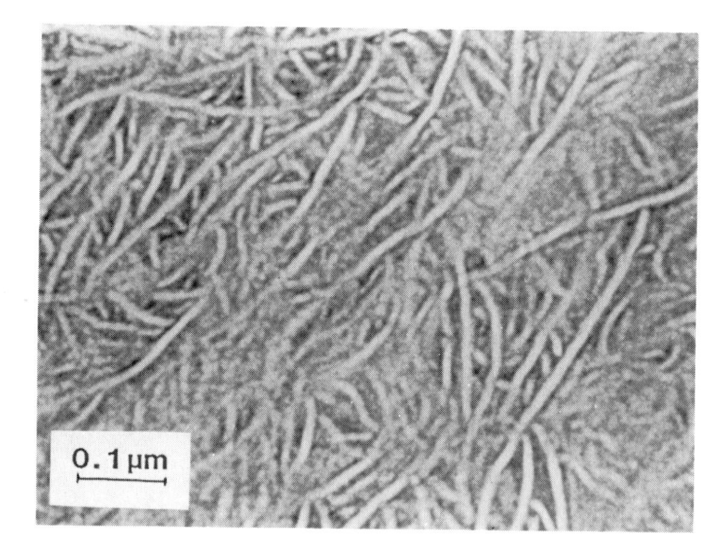

Figure 10.16 Structural details revealed by staining, in this case showing lamellar detail in cross-linked polyethylene (chlorosulphonation).

Of all the available staining reagents, osmium tetroxide (OsO_4) is of particular importance, being one of the most widely used chemicals for staining polymers. OsO_4 reacts with carbon–carbon double bonds and is therefore widely used as a means of enhancing contrast in unsaturated systems [146–149]. During this reaction cross-links are formed and the morphology present is therefore simultaneously stained and fixed. This technique has been used extensively to study unsaturated rubbers, where the topics investigated range from phase segregation in copolymers [150–154] to the crystallisation and growth of spherulites in polyisoprene [155, 156]. In unsaturated polymers, staining by OsO_4 involves a chemical reaction. However, this reagent has also been used in the study of saturated systems. In the simplest case, amplitude contrast may be induced in the TEM image by the preferential absorption of OsO_4, and such an approach has been employed to study PET [157]. However, a more subtle application of OsO_4 staining has been developed for the study of unsaturated materials that are porous, the so-called reactive inclusion technique [157]. In this approach the sample is first treated with a suitable unsaturated surfactant and then exposed to OsO_4. In this two-stage process the pores are first coated and then stained, so that the porous regions appear dark. The use of OsO_4 in this way as a negative stain, although initially developed for the study of microporous polypropylene membranes, has general utility for revealing the structure of voided materials. OsO_4 has been used in a conceptually similar way to study poly(tetrahydrofuran) [158]. After diffusion of *N*-vinylcarbazole into the polymer, the sample is then exposed to OsO_4. The *N*-vinylcarbazole within the amorphous regions reacts with the OsO_4 to give contrast in the final image.

Other staining reagents function in a similar manner to that described above. Phosphotungstic acid (PTA) reacts with functional groups such as hydroxyl, carbonyl and amino groups, and has been used in the study of, for example, polyoxymethylene [159] and polyamides [160]. A technique has also been developed for PTA staining of nitric acid-digested polypropylene [161]. However, the penetration of PTA into bulk polymers is limited to $\sim 100\,\mu m$ and for this reason, unlike the situation with OsO_4, samples are first sectioned and then stained.

Ruthenium tetroxide (RuO_4) has much in common with OsO_4 and it has been used in a similar way to study unsaturated systems [162, 163]. However, RuO_4 is a more strongly oxidising reagent than OsO_4 and appears to react with groups that include ethers, aromatic rings, alcohols and amines. Indeed, it has been reported that even linear polyethylene [164] may be lightly stained with RuO_4. Other examples of systems that have been successfully studied using RuO_4 staining are polystyrene [165, 167], cross-linked polyethylene [166, 167] and many blends and copolymers [168–170].

Unlike OsO_4, PTA and RuO_4, which have each been applied with success to many different types of polymers, staining with chlorosulphonic acid is a technique that was developed by Kanig [171] specifically to enhance lamellar contrast in polyethylene; it has not been applied to other materials. This technique has been used to reveal lamellar morphologies in cross-linked and low-density polyethylene [172], linear polyethylene [133, 134] and polyethylene blends [135, 136]. In addition to the study of the morphology of polyethylene itself, chlorosulphonation has also been used in the study of water-tree growth in cable insulation materials [173, 174]. This technique provides detailed, high-resolution images that reveal small portions of water trees, on the basis of which a mechanism of growth has been proposed. However, in line with general comments concerning staining made in section 10.4.2.1, this approach is less suitable for the study of overall water-tree morphologies than is the fixation–etching technique described in section 10.3.2.2. Therefore, in this case, as in the more general study of polymer morphology, staining and etching constitute complementary techniques, each of which has its particular forte.

Although the four staining reagents discussed above are probably the most widely used in connection with polymeric materials, many other chemicals have been used as a means of enhancing contrast in the TEM. For information concerning alternative reagents and details of the selection and application of staining to particular polymeric types, reference to *Polymer Microscopy* by Sawyer and Grubb [157] is recommended.

10.4.2.3 Permanganic etching. The first etchants to be devised which approached the ideal requirements were based upon oxidation by chromium, either in the form of chromic acid proper or else in various solutions containing sulphuric acid, phosphoric acid and water [175–177]. Recently, a similar etchant has been devised for poly(vinylidene fluoride) (PVF_2), a material that is not affected by the permanganic reagent [178]. Figure 10.17 shows an etched surface of PVF_2 crystallised at 163°C. Although good structural detail is evident, in regions where lamellae are viewed edge-on (A) the observed

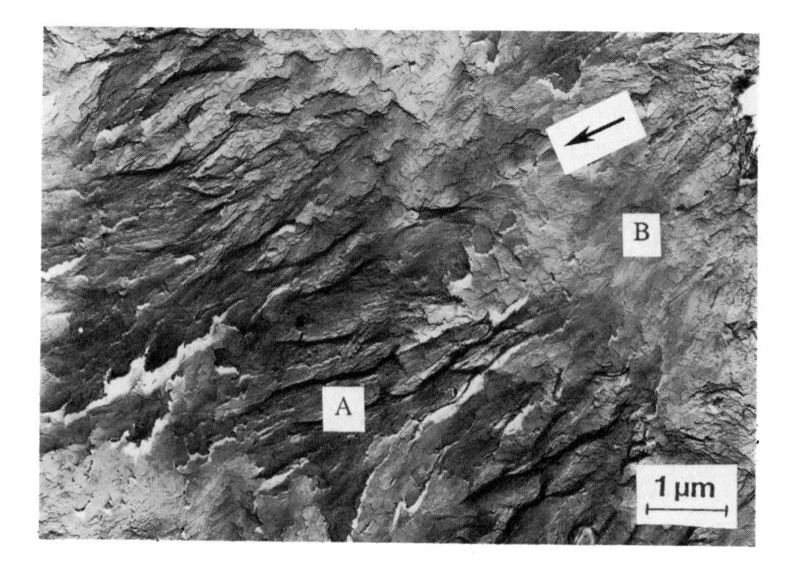

Figure 10.17 Etched surface showing lamellar detail in a banded spherulite of poly(vinylidene fluoride). Moving outwards along the radius (arrowed) the lamellar orientation changes from being seen close to flat on [B] to being edge on [A]. In region [A] the very dark radial features are a consequence of material being stripped from the specimen surface during replication.

contrast clearly indicates that some polymer has been stripped from the surface during replication. In a sense this indicates that the etching conditions employed were not ideal, that is, the etched surface has become friable as a consequence of amorphous regions being attacked too vigorously. However, mechanically removed polymer can be useful in certain circumstances, as discussed below.

Although several specific etchants have been devised [179–182], by far the most successful polymeric etching technique is permanganic etching [143, 183]. Initially reported in 1979, this approach to the study of polymer

(a)

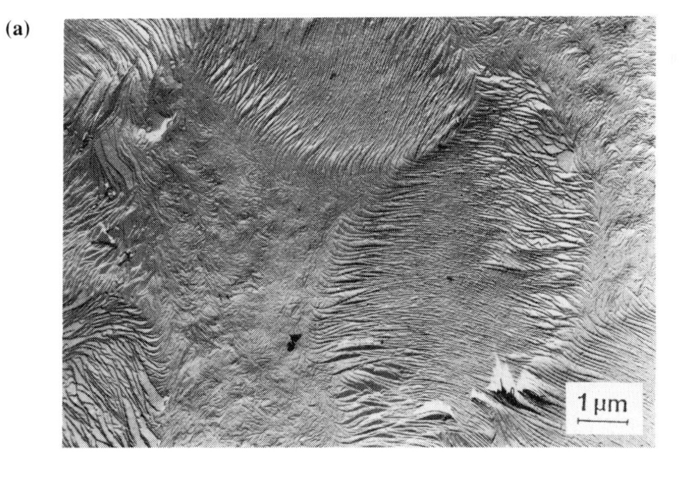

(b)

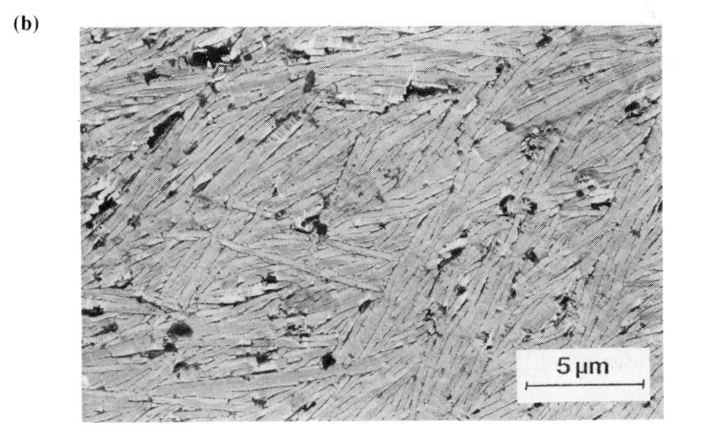

Figure 10.18 Samples etched with the permanganic reagent; (a) a blend of linear and low-density polyethylene; and (b) a main-chain liquid-crystal polymer.

morphology has, over the last thirteen years, been constantly refined and modified, so that there is now a wide spectrum of reagents that are all based upon potassium permanganate, the composition of the reagent being adjusted to optimise the resulting detail. This approach has been used to study morphology in systems which include polyethylene [184–189], polypropylene [190, 191], isotactic polystyrene [192–194], polybutene [195], PEEK [196, 197] and some main-chain liquid-crystal copolymers [198] (see Figure 10.18). The procedure for permanganic etching generally involves three steps. Firstly, the surface to be etched is prepared; this is usually a non-specific internal surface prepared by cutting the sample. It has been demonstrated that the technique of permanganic etching is highly selective and depends upon both structural factors and material deformation [199]. It is therefore important to prepare internal surfaces in such a way that damage introduced into the exposed surface by the preparation procedure is minimised. With this aim in mind, microtomy is ideal for this application. In addition, it also enables the optical appearance of sections removed from immediately adjacent to the exposed surface to be compared with the surface relief induced by etching, to check that the topography is indeed genuine and not merely an artefact of the chemical treatment [178, 196].

The exposed surface is then ready to be etched; a typical procedure is as follows. The etching reagent is made up by adding potassium permanganate, with stirring, to a 2:1 mixture of concentrated sulphuric acid and orthophosphoric acid to give a 0.7% solution. Initially, sulphuric acid was used alone, but this reagent is not without its problems. During etching, crystals precipitated from solutions on to the specimen surface and locally inhibited etching, with the result that artefacts were formed [183]. Typically, etching times are of the order of 2 h at room temperature, after which time etching is terminated by decanting the excess reagent into water and washing the samples first in dilute sulphuric acid and then in a solution of hydrogen peroxide in dilute sulphuric acid. Finally, the specimens are washed about six times in distilled water and twice in methanol, to remove all traces of the acid. This procedure is described in more detail in the initial publications [143, 183], as are the etchant variations that have been developed to meet particular requirements.

Etched surfaces, as previously noted, may be examined directly by SEM, but for TEM study it is necessary to produce a replica [200, 201]. Replication techniques for etched polymers are entirely in line with established metallurgical practices. Replication techniques can be categorised under two headings, indirect (two-stage) replication and direct replication. In two-stage replication, an intermediate replica is first formed in some material such as cellulose acetate, and it is this that is coated with the shadowing material and carbon. The acetate is finally dissolved away to leave the shadowed carbon replica sitting on a TEM grid. In direct replication, the sample surface is shadowed and carbon-coated directly. The problem then is how to remove the fragile replica from the sample surface. This may be achieved either chemically or mechanically. The most straightforward approach is to simply dissolve the sample away in an analogous approach to that used in the two-stage technique.

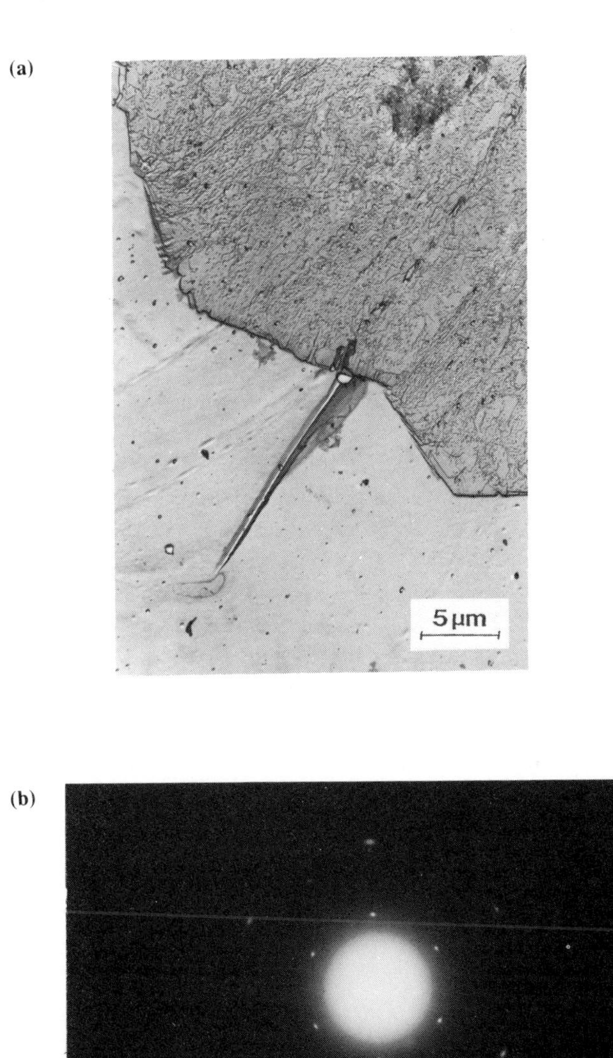

Figure 10.19 (a) Replica of an etched surface of isotactic polystyrene crystallised at 215°C in the form of a thin film. Replication involved an extraction procedure whereby a thin film of crystalline material remained adhering to the carbon replica. (b) Electron diffraction pattern from the replica shown in (a).

This route has the disadvantages that it may be slow, the sample itself is destroyed in the process and also, if the polymer swells prior to dissolution, the accompanying dimensional changes will shatter the replica. Thus, particularly with bulk samples, it is possible to wait a considerable time for dissolution and still to end up with no sample and little or no replica. However, sample dissolution can be a very effective approach. Consider the thin film of isotactic polystyrene shown in Figure 10.19; this was coated directly and then exposed to refluxing dichloromethane which removed all the non-crystalline polymer. The specimen was then, very briefly, exposed to refluxing chlorobenzene which disrupted most of the remaining crystalline material (see extraction replication, ref. 127). Finally, this disrupted material was removed by a third refluxing stage, again with dichloromethane. As previously noted, a major disadvantage of replication techniques is that they generally only provide information relating to the architecture present in the sample. However, by leaving a small amount of the sample material itself attached to the replica, diffraction experiments can also be performed, so that both crystal habits and the underlying crystallography or molecular orientation can be explored. A similar result can be achieved by mechanically stripping the replica together with small fragments of polymer from the etched surface. This is usually achieved by placing a droplet of poly(acrylic acid) (PAA) in solution in water on to the shadowed and carbon-coated etched surface. Once the water has evaporated and the PAA has dried, the backed replica may be prized off the etched surface. The shadowed carbon replica is finally recovered by dis-

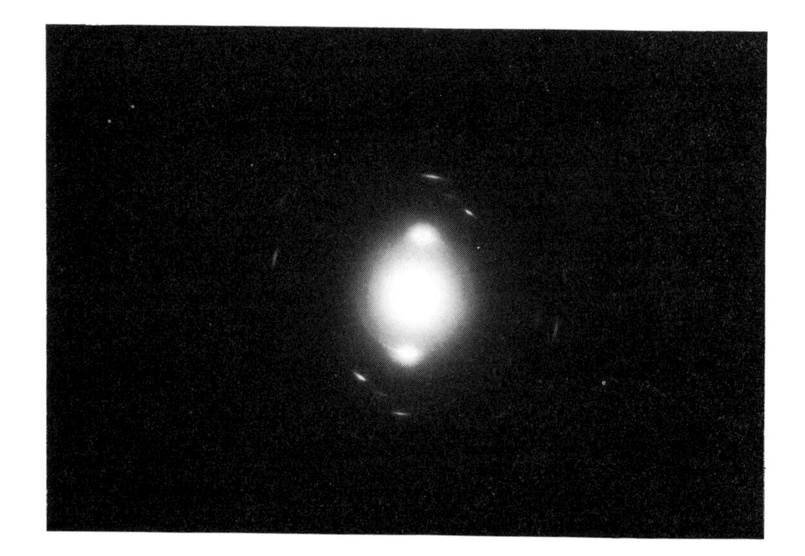

Figure 10.20 Electron diffraction pattern from a replica of a spherulite grown in a blend of linear and low-density polyethylene.

solving away the PAA in much the same way as for a cellulose acetate replica. On occasions, some polymer may be removed from the sample surface during two-stage replication. This is not generally desirable, since it interferes with the shadowing process and so disrupts the contrast that is seen in the final image. However, in systems that are already well characterised morphologically, etching conditions may be deliberately adjusted to produce a friable surface that will break up during replication. In this way complementary diffraction data may be obtained. Figure 10.20 shows an electron diffraction pattern obtained from an etched blend of linear and low-density polyethylene. Prolonged etching with a vigorous reagent has produced a friable surface, enabling this selected area diffraction pattern to be obtained.

Although etching and replication is generally considered to be a technique for the study of surface topography via inert, beam-stable replicas, it is clear from the above examples that by choosing the etchants specifically to develop surface relief or to disrupt the subsurface structure, many different effects may be achieved. Indeed, by adopting this wider view of etching, this approach to the study of structure may develop into new areas. Initially, permanganic etching was developed primarily for the study of lamellar morphologies in polyethylene. During the 1980s it was expanded into other polyolefins, aromatic systems, and liquid-crystal polymers, and now even non-crystalline conducting polymers have been studied by using this approach. Now, many of the challenges necessitate obtaining information concerning molecular orientation in addition to supermolecular data. Although etching has not been extensively employed with these aims in the past, the potential is clearly there and as understanding of the etching process develops the experimental possibilities will increase.

10.5 Conclusions

Although the preceding account has concentrated on particular aspects of microscopy, it is clear that, even in the more traditional areas that have been addressed here, microscopy of polymers is a topic that is both varied and rewarding. Limitations of space have inevitably prevented as full a discussion as would be ideal, but it is to be hoped that through reference to the cited works, both additional depth and breadth of information may be obtained.

Ultimately, microscopy is only one out of a very wide range of analytical techniques that can be used to characterise polymeric materials. In general, the strength of microscopy is that it enables the microstructure present to be visualised. Although scattering techniques which sample a much larger specimen volume are generally superior as far as the quantification of microstructure is concerned, it is nevertheless microscopy that initially enables realistic models to be developed by which scattering data may be interpreted. As such, the microscope is a potent instrument for the study of polymers, particularly when used in conjunction with other complementary analytical techniques.

Acknowledgements

The author is indebted to Professor D. C. Bassett, Dr D. Patel, Dr S. M. Moody and BICC, Dr G. C. Stevens and National Power, Ms J. C. Cooper and the National Grid Company, Mr S. J. Sutton, Mr P. Butler and Mr A. Forgham for providing the material without which this chapter could not have been produced.

References

1. P. J. Goodhew and F. J. Humphreys (1988). *Electron Microscopy and Analysis*, Taylor and Francis, London.
2. D. Pitt (1991). *Physics World*, **4**(10), 19–20.
3. P. Fischer (1982). In *Electrical Properties of Polymers*, D. A. Seanor (ed.), Academic Press, New York, pp. 319–367.
4. D. A. Hemsley (1989). In *Applied Polymer Light Microscopy*, D. A. Hemsley (ed.), Elsevier, London, pp. 39–72.
5. B. Bridge, M. J. Folkes and H. Jahankhani (1990). *J. Mater. Sci.*, **25**, 3061–66.
6. W. Gölz (1985). *Colloid Polym. Sci.*, **263**, 286–92.
7. R. H. Olley, D. C. Bassett, A. S. Vaughan, V. A. A. Banks, P. B. McAllister and S. M. Moody (1992). *J. Mater. Sci.*, **27**, in the press.
8. W. G. Hartley (1979). *Hartley's Microscopy*, Senecio Publishing, Charlbury.
9. S. Z. D. Cheng, S. K. Lee, J. S. Barley, S. L. C. Hsu and F. W. Harris (1991). *Macromolecules*, **24**, 1883–89.
10. H. Hakemi and H. A. A. Rasoul (1990). *Polym. Commun.*, **31**, 82–85.
11. G. R. Mitchell and A. H. Windle (1988). In *Developments in Crystalline Polymers—2*, D. C. Bassett (ed.), Elsevier, London, pp. 115–175.
12. R. Muller and D. Froelich (1985). *Polymer*, **26**, 1477–82.
13. H. Saito, M. Takahashi and I. Inoue (1991). *Macromolecules*, **24**, 6536–38.
14. T. E. Strzelecka and R. L. Rill (1991). *Macromolecules*, **24**, 5124–33.
15. S. J. Pichen, J. Aerts, H. L. Doppert, A. J. Reuvers and M. G. Northolt (1991). *Macromolecules*, **24**, 1366–75.
16. R. S. Farinato (1988). *Polymer*, **29**, 2182–90.
17. T. Inoue, H. Okamoto and K. Osaki (1991). *Macromolecules*, **24**, 5670–75.
18. S. L. Bazhenov and A. A. Berlin (1990). *J. Mater. Sci.*, **25**, 3941–49.
19. B. P. Saville (1989). In *Applied Polymer Light Microscopy*, D. A. Hemsley (ed.), Elsevier, London, pp. 73–109.
20. A. F. Hallimond (1970). *The Polarizing Microscope*, Vickers, York.
21. H. D. Keith and F. J. Padden (1964). *J. Appl. Phys.*, **35**, 1270–85.
22. G. S. Y. Yeh and S. L. Lambert (1972). *J. Polym. Sci. A-2*, **10**, 1183–91.
23. Z. Bartczak, A. Galeski and M. Pracella (1986). *Polymer*, **27**, 537–43.
24. J. G. Lim, W. George and B. S. Gupta (1987). *J. Appl. Polym. Sci.*, **33**, 989–96.
25. H. Awaya (1988). *Polymer*, **29**, 591–96.
26. A. J. Lovinger and T. T. Wang (1979). *Polymer*, **20**, 725–32.
27. C.-M. Chu and G. L. Wilkes (1974). *J. Makromol. Sci. Phys.*, **B10**, 231–54.
28. E. Devaux and B. Chabert (1991). *Polym. Commun.*, **32**, 464–68.
29. K. Shimamura (1983). *Makromol. Chem., Rapid Commun.*, **4**, 107–11.
30. D. C. Bassett and B. Turner (1974). *Phil. Mag.*, **29**, 285–307.
31. P. H. Lindenmeyer and V. F. Holland (1964). *J. Appl. Phys.*, **35**, 55–58.
32. H. D. Keith and F. J. Padden (1959). *J. Polym. Sci.*, **39**, 101–22.
33. A. Keller (1959). *J. Polym. Sci.*, **39**, 151–73.

34. D. C. Bassett (1981). *Principles of Polymer Morphology*, Cambridge University Press, Cambridge.
35. W. J. Patzelt (1985). *Polarized Light Microscopy: Principles, Instruments, Applications*, Leitz, Wetzlar.
36. A. H. Bennett, H. Osterberg, H. Jupnik and O. W. Richards (1951). *Phase Microscopy: Principles and Applications*, Wiley, New York.
37. D. C. Bassett and A. Keller (1962). *Phil. Mag.* **7**, 1553–84.
38. H. D. Keith (1964). *J. Appl. Phys.*, **35**, 3115–26.
39. A. J. Lovinger, D. D. Davis and B. Lotz (1991). *Macromolecules*, **24**, 5552–60.
40. W. Salomons, G. ten Brinke and F. E. Karasz (1991). *Polym. Commun.*, **32**, 185–87.
41. H. D. Keith (1964). *J. Polymer Sci. A*, **2**, 4339–60.
42. R. M. Briber and F. Khoury (1987). *Polymer*, **28**, 38–46.
43. A. J. Hale (1958). *The Interference Microscope in Biological Research*, Livingstone, Edinburgh.
44. D. Hemsley (1989). In *Comprehensive Polymer Science*, Vol. 1, *Polymer Characterization*, C. Booth and C. Price (eds), Pergamon Press, Oxford, pp. 765–784.
45. R. Hoffman (1989). In *Applied Polymer Light Microscopy*, D. A. Hemsley (ed.), Elsevier, London, pp. 151–184.
46. R. Zbinden (1964). *Infrared Spectroscopy of High Polymers*, Academic Press, New York.
47. P. Calvert and N. C. Billingham (1989). In *Applied Polymer Light Microscopy*, D. A. Hemsley (ed.), Elsevier, London, pp. 233–271.
48. C. W. Oatley (1972). *The Scanning Electron Microscope*, Cambridge University Press, Cambridge.
49. J. I. Goldstein and H. Yakowitz (eds) (1975). *Practical Scanning Electron Microscopy: Electron and Ion Microprobe Analysis*, Plenum, New York.
50. D. B. Holt, M. D. Muir, P. R. Grant and I. M. Boswarva (eds) (1974). *Quantitative Scanning Electron Microscopy*, Academic Press, London.
51. Y. Lu, J. Li and W. Wu (1989). *Synth. Meth.*, **30**, 87–95.
52. A. Charlesby (1960). *Atomic Radiation and Polymers*, Pergamon Press, Oxford.
53. A. Chapiro (1962). *Radiation Chemistry of Polymeric Systems*, Interscience, London.
54. D. Campbell and J. R. White (1989). *Polymer Characterization*, Chapman and Hall, London.
55. A. Chenite and F. Brisse (1991). *Macromolecules*, **24**, 2221–25.
56. W. H. Jo, H. C. Kim and D. H. Baik (1991). *Macromolecules*, **24**, 2231–35.
57. A. Molnár and A. Eisenberg (1991). *Polym. Commun.*, **32**, 370–73.
58. H. Kanazawa, J. Stejny and A. Keller (1990). *J. Mater. Sci.*, **25**, 3838–42.
59. S. Saiello, J. Kenny and L. Nicolais (1990). *J. Mater. Sci.*, **25**, 3493–96.
60. S.-A. Chen and T.-S. Lee (1987). *J. Polym. Sci. C, Polym. Lett.*, **25**, 455–60.
61. J. H. Hwang and S. C. Yang (1989). *Synth. Meth.*, **29**, E271–E276.
62. K. M. Cheung, D. Bloor and G. C. Stevens (1990). *J. Mater. Sci.*, **25**, 3814–37.
63. R. H. Olley (1986). *Sci. Prog., Oxford*, **70**, 17–43.
64. N. J. Tyrer and P. R. Sundararajan (1985). *Macromolecules*, **18**, 511–18.
65. L. Bartosiewicz and Z. Mencik (1974). *J. Polym. Sci., Polym. Phys. Ed.*, **12**, 1163–75.
66. F. P. Reding and E. R. Walter (1959). *J. Polym. Sci.*, **38**, 141–55.
67. L. J. Mao, Z. P. Zhang and S. K. Ying (1991). *Polym. Commun.*, **32**, 242–44.
68. D. C. Bassett, R. H. Olley and I. A. M. Al Raheil (1988). *Polymer*, **29**, 1539–43.
69. J. A. Di Corleto and D. C. Bassett (1990). *Polymer*, **31**, 1971–77.
70. W.-J. Ho and R. Salovey (1981). *Polym. Eng. Sci.*, **21**, 839–43.
71. F. Khoury, personal communication.
72. A. Miyamoto, H. Kikuchi, S. Kobayashi, Y. Morimura and T. Kajiyama (1991). *Macromolecules*, **24**, 3915–20.
73. Y. S. Yang and L. J. Lee (1988). *Polymer*, **29**, 1793–1800.
74. R. P. Palmer and A. J. Cobbold (1964). *Makromol. Chem.*, **74**, 174–89.
75. A. Keller, E. Martuscelli, D. J. Priest and Y. Udagawa (1971). *J. Polym. Sci. A-2*, **9**, 1807–37.
76. R. H. Olley and D. C. Bassett (1977). *J. Polym. Sci. Polym. Phys. Ed.*, **15**, 1011–27.
77. T. J. Pecorini, R. W. Hertzberg and J. A. Mason (1990). *J. Mater. Sci.*, **25**, 3385–95.
78. C. W. Hock (1966). *J. Polym. Sci. A-2*, **4**, 227–42.
79. J. Rhee amd B. Crist (1991). *Macromolecules*, **24**, 5663–69.
80. J. T. Muellerleile, G. L. Wilkes and G. A. York (1991). *Polym. Commun.*, **32**, 176–79.

81. T. A. Skotheim (ed.) (1986). *Handbook of Conducting Polymers*, Vols. 1 and 2, Marcel Dekker, New York.
82. A. Garton, P. Z. Sturgeon, D. J. Carlsson and D. M. Wiles (1978). *J. Mater. Sci.*, **13**, 2205–10.
83. T. Tagawa and J. Mori (1978). *J. Electron Microsc.*, **27**, 267–74.
84. J. G. M. van Gisbergen, C. P. J. H. Borgmans, M. C. M. van der Sanden and P. J. Lemstra (1990). *Polym. Commun.*, **31**, 162–64.
85. F. D. Egitto, V. Vukanovic and G. N. Taylor (1990). In *Plasma Deposition, Treatment and Etching of Polymers*, R. d'Agostino (ed.), Academic Press, New York, pp. 321–422.
86. P. H. Till (1957). *J. Polym. Sci.*, **14**, 301–06.
87. V. F. Holland and R. F. Miller (1964). *J. Appl. Phys.*, **35**, 3241–48.
88. A. J. Lovinger and D. D. Davis (1985). *Polym. Commun.*, **26**, 322–24.
89. A. Toda (1989). *J. Polym. Sci. B, Polym. Phys.*, **27**, 53–70.
90. A. J. Lovinger (1980). *J. Polym. Sci. Polym. Phys. Ed.*, **18**, 793–809.
91. C. C. Hsu and P. H. Geil (1984). *J. Appl. Phys.*, **56**, 2404–11.
92. A. J. Lovinger and D. D. Davis (1985). *J. Appl. Phys.*, **58**, 2843–53.
93. H. D. Keith, F. J. Padden, B. Lotz and J. C. Wittmann (1989). *Macromolecules*, **22**, 2230–38.
94. B. C. Edwards and P. J. Phillips (1974). *Polymer*, **15**, 351–56.
95. D. T. Grubb (1974). *J. Mater. Sci.*, **9**, 1715–36.
96. J. E. Breedon, J. F. Jackson, M. J. Marcinkowski and M. E. Taylor (1973). *J. Mater. Sci.*, **8**, 1071–82.
97. E. S. Kempner, R. Wood and R. Salovey (1986). *J. Polym. Sci. B, Polym. Phys.*, **24**, 2337–43.
98. P. B. Hirsch, A. Howie, R. B. Nicholson, D. W. Pashley and M. J. Whelan (1965). *Electron Microscopy of Thin Crystals*, Butterworths, London.
99. R. S. Saunders, R. E. Cohen and R. S. Schrok (1991). *Macromolecules*, **24**, 5599–605.
100. J. M. DeSimone, G. A. York, J. E. McGrath, A. S. Gozdz and M. J. Bowden (1991). *Macromolecules*, **24**, 5330–39.
101. V. Sankaran, R. E. Cohen, C. C. Cummins and R. R. Schrok (1991). *Macromolecules*, **24**, 6664–69.
102. A. I. Schneider, J. Blackwell, H. Pierlartzik and A. Karbach (1991). *Macromolecules*, **24**, 5676–82.
103. C. Creton, E. J. Kramer and G. Hadziioannon (1991). *Macromolecules*, **24**, 1846–53.
104. P. Blais and R. St John Manley (1960), *J. Polym. Sci. A-2*, **4**, 1022–24.
105. H. D. Keith, R. G. Vadimsky and F. J. Padden (1970). *J. Polym. Sci. A-2*, **8**, 1687–96.
106. A. Thiery, C. Straupé, B. Lotz and J. C. Wittmann (1990). *Polym. Commun.*, **31**, 299–301.
107. A. J. Waddon, L. C. Brookes, L. J. Heyderman and M. J. Hill (1990). *Polym. Commun.*, **31**, 5–7.
108. L. Leemans, R. Fayt, Ph. Teyssie and N. C. de Jaeger (1991). *Macromolecules*, **24**, 5922–25.
109. D. T. Grubb (1982). In *Developments in Crystalline Polymers—1*, D. C. Bassett (ed.) Applied Science, London, pp. 1–35.
110. P. H. Geil (1963). *Polymer Single Crystals*, Interscience, New York.
111. J. Petermann and Y. Xu (1990). *Polym. Commun.*, **31**, 428–30.
112. R. J. Spontak and A. H. Windle (1990). *J. Mater. Sci.*, **25**, 2727–36.
113. D. C. Bassett, F. C. Frank and A. Keller (1959). *Nature*, **184**, 810–11.
114. D. C. Bassett (1964). *Phil. Mag.*, **10**, 595–615.
115. J. R. White, personal communication.
116. H. Chanzy, T. Folda, P. Smith, K. Gardner and J.-F. Revol (1986). *J. Mater. Sci. Lett.*, **5**, 1045–47.
117. J.-F. Revol and R. St. John Manley (1986). *J. Mater. Sci. Lett.*, **5**, 249–51.
118. D. C. Martin, L. L. Berger and K. H. Gardner (1991). *Macromolecules*, **24**, 3921–28.
119. W. Zhang and E. L. Thomas (1991). *Polym. Commun.*, **32**, 482–85.
120. J. Mayer and W. Probst (1992). *Microscopy and Analysis*, **28**, 27.
121. M. Kunz, M. Möller and H.-J. Cantow (1987). *Makromol. Chem., Rapid Commun.*, **8**, 401–10.
122. J. A. N. Zasadzinski, A. Chu and R. K. Prud'homme (1986). *Macromolecules*, **19**, 2960–64.
123. D. C. Bassett and R. Davitt (1974). *Polymer*, **15**, 721–28.
124. F. Khoury and E. Passaglia (1976). In *Treatise on Solid State Chemistry*, Vol. 3, *Crystalline and Non-crystalline Solids*, N. B. Hannay (ed.), Plenum, New York, pp. 335–496.
125. D. C. Bassett and D. R. Carder (1973). *Phil. Mag.*, **28**, 535–45.
126. D. C. Bassett and A. M. Hodge (1978). *Proc. Roy. Soc., London*, **A359**, 121–32.

127. D. C. Bassett (1961). *Phil. Mag.*, **6**, 1053–56.
128. A. J. Lovinger (1981). *Macromolecules*, **14**, 322–25.
129. A. J. Lovinger (1983). *J. Polym. Sci. Polym. Phys. Ed.*, **21**, 97–110.
130. B. Lotz, S. Graff and J. C. Wittmann (1986). *J. Polym. Sci. Part B, Polym. Phys.*, **24**, 2017–32.
131. J. C. Wittmann and B. Lotz (1985). *J. Polym. Sci. Polym. Phys. Ed.*, **23**, 205–26.
132. R. J. Spontak, M. C. Williams and C. N. Schooley (1986). *J. Mater. Sci.*, **21**, 3173–78.
133. I. G. Voigt-Martin and L. Mandelkern (1989). *J. Polym. Sci. B, Polym. Phys.*, **27**, 967–91.
134. I. G. Voigt-Martin, A. J. Peacock and L. Mandelkern (1989). *J. Polym. Sci. B, Polym. Phys.*, **27**, 957–65.
135. M. T. Conde Braña, J. I. Iragorri Sainz, B. Terselius and U. W. Gedde (1989). *Polymer*, **30**, 410–15.
136. M. Rego Lopez, M. T. Conde Braña, B. Terselius and U. W. Gedde, (1988). *Polymer*, **29**, 1045–51.
137. K. C. Dauzinas, R. E. Cohen and A. F. Halasa (1991). *Macromolecules*, **24**, 4457–59.
138. L. Cazzaniga and R. E. Cohen (1991). *Macromolecules*, **24**, 5817–22.
139. K. Ishizu, A. Omote and T. Fukutomi (1991). *Polymer*, **31**, 2135–40.
140. L. A. Lay (1991). *Metals and Materials*, **7**, 543–47.
141. E. Jakopić (1960). In *Proc. European Regional Conf. Electron Microscopy*, Delft, 1960, Vol. 1, A. C. Houwink and B. J. Spit (eds.), pp. 559–563.
142. B. J. Spit (1960). In *Proc. European Regional Conf. Electron Microscopy*, Delft, 1960, Vol. 1, A. C. Houwink and B. J. Spit (eds) pp. 564–567.
143. R. H. Olley, A. M. Hodge and D. C. Bassett (1979). *J. Polym. Sci. Polym. Phys. Ed.*, **17**, 627–43.
144. T. M. Babchinitser, A. E. Shworak, C. M. Frenkel and D. V. Smelianski (1991). *Polym. Commun.*, **32**, 409–11.
145. Y.-C. Chen, V. Dimonie and M. S. El-Aasser (1991). *Macromolecules*, **24**, 3779–87.
146. R. A. Weiss, S. Sasongko and R. Jerome (1991). *Macromolecules*, **24**, 2271–77.
147. J. Csernica, D. H. Rein, R. F. Baddour and R. E. Cohen (1991). *Macromolecules*, **24**, 3612–17.
148. O. S. Gebizlioglu, A. S. Argon and R. E. Cohen (1985). *Polymer*, **26**, 519–28.
149. R. A. Mendelson (1985). *J. Polym. Sci. Polym. Phys. Ed.*, **23**, 1975–95.
150. T. Hashimoto, M. Shibayama and H. Kawai (1980). *Macromolecules*, **13**, 1237–47.
151. T. Hashimoto, Y. Tsukahara, K. Tachi and H. Kawai (1983). *Macromolecules*, **16**, 648–57.
152. T. Hashimoto, K. Kimishima and H. Hasegawa (1991). *Macromolecules*, **24**, 5704–12.
153. K. I. Winey, E. L. Thomas and L. J. Fetters (1991). *Macromolecules*, **24**, 6182–88.
154. J. A. Odell, J. Dlugosz and A. Keller (1976). *J. Polym. Sci. Polym. Phys. Ed.*, **14**, 861–67.
155. C. K. L. Davies and O. E. Long (1977). *J. Mater. Sci.*, **12**, 2165–83.
156. G. J. Rensch, P. J. Phillips, N. Vatansever and A. Gonzalez (1986). *J. Polym. Sci. B, Polym. Phys.*, **24**, 1943–59.
157. L. C. Sawyer and D. T. Grubb (1987). *Polymer Microscopy*, Chapman and Hall, London.
158. S. Ishikawa, K. Ishizu and T. Fukutomi (1990). *Polym. Commun.*, **31**, 407–08.
159. A. Peterlin, P. Ingram and H. Kiho (1965). *Makromol. Chem.*, **86**, 294–97.
160. J. A. Rusnock and D. T. Hansen (1965). *J. Polym. Sci. A*, **3**, 647–58.
161. C. W. Hock (1967). *J. Polym. Sci. A-2*, **5**, 471–78.
162. R. Vitali and E. Montani (1980). *Polymer*, **21**, 1220–22.
163. B. A. Wood (1989). *Soc. Plast. Eng., Tech. Pap.*, **35**, 1859–61.
164. Y. M. T. Tervoort-Engelen and P. J. Lemstra (1991). *Polym. Commun.*, **32**, 343–45.
165. K. Ishiza, S. Yukimasa and R. Saito (1991). *Polym. Commun.*, **32**, 386–89.
166. N. Hozumi, T. Okamoto and H. Fukagawa (1988). *Jap. J. Appl. Phys.*, **27**, 1230–33.
167. T. Okamoto, M. Ishida and N. Hozumi (1989). *IEEE Trans. Electrical Insulation*, **EI-24**, 599–607.
168. T. Yamaguchi, S. Nakao and S. Kimura (1991). *Macromolecules*, **24**, 5522–27.
169. J. S. Trent, J. I. Scheinbeim and P. R. Couchman (1983). *Macromolecules*, **16**, 589–98.
170. Y. M. T. Tervoort-Engelen and J. van Gisbergen (1991). *Polym. Commun.*, **32**, 261–63.
171. G. Kanig (1973). *Kolloid-Z. Z. Polym.*, **251**, 782–83.
172. G. C. Stevens, personal communication.

173. G. Capaccio, W. Goltz and L. J. Rose (1985). In *ETG Conf. Proc. 16, Long-term Performance of High-voltage Insulations*, M. Beyer (ed.), VDE, Berlin, pp. 123–126.
174. J. J. de Bellet, G. Matey, L. Rose, V. Rose, J. C. Filippini, Y. Poggi and V. Raharimalala (1987). *IEEE Trans. Electrical Insulation*, **EI-22**, 211–17.
175. V. J. Armond and J. R. Atkinson (1969). *J. Mater. Sci.*, **4**, 509–17.
176. C. B. Bucknall and I. C. Drinkwater (1974). *Polymer*, **15**, 254–55.
177. D. R. Fitchmun and Z. Menick (1973). *J. Polymer Sci. Polym. Phys. Ed.*, **11**, 951–71.
178. A. S. Vaughan (1992). *J. Mater. Sci.* In the press.
179. H. S. Bu, S. Z. D. Cheng and B. Wunderlich (1987). *Polym. Bull.*, **17**, 567–71.
180. H. S. Bu, S. Z. D. Cheng and B. Wunderlich (1988). *Polymer*, **29**, 1603–07.
181. T.-J. Lemmon, S. Hanna and A. H. Windle (1989). *Polym. Commun.*, **30**, 2–4.
182. D. P. Heberer, S. Z. D. Cheng, J. S. Barley, S. H.-S. Lien, R. G. Bryant and F. W. Harris (1991). *Macromolecules*, **24**, 1890–98.
183. R. H. Olley and D. C. Bassett (1982). *Polymer*, **23**, 1707–10.
184. D. C. Bassett and A. M. Hodge (1981). *Proc. Roy. Soc. London*, **A377**, 25–37.
185. D. C. Bassett, A. M. Hodge and R. H. Olley (1981). *Proc. Roy. Soc., London*, **A377**, 39–60.
186. D. C. Bassett and A. M. Hodge (1981). *Proc. Roy. Soc., London*, **A377**, 61–71.
187. R. M. Gohil (1986). *Colloid Polym. Sci.*, **264**, 951–64.
188. D. R. Norton and A. Keller (1984). *J. Mater. Sci.*, **19**, 447–56.
189. K. L. Naylor and P. J. Phillips (1983). *J. Polym. Sci. Polym. Phys. Ed.*, **21**, 2011–26.
190. D. C. Bassett and R. H. Olley (1984). *Polymer*, **25**, 935–43.
191. R. H. Olley and D. C. Bassett (1989). *Polymer*, **30**, 399–409.
192. D. C. Bassett and A. S. Vaughan (1985). *Polymer*, **26**, 717–25.
193. A. S. Vaughan, D. C. Bassett and R. H. Olley (1986). In *Morphology of Polymers*, B. Sedlacek (ed.), Walter de Gruyter, Berlin, pp. 387–398.
194. A. S. Vaughan and D. C. Bassett (1988). *Polymer*, **29**, 1397–1401.
195. T.-C. Hsu and P. H. Geil (1990). *Polym. Commun.*, **31**, 105–08.
196. R. H. Olley, D. C. Bassett and D. J. Blundell (1986). *Polymer*, **27**, 344–48.
197. D. C. Bassett, R. H. Olley and I. A. M. Al Raheil (1988). *Polymer*, **29**, 1745–54.
198. R. N. Dutton, D. C. Bassett and G. R. Mitchell, unpublished work.
199. A. M. Freedman, D. C. Bassett, A. S. Vaughan and R. H. Olley (1986). *Polymer*, **27**, 1163–69.
200. D. E. Bradley (1965). In *Techniques for Electron Microscopy*, D. H. Kay (ed.), Blackwell, Oxford, pp.98–152.
201. M. Tsuji (1989). In *Comprehensive Polymer Science*, Vol. 1, *Polymer Characterization*, C. Booth and C. Price (eds), Pergamon Press, Oxford, pp. 785–840.

11 The characterisation of polymer surfaces by XPS and SIMS

H. S. MUNRO and S. SINGH

11.1 Introduction

The availability of an increasing number of techniques for the characterisation of polymeric surfaces over the last twenty years or so has lead to many important improvements in the understanding of interfacial phenomena (e.g. adhesion, wetting, printing, biocompatability) in these materials. In particular the development of X-ray photoelectron spectroscopy (XPS), as applied to polymers, has allowed a wealth of information regarding the surface elemental and functional-group composition to be obtained. Many interesting correlations between the surface composition, as determined by XPS, and interfacial phenomena have been noted. These observations have led to the realisation that the structure and properties of the outermost 5 nm or so of a polymer can control the performance of the material in many applications. As with many characterisation techniques, there are limitations on the information obtained by XPS. This has, in part, provided the impetus for developing other techniques which would complement XPS, in terms of both compositional data and surface sensitivity. Static secondary ion mass spectrometry (SSIMS) has emerged in the last ten years as one of the most powerful techniques complementary to XPS. As implied by the name of the technique, a mass spectrum representative of the surface is obtained. This chapter will give brief outlines of the two techniques and some of their applications.

Before commencing on the main points of this chapter, it is worth emphasising that both of these techniques require high-vacuum conditions in order to operate. At first sight, many might expect that on introducing a polymeric material into a high-vacuum system serious outgassing and contamination problems would ensue. However, the continued application of XPS and SSIMS to polymeric materials is evidence that these are not major problems if due care and attention is given to sample handling.

11.2 X-ray photoelectron spectroscopy (XPS)

This section will only give the brief essentials of the technique, as it has been covered in detail elsewhere [1]. An example of the configuration of a typical commercial XPS instrument is shown in Figure 11.1. The main features are a vacuum chamber, an X-ray gun, a sample manipulator and an electron-energy analyser. When a sample is irradiated with soft X-rays, photoelectrons are emitted. Depending on the energy of the X-ray, the photoelectrons can arise from valence or core levels. A schematic diagram of the processes involved is given in Figure 11.2. In this case the electron is shown to to arise from the 1s level. On removal of this electron, a core-level vacancy is left, i.e. the atom from which the electron is emitted is in a core-ionised state. The vacancy may be filled by an electron from a higher level. The energy released in filling the core vacancy may result in the emission of an X-ray or may be transferred to a third, more weakly bound electron. The latter is then emitted as an Auger electron. Typical X-ray sources are Mg Kα (1253.6 eV) and Al Kα (1486.6 eV). These are energetic enough to remove the 1s electrons from, for example, carbon, oxygen, nitrogen and fluorine.

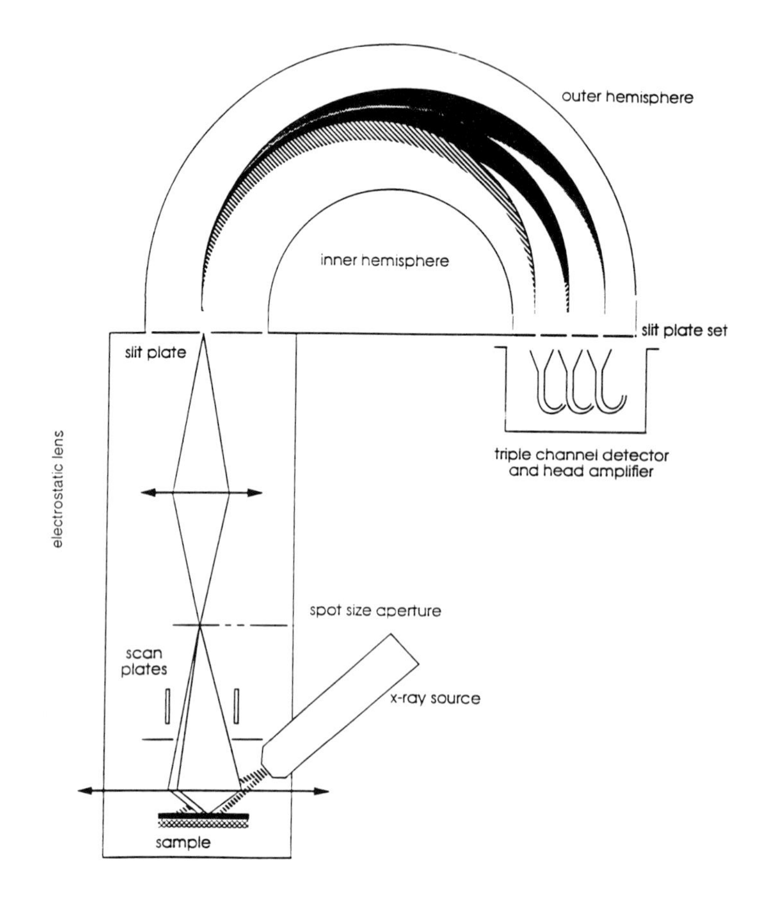

Figure 11.1 Schematic diagram of a commercial photoelectron spectrometer.

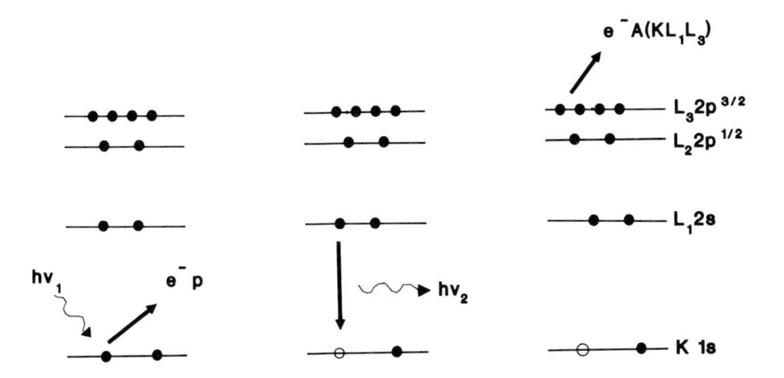

Figure 11.2 Schematic diagram of the processes involved in core-level photoemission.

11.2.1 Spectra

The photoelectrons are collected by a lens system and focussed into the energy analyser. The analyser effectively counts the number of electrons with a given kinetic energy. By allowing the analyser to sweep across a wide energy scale a survey spectrum of the type shown in Figure 11.3 is obtained. The energy scale is displayed as binding energy. This is obtained from the Einstein relation:

$$\text{B.E.} = h\nu - \text{K.E.} - \phi$$

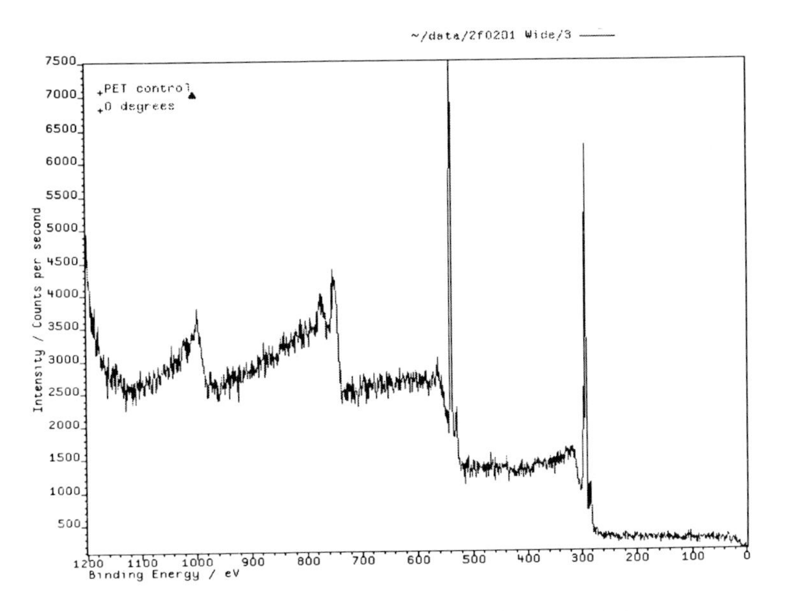

Figure 11.3 Survey spectrum of poly(ethylene terephthalate).

where $h\nu$ is the X-ray energy, ϕ the sample work function, K.E. the kinetic energy of the photoelectron and B.E. the binding energy. The binding energies are characteristic of the atomic core level from which the photoelectron was emitted, and enable the elemental composition of the surface to be obtained. All elements except hydrogen can be detected.

The main features in the spectrum are a secondary electron background on which a number of peaks are superimposed. The main peaks are due to the photoelectrons e.g. C 1s at ~ 290 eV and O 1s at ~530 eV. Weaker, broader peaks at higher binding energy arise from oxygen and carbon Auger electrons. Although not evident in this spectrum, photoelectrons arising from the valence levels would appear in the 0–30 eV region. They are present in the spectrum in Figure 11.3 but are weak in intensity and hence lost in the background. It is in part due to the low intensity of the valence-level relative to the core level-electrons that XPS is primarily a core-level technique. This apparent restriction would appear to restrict the application of XPS to elemental analysis only. However, as stated in the introduction, some form of functional-group analysis is also possible, i.e. what is bonded to what.

11.2.2 Chemical shifts

Chemical bonding directly involves the valence electrons of the atoms and it might be thought that the core-level electrons are too inaccessible to have their binding energies perturbed. However, the core-level binding energies in an atom are influenced by the local electronic environment. As a result an atom in a molecule may exhibit a small range of binding energies (chemical shifts). Figure 11.4 is the C 1s region for poly(ethylene terephthalate) (PET) recorded at a higher energy resolution than in Figure 11.3. The most intense peak at 292.3 eV arises from carbons bonded to carbon and hydrogen. The shoulder to higher binding energy is due to carbon singly bonded to oxygen, and the peak at 296.4 eV is due to a carboxylate group.

There are two further environments identifiable from the C 1s envelope. The first is the weak peak about 3 eV to the higher binding-energy side of the carboxylate peak. This is a shake-up satellite and is diagnostic of the presence of unsaturation. For PET, the source of the unsaturation is the aromatic ring. As can be seen from Figure 11.4 a number of Gaussian peaks have been fitted to the C 1s envelope. The peak-fitting procedure is based on a chemical fit, not a mathematical one. The relative positions for carbon–oxygen environments are well documented (as they are for others) [1, 2]. In the example shown, the full widths at half maximum of the peaks are assumed to be constant. Thus, the only parameter to be optimised is the height of the peak. Other peak shapes can be used; however, for most applications the approximation to a Gaussian is perfectly adequate. The use of good peak-fitting procedures allows important insights to the nature of the functional groups present at the surface of the polymer.

From the peak fitting, the final carbon environment becomes apparent. It arises from a carbon atom attached to an ester group (i.e. the aromatic carbons

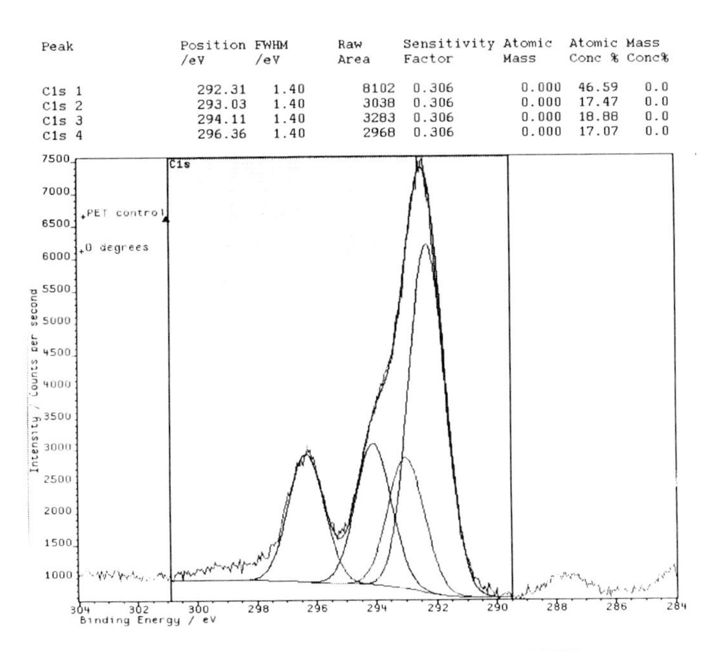

Peak	Position /eV	FWHM /eV	Raw Area	Sensitivity Factor	Atomic Mass	Atomic Conc %	Mass Conc%
C1s 1	292.31	1.40	8102	0.306	0.000	46.59	0.0
C1s 2	293.03	1.40	3038	0.306	0.000	17.47	0.0
C1s 3	294.11	1.40	3283	0.306	0.000	18.88	0.0
C1s 4	296.36	1.40	2968	0.306	0.000	17.07	0.0

Figure 11.4 High-resolution C 1s spectrum of PET.

bonded to the ester groups). These carbons experience a secondary shift [3]. Carbon attached to a nitrile group is also secondary-shifted [1].

The binding energies quoted above do not correspond to the ones quoted in the literature. They are approximately 8 eV too high. As the sample is irradiated and photoelectrons are emitted, the surface of an insulator will become positively charged. As a result, the electrons being emitted experience a retarding field and lose kinetic energy. This is referred to as sample charging. When non-monochromated X-rays are used, there is a sufficient supply of secondary electrons from the window of the X-ray gun to reduce the charging to a few eV. Monochromated sources can give rise to sample charging of ~ 100 eV or greater. By exposing the sample to a source of electrons from a low-energy electron gun the charging can be neutralised. All the core levels will be shifted by sample charging to the same degree. An internal energy reference is used to recalibrate the spectra once the sample charging has been minimised and has reached equilibrium. For polymers, the internal reference is the hydrocarbon peak (quoted as 285 or 284.6 eV in the literature).

11.2.3 Surface sensitivity

The surface sensitivity of XPS arises not from the penetration depth of the X-rays but from the limited distance an electron of a given kinetic energy can travel through a material before undergoing an inelastic collision. For an electron with a kinetic energy of 965 eV (e.g. a C 1s electron photoemitted with Mg Kα radiation), the inelastic mean free path is about 1.5–2 nm. The sampling depth of the technique refers to depth from which 95% of the signal intensity arises:

$$\text{Sampling depth} = 3\lambda \cos \theta$$

where λ is the inelastic mean free path and θ is the angle between the energy analyser and the sample surface (Figure 11.5). The sampling depth for the 965 eV electron in a polymeric material is about 4.5–6 nm. The values of the mean free paths are still open to some debate. However, in the polymer literature for core levels ionised by Mg Kα radiation, the mean free paths for C 1s, O 1s, N 1s and F 1s electrons are often given as 1.5 nm, 1.1 nm, 1.0 nm, and 0.7 nm, respectively.

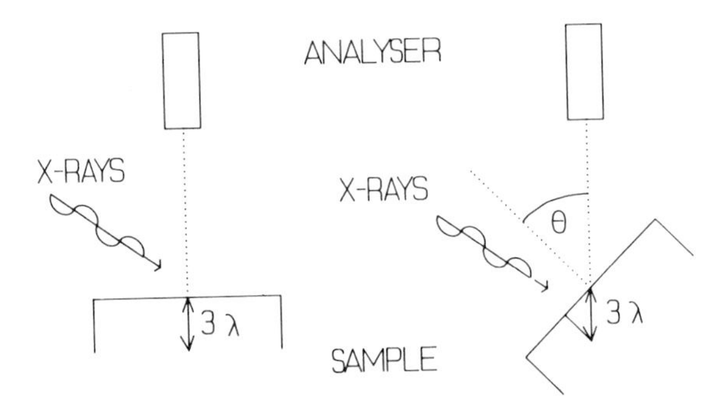

Figure 11.5 Schematic diagram for variable take-off angles.

By varying the angle θ, the surface sensitivity of the technique can be altered. The lateral and in some cases vertical homogeneity of the sample can be checked. Models for the interpretation of this type of data have been discussed, for example by Dilks [2].

11.2.4 Quantification

In addition to the identification of the elemental and functional groups within the surface, the XPS data can also be quantified. From a knowledge of the photoionisation cross-sections, the kinetic-energy dependence of the inelastic mean free path and the electron-analyser transmission function, instrument sensitivity factors can be calculated. By measuring the area under the core-level envelopes and multiplying by the appropriate sensitivity factors, the relative atomic composition can be calculated. As there is still some dispute regarding the kinetic-energy dependance of the mean free paths, it is advisable to determine the sensitivity factors experimentally.

On peak-fitting a core level, the areas of the individual components can be calculated and hence the functional group concentration. These data can then be used to estimate the intensity of another core level. In the case of PET, the C 1s components can be used to calculate the intensity of the O 1s core level. Comparison of the two sets of data can then allow (a) the sensibility of the peak fit to be checked, and (b) the homogeneity of the surface to be examined. In the latter case, the differences in the mean free paths between, for example, C 1s and O 1s photoelectrons means that the sampling depths of the two core levels are different. If the predicted O 1s intensity derived from the C 1s peak is in good agreement with the measured O 1s core level intensity, then the surface is probably vertically homogeneous. However, if the predicted value is less than the measured O 1s intensity then the surface is probably not vertically homogeneous.

Reference has been made to the identification of surface functional groups by XPS. Some caution is needed in the extent of interpretation of the data for two main reasons. The first relates to the limited dynamic range of chemical shifts in most core levels of interest to the polymer scientist. C 1s core levels have a chemical shift range of ~ 10 eV. For example, the binding energy of trifluoromethyl groups is shifted by ~ 10 eV relative to hydrocarbon. The small range of chemical shifts can result in several functional groups appearing at the same binding energy. Carbon singly bonded to oxygen can be present either as an ether or as a hydroxyl functionality. The chemical shift for both groups is 1.6 eV. Similarly, acids and esters have very similar chemical shifts (~ 4.2 eV). As a result, unambiguous identification of these functional groups is not possible. There have been attempts to overcome this problem by the use of derivatising agents [1]. The approach is to derivatise the functionality of interest, so that a new element is introduced (e.g. fluorine). While derivatisation allows further information to be gained about the surface, problems with the surface reorganising on exposure to derivatising agent means that quantification becomes ambiguous.

The second reason for caution is the concentration-sensitivity of the technique. Within a given core level, detection of a particular functional group is not possible below 0.5%. In most practical situations, if the area of the component peak is less than 1%, then it will not be detected. For example, in monitoring the oxidation of a polyolefin, the C 1s envelope will not exhibit discernable carbon–oxygen functionality until at least 1 in 100 of the carbon atoms present in the surface has been oxidised. When comparing this level of oxidation with that monitored in the bulk of the polymer, it is found that observable chemical changes in the surface correspond to extensive bulk oxidation. The moral of the above description is that if a species is not detected by XPS, it does not mean that it is not present.

11.2.5 Imaging

Since the first applications of XPS to polymers appeared in the 1970s, several advances in instrumentation have taken place, the most obvious being the use of computers to control data aquisition and to manipulate data. Improvements in electron optics and electron detection have lead to reduced acquisition times. Recent developments in the use of monochromated X-rays have lead to significant improvements in the energy resolution within a core level [1]. This latter advance may help to remove some of the ambiguities arising from the small range of chemical shifts. However, to the authors one of the most exciting new features of XPS is the development of imaging.

Over the last decade there has been a drive towards reducing the sampling area of XPS. It is now possible to analyse small spots (typical width 150–200 μm). As a result of this ability to obtain information from small areas, the logical next step, to image the sample surface, has now been achieved. A number of approaches to this problem have been successful [1]. At present it is possible to image at resolution of ~ 10 μm. An example of XPS imaging at low spatial resolution is shown in Figure 11.6. A spot of silicon grease was placed on to a PET film. The sample was imaged over an area of approximately 3 mm × 3 mm for silicon (Si 2p) and the ester peak from the C 1s envelope. These were chosen so as to be unambiguously representative of the silicon grease and the PET substrate. As can be seen from Figure 11.6, the silicon and ester maps complement one another. Although this is a relatively crude example of XPS imaging, it does indicate the potential of the technique. In the past, XPS allowed for measurement of vertical homogeneity. In the future, it will be routinely possible to check for lateral homogeneity. XPS imaging will become increasingly important in problem-solving applications. However, some technical progress is still required before it effectively complements imaging SSIMS.

(a)

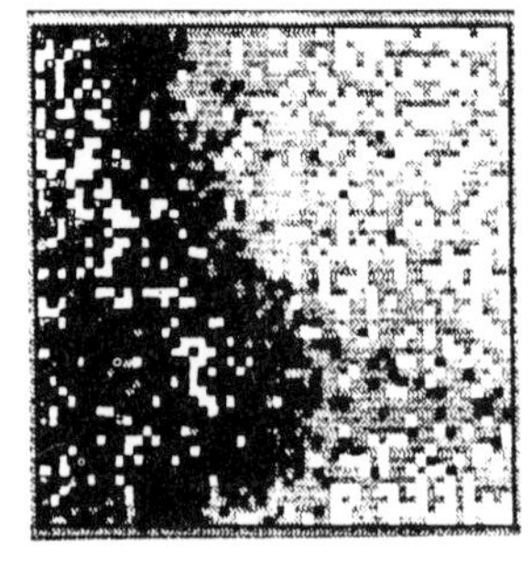

(b)

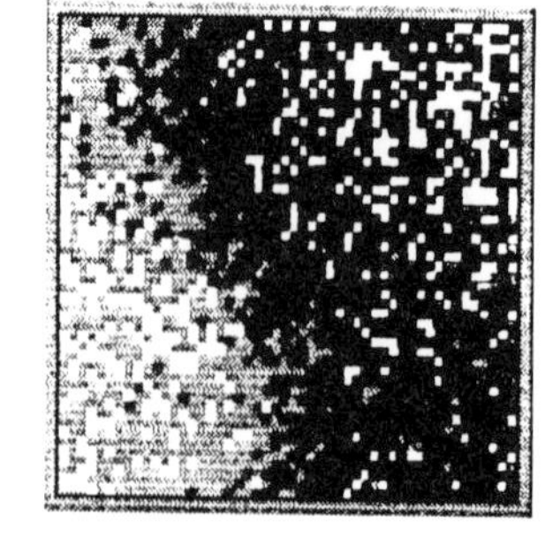

Figure 11.6 A 3 mm × 3 mm imaged area showing the distribution of silicon, Si 2p (a) and the ester peak from the C 1s envelope (b) to represent the silicon grease and PET respectively. White represents low signal and black represents high signal.

11.2.6 Applications

Typical applications of XPS include:

(a) Detection of surface contamination.
(b) Surface modification.

(c) Photo-oxidation and weathering.
(d) Polymer blends.
(e) Polymer–metal interactions.
(f) Biomaterials.

A bibliography of papers pertaining to these areas appears in ref. 1. It is not the intention of this section to give a review of the above areas when they have already been covered. It is the intention, however, to stress the importance of obtaining surface-compositional data in the above applications. Attempts to extrapolate what is known about the bulk of the material to the surface are fraught with danger. The surface is not always the same as the bulk. Hence, in trying to understand why a polymeric material behaves in a particular way, due consideration of the role that the surface plays is essential.

One of the most significant impacts on polymer technology that XPS has had to date is that on the surface modification of polymers. Polymers are frequently surface-treated to permit enhanced wetting, printing, adhesion, etc. Understanding the changes in the surface composition as a result of these treatments is the key to the development of the right method and extent of treatment. XPS has allowed considerable advances in this area to be made. For example, in the corona treatment of polyolefins, a combination of monitoring elemental and functional-group composition as functions of treatment parameters brought about a new model for explaining auto-adhesion in these materials. The interesting feature of the model is the proposal that keto–enol tautomerism across the interface was important.

XPS can be a rich source of information about the surface of a polymer. It is essential that all the available information levels that have been outlined above are taken into account in the spectral interpretation. On many occasions data interpretation could have been enhanced if a fuller appreciation of the technique's capabilities had been realised.

11.3 Secondary ion mass spectrometry (SIMS)

Bombardment of a surface of interest with a beam of primary projectiles (e.g. ions, electrons, atoms, etc.) results in the sputtering of material characteristic of that surface (see Figure 11.7). A small fraction of this sputtered material will be positively or negatively ionised species, termed secondary ions, the majority being neutral species and electrons. With the aid of purpose-designed ion-extraction optics, these secondary ions can be extracted and mass-analysed to yield a mass spectrum of the surface. This characterisation technique is known as secondary ion mass spectrometry, SIMS [4–6]. The detection of sputtered material is not limited to ionised species. Recent instrumentation developments [7–9] have shown that sputtered neutral species can also be

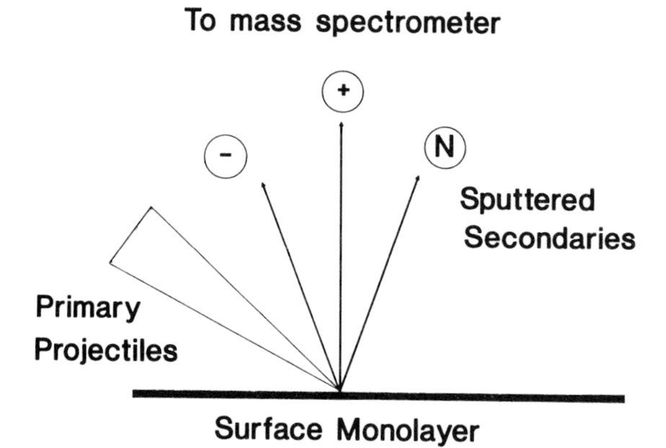

Figure 11.7 A schematic representation of the SIMS process.

subsequently ionised to give a mass spectrum. The emission of secondary ions after primary ion bombardment was first observed by J. J. Thompson [10] in 1910, but it was not until the 1950s that instrumentation was developed to exploit this phenomenon [11, 12].

The process of SIMS is, by definition, a destructive one and was therefore limited in its application until the early 1970s, when Benninghoven [13] and others demonstrated the possibility of obtaining chemical information from organic species with minimal specimen degradation. This involves reducing the primary beam flux to such a level that the probability of sampling a previously examined region on the specimen surface is considerably reduced. For a typical primary beam dose of 10^{10} projectiles cm^{-2}, there is approximately 1 in 10^5 [13, 14] chance of deriving data from an already investigated region. This assumes that the primary particle is approximately 10 Å in diameter. Under these instrumental parameters the technique is termed static secondary ion mass spectrometry, SSIMS [13, 15, 16]. Dynamic secondary ion mass spectrometry [17] (DSIMS) describes the condition where appreciable specimen degradation or modification occurs during characterisation. This usually takes place at primary beam doses of greater than 10^{14} projectiles cm^{-2}. In the DSIMS mode of operation, a high flux of primary ions is directed at the material's surface to obtain a very high yield of secondary ions. The surface is eroded very rapidly, it is possible to monitor changes in elemental composition with depth, and thus a depth profile can be generated [18]. The major application of DSIMS has been in the semiconductor industry [19].

11.3.1 The static SIMS experiment

The yield of secondary ions can be simply written as:

$$Y_s = I_p S R_{+/-} T \tag{11.1}$$

where Y_s = secondary ion current of mass species Y, I_p = primary beam flux, $R_{+/-}$ = ionisation probability of species Y, T = transmission of mass spectrometer, inclusive of fraction of Y collected, and S = sputter rate of Y.
Also, the monolayer lifetime is given by:

$$T_m = 10^{15}/S \quad (11.2)$$

where 10^{15} = density of atoms in the surface.

To increase monolayer lifetime and hence achieve SSIMS conditions, the sputter rate needs to be reduced (eqn (11.2)), and this can be realised by reducing the primary beam flux (eqn (11.1)). However, this also has the effect of reducing the secondary ion current. The choice of mass analyser is, therefore, most important for optimum spectrometer performance and to maximise the chemical information available from the material's surface.

11.3.2 Vacuum system

Due to the very low number of secondary ions produced in a SSIMS experiment, the first requirement for any mass spectrometer is a good vacuum system. This ensures that ions travelling through the instrument will not suffer any collisions with gas molecules which will cause them to be scattered or change their chemical state. To ensure this, the mean free path of the ions at the pressure within the vacuum system must be long, compared with the path length as governed by the spectrometer design.

The mean free path in any gas increases as the pressure is reduced. A simple approximation [20] from the kinetic theory of gases shows that the relationship between the mean free path, L, and the pressure, P, at a fixed temperature, T, and for a given type of gas, is given by:

$$L = k/P \quad (11.3)$$

where k is a constant. For nitrogen (and hence approximately for air) eqn (11.3) takes on the practical form of:

$$L = 5 \times 10^{-5}/p \quad (11.4)$$

Thus, if vacua of 10^{-6} or better can be produced, then the mean free path is greater than 50 m.

11.3.3 Mass analyser

Another requirement for a mass spectrometer is to use the most appropriate mass analyser. Several mass analysers with varying performance characteristics are available. However, the most commonly used in the SIMS experiment are magnetic-sector [21–23], quadrupole [24–26] and, recently, time-of-flight (TOF) [27–30] analysers.

11.3.3.1 Quadrupole mass analyser. A large number of SSIMS systems presently available incorporate the quadrupole mass analyser. This consists of four parallel metal rods of circular cross-section arranged as shown in Figure 11.8. Opposite pairs are then connected together and superimposed DC and RF voltages are applied. The action of the generated electric field is to mass-select low-energy (approximately 10 eV) ion species by varying either the RF/DC voltage amplitudes or the frequency of the RF voltage at a fixed DC value. The former method has the advantage of yielding a mass spectrum on a linear scale. Quadrupole mass analysers are generally of a small size; however, longer rod lengths have been used to obtain mass ranges in excess of 1000 atomic mass units (amu). The transmission, T, of secondary ions through this mass analyser is of the order of 0.1%.

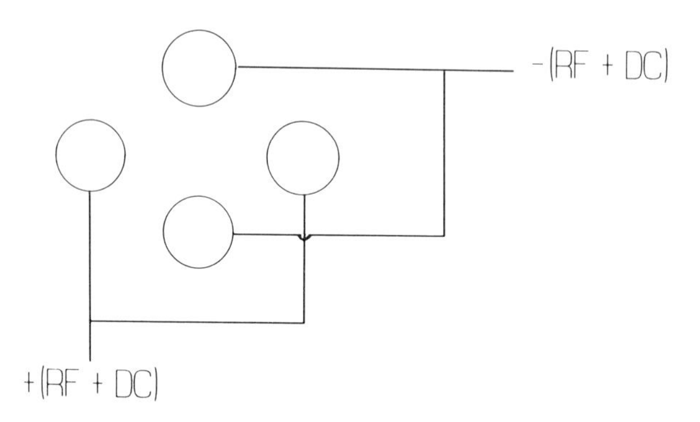

Figure 11.8 Schematic diagram showing the cross-sectional view of a quadrupole mass analyser.

11.3.3.2 Magnetic mass analyser. Where high mass resolution is a requirement in the SIMS experiment, then the traditional magnetic-sector analyser may be utilised. Presently, however, these magnetic mass analysers are used in the DSIMS mode of characterisation. The transmission of secondary ions is much higher than in the quadrupole system, approximately 1%. Due to its lack of usage within the SSIMS area it will not be described further.

11.3.3.3 Time-of-flight (TOF) mass analyser. The TOF mass analyser works on the principle that ions of differing mass/charge (m/e) ratios, when accelerated through the same potential, will travel a fixed length in different times [30]. Accurate time measurements can then be made to yield a mass spectrum through the utilisation of the relationship:

$$m/e = 2Vt^2l^{-2} \tag{11.5}$$

where t = measured time of ion travel, l = ion flight-path length, and V = accelerating potential.

The majority of TOF mass analysers used for SSIMS experiments are of the design put forward by Poschenrieder [27] or Mamyrin [28]. The main advantages of this type of analyser are its greater transmission capability (greater than 30%) and its quasi-simultaneous detection of all masses [31].

11.3.4 Ionisation process

An important feature of SIMS is the fact that considerable chemical information can be obtained through the procedurally simple physical technique of sputtering. While there has been rapid progress in many areas of SIMS, the generation of organic ionic species is still being debated and investigated. Several theories [6, 32, 33] do exist, with some experimental data. The most pertinent one with some relevance to organics is known as the desorption ionisation (DI) model [32], and is fundamentally based on the following criteria:

(a) Direct conversion of primary projectile kinetic energy into nett translational energy sufficient for a secondary ion to leave the surface, i.e. direct desorption of an ion.
(b) Ion–molecule and other chemical reactions that occur at the surface and in the 'selvedge' [34], the plasma formed at and immediately above the surface during sputtering, and lead to the transformation of molecules of interest into gas-phase ions.
(c) Secondary processes such as unimolecular dissociation of an excited secondary ion.

Three types of ionisation process can be distinguished as contributing to the ions generated in the SSIMS experiment. These are illustrated very simplistically by eqns (11.6)–(11.8) and describe: direct desorption of pre-charged material (eqn (11.6)), cationisation/anionisation (eqn (11.7)) and electron ionisation (eqn (11.8)):

$$C^+A^- \text{ (s)} \longrightarrow C^+ \text{ (g)} + A^- \text{ (g)} \qquad (11.6)$$

$$M \text{ (s)} \longrightarrow M \text{ (g)} \xrightarrow{+C^+} (M + C)^+ \text{ (g)} \qquad (11.7)$$

$$M \text{ (s)} \longrightarrow M \text{ (g)} \xrightarrow{-e^-} M^+ \text{ (g)} \qquad (11.8)$$

where s = solid phase, g = gas phase, C^+ = cation, A^- = anion, M = neutral parent molecule, and M^+ = molecular ion.

Desorption of precharged material (i.e. salts) is a highly efficient process, since energy is not translated into both an ionisation step and a desorption step; existing ions in the solid specimen are simply transferred to the gaseous phase. This effect can be seen in the SIMS spectra of quaternary ammonium salts. Cationisation or anionisation of neutral molecules by attachment [35, 36] of metal ions, protons and other charged species is the second most commonly observed ionisation process in SIMS. The third process, formation of radicals [37], is not such an efficient process but has been observed.

11.3.5 Surface charging during the SSIMS experiment

SIMS, in common with other surface characterisation methods that involve charged particles, can lead to surface charge build-up on insulating materials [38–40]. As most polymers are electrically insulating to varying degrees, this problem can lead to unreliable spectral interpretation due to induced instrumental aberrations. During bombardment in the SIMS experiment, the major sputtered species will be neutral species and electrons. With positive-ion bombardment, insulating surfaces will usually become positively charged [40]. The consequence of this charging is that positive ions are further accelerated during extraction to a point beyond the acceptance window of the mass analyser. This is a particular problem with quadrupole systems, which have a very narrow energy-acceptance window. In the case of negative ions, a positive surface charge can suppress their emission. Another consequence is that the surface potential will perturb the well-defined secondary-ion extraction optics, causing ions to traverse in directions other than the desired flight path into the mass analyser.

Various methods are employed to overcome this charging problem, including mixing graphite with the specimen, placing a conducting grid over the surface, and electrically biasing the specimen. Although improving the situation somewhat, these methods do not always provide a uniform surface potential. Another method, most commonly employed, is charge compensation by the use of a low-energy electron-flood gun [41, 42]. An important consideration in this type of charge neutralisation is that the correct flux of electrons needs to be employed, with the right kinetic energy. Although a suitable method of charge neutralisation, it can, if not used correctly, produce electron-stimulated ion emission (ESIE) in polymers [43]. Over-compensation can lead to the loss of signal in positive-ion spectra.

An example of an effective use of low-energy electrons to compensate for surface charging can be seen in Figure 11.9. Figure 11.9(a) shows a negative-ion mass spectrum derived from polytetrafluoroethylene (PTFE) with no compensation. No useful information is observed above 20 amu. This should be compared with Figure 11.9(b), where charge compensation has been optimised, and the high-mass peaks characteristic of PTFE are clearly identifiable.

Surface-charging problems can also be minimised by the use of neutral beams to bombard surfaces instead of primary ions, a technique known as fast atom bombardment mass spectrometry (FABMS) [44–47]. A disadvantage of this type of process is that chemical information is limited to a spatial resolution of a few millimetres only, and no information is available on the spatial distribution of specific chemical moieties.

11.3.6 Spectral interpretation

Mass-spectral peaks correspond to both the original molecule and fragments derived from it. In addition to the masses of these peaks, the empirical formula

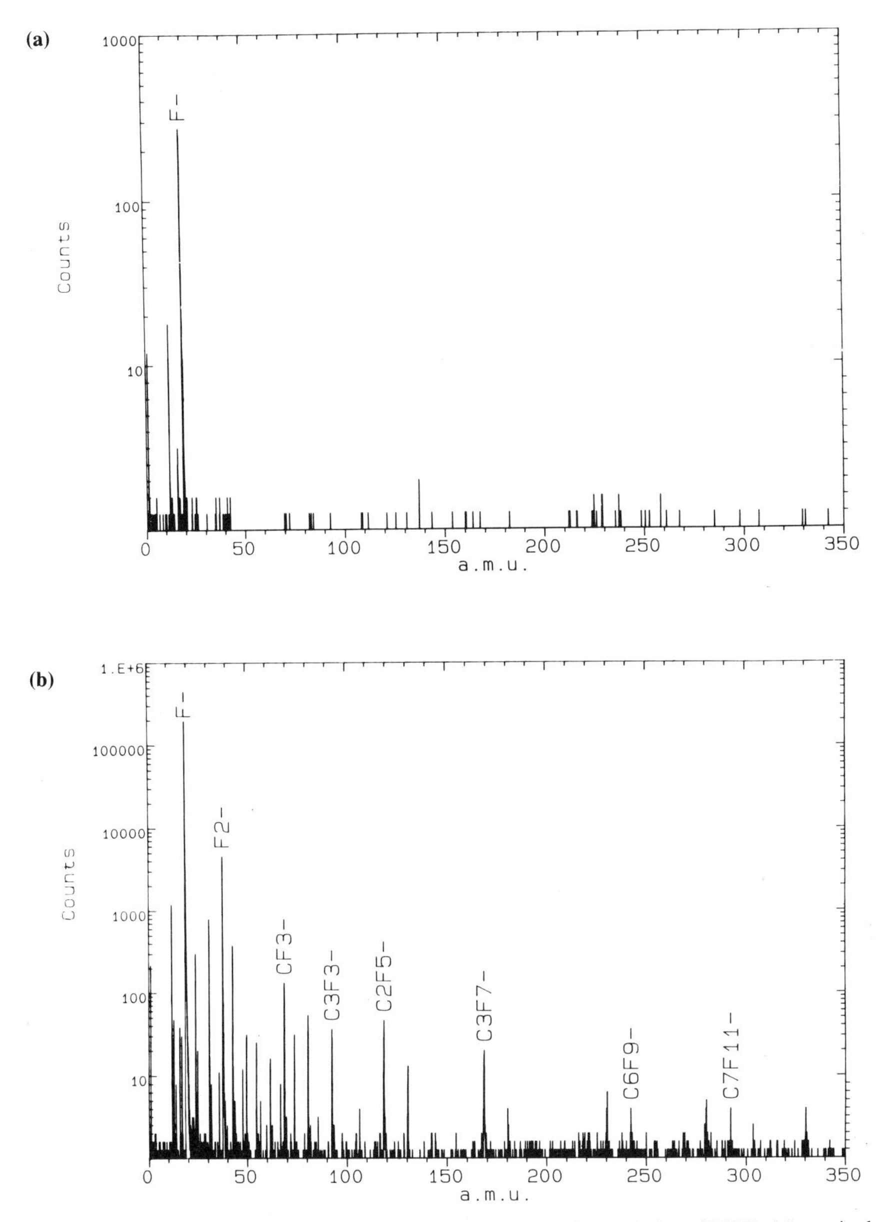

Figure 11.9 Negative-ion mass spectra derived from polytetrafluoroethylene (PTFE); (a) acquired after no charge compensation; (b) optimised charge compensation through the use of low-energy electrons. The ion dose was approx. 10^{12} ions cm^{-2}.

can also be deduced. A powerful method to achieve this is to use accurate mass-measurement techniques coupled with a high mass-resolution capability. These developments are currently underway within the SSIMS area and some progress has been made. However, even with instrumentation of unit-mass resolution, the presence of isotopes of known natural abundance makes possible a useful and simple method for deducing the empirical formulae.

The isotopic abundances of the elements can be classified into three general categories: 'A', elements with only one natural isotope: 'A + 1', elements that have two isotopes, the second of which is one mass-unit heavier than the most abundant isotope; 'A + 2', elements that have an isotope which is two mass-units heavier than the most abundant isotope. It is clear that various combinations of elements will result in specific isotopic-peak ratios in a mass spectrum. Statistically calculated relative isotopic abundances in various combinations, including carbon, hydrogen, oxygen, nitrogen, etc., exist, and provide a ready source of information for empirical formulae derivation [48, 49]. If, after the utilisation of exact mass calculation, fingerprint spectra (fragmentation pathway identification) and relative isotopic determination, further interpretation is still required, then derivatisation can offer complimentary information [50].

11.3.7 Application of SSIMS to polymer characterisation

It has been demonstrated that SIMS gives different fragmentation patterns for polymers that are structurally different, if the experiments are carried out in conditions of extremely low primary-ion current densities [51–55]. The possibility of obtaining structural information from polymer surfaces, not previously obtainable through utilisation of XPS, has stimulated research in this area since the late 1970s. Systematic studies [56–63] have been carried out by various research groups using mass spectrometers with high mass range and sensitivity. Optimum conditions for polymer surface characterisation were determined, in terms of primary-beam and extraction conditions, which resulted in lower beam damage, higher mass-range spectra and much improved negative-ion spectra.

The potential of SIMS has already been amply demonstrated within the scientific literature. The following series of practical examples serves to give an insight into the applicability of SSIMS to polymer surface characterisation.

11.3.7.1 Identification of plasticisers at polymer surfaces. Plasticisation, in general, refers to a change in the thermal and mechanical properties of a given polymer. This is normally achieved by either (a) compounding the given polymer with a low-molecular weight material, or (b) introducing into the

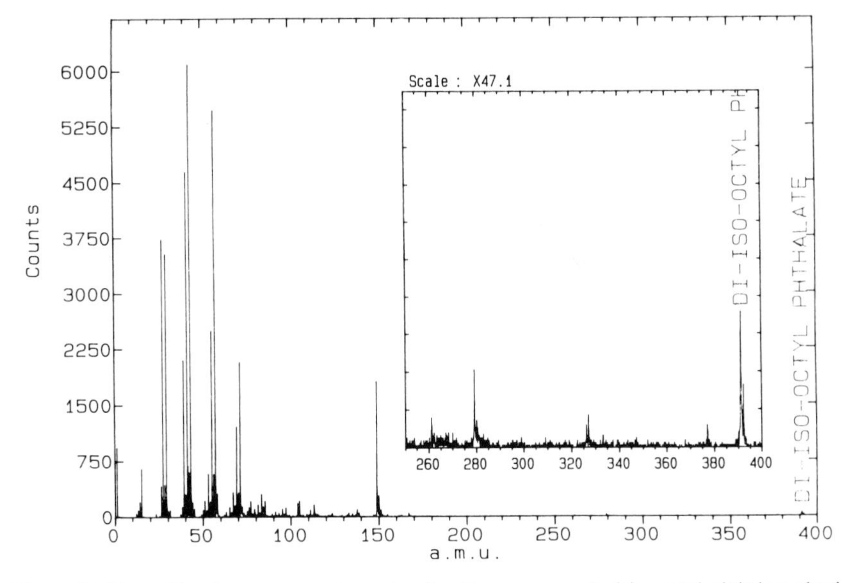

Figure 11.10 Positive-ion mass spectrum showing the presence of di-iso-octyl phthalate plasticiser on the surface of a poly(vinyl chloride)-based material. The ion dose was approx. 10^{12} ions cm^{-2}.

original polymer a comonomer that reduces crystallisability and increases chain flexibility. Plasticiser migration can lead to adverse surface-chemistry effects, resulting in loss of performance, for example, in areas such as adhesion and tribology. In these situations, SSIMS provides a means of monitoring the surface for any possible migration. Figure 11.10 is a positive-ion mass spectrum derived from the surface of a thick (5 mm) poly(vinyl chloride)-based material. The mass species at 391 amu is due to di-iso-octyl phthalate plasticiser. The lower masses are fragments of the plasticiser molecule and the polymer. Clearly, the presence of this plasticiser at the surface must be taken into consideration when assessing the applicability of this material. This type of migration can lead to undesirable surface chemistry effects and ultimately to impaired performance.

11.3.7.2 Surface distribution of chemical moieties. By blending polymers it is possible to obtain material properties not available through the utilisation of single constituents [64]. It has already been shown [65–67] that the surface composition of polymer blends may not necessarily be the same as that of the bulk. Models of the surface morphology have been proposed for various blend systems. In order to investigate the validity of these models chemical mapping of the surface morphology is required. The following example deals with the application of imaging SSIMS to investigate the surface morphology of a 40:60 ratio blend of poly(vinyl chloride)–poly(methyl methacrylate) (PVC–

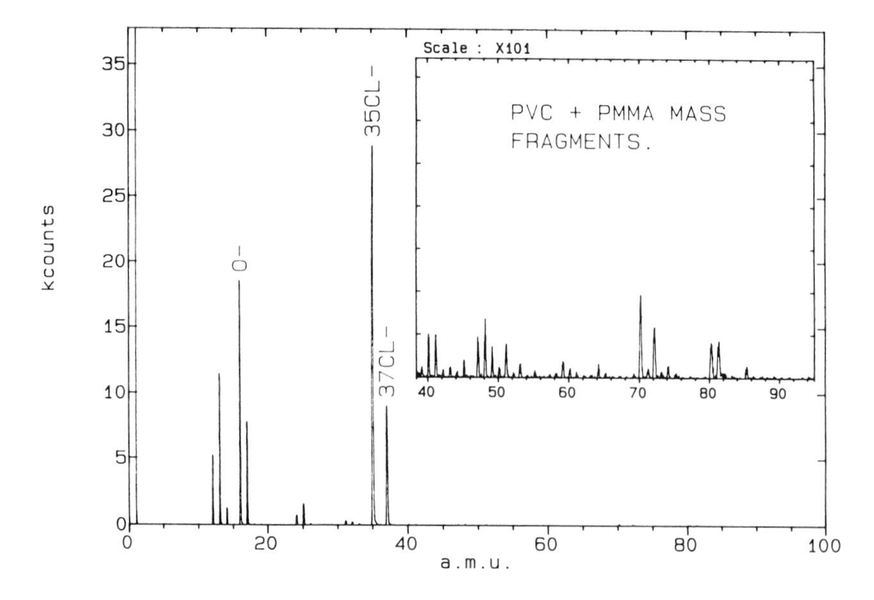

Figure 11.11 Negative-ion mass spectrum derived from a 40:60 ratio blend of PVC–PMMA. cast from THF. The ion dose was approx. 10^{12} ions cm^{-2}.

PMMA) cast from tetrahydrofuran (THF). This system has previously been reported to be phase-separated in the bulk [68, 69].

Chemical imaging is achieved by digitally raster-scanning the highly focused primary-ion beam across an area of the specimen's surface, and recording the sputtered secondary ions from each digital step (pixel). Specific ion species can be searched for and recorded in a pixel array to build up a chemical image. Figure 11.11 is a negative-ion mass spectrum; the ions O^- and OH^- are derived from PMMA, and Cl^- is derived from PVC. Figure 11.12 is a surface 'chemical picture' showing the distribution of PMMA and PVC. This clearly shows that this blend has phase-separated in the surface as well as in the bulk.

11.3.7.3 Characterisation of surfactants. Surfactants are surface-active species applied to a material to bring about a change in that material's surface energy [70]. This surface energy change can be specifically orchestrated to yield desirable material-performance characteristics [71]. As a consequence, surfactants have found applications in many chemical industries. Therefore, an understanding of the basic phenomenon involved in the application of surface-active components, for example in the preparation of emulsions [72] and suspensions [73], in microencapsulation [74], and in wetting and adhesion

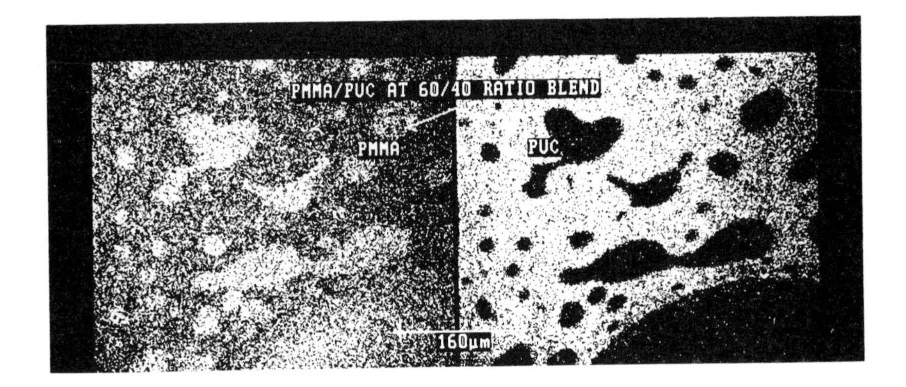

Figure 11.12 The surface chemical distribution of a 40:60 ratio blend of PVC–PMMA. cast from THF. Data were acquired by recording signals due to oxygen and chlorine to represent PMMA and PVC respectively. The ion dose was approx. 5×10^{12} ions cm^{-2}; the acquisition time was approx. 2 min.

[75], is important in arriving at the right formulation and control of the systems involved. A prerequisite for a fundamental understanding of the work of surfactants is to have the ability to characterise them. Since the applied surfactant resides on the surface, at very low concentrations, this characterisation needs to be undertaken with extreme surface and detection sensitivity. SSIMS offers this capability.

Fatty acids and fatty acid–poly(ethylene glycol)-based esters (PEG esters) are commonly used as surfactants because of their complementary hydrophilic and hydrophobic nature. Figure 11.13(a) is a positive-ion mass spectrum showing initially what appears to be the distribution of the fatty acids [4]: *n*-octanoic, *n*-decanoic, *n*-dodecanoic, *n*-tetradecanoic, *n*-hexadecanoic and *n*-octadecanoic acids, derived from the surface of a polymeric fibre specimen. However, this should be considered with reference to the negative-ion spectrum, Figure 11.13(b), in which an identical peak pattern is observed, but shifted down in mass by 28 amu. Also, the peak integrals associated with these fatty acids agree with the quoted mix ratios and may be used as a specimen purity check. Considering both the positive- and the negative-ion spectra, it can be inferred that the positive ions are derived from the PEG ester, leaving the $-CH_2CH_2$ fragment from the PEG attached to the fatty acid end. The result is a shift of 28 amu in the masses of the fatty acids (eqn (11.9), reaction *a*). The negative ions can be formed via two mechanisms: (i) through the fragmentation of the PEG ester (eqn (11.9), reaction *b*), or (ii) loss of a proton from the free fatty acid (eqn (11.10), reaction *c*). It can be shown that by varying the concentration of fatty acid in PEG ester–fatty acid mixture, the

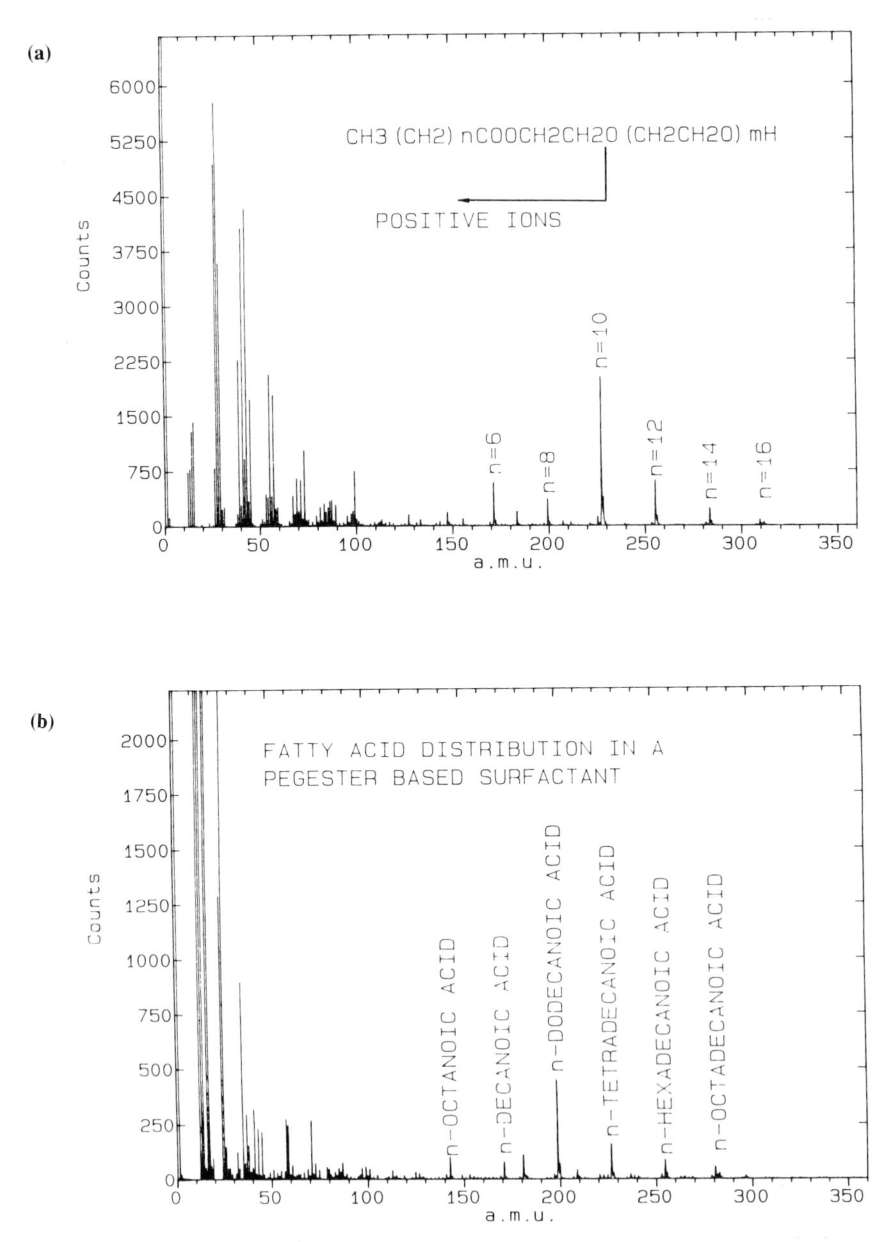

Figure 11.13 SSIMS spectra derived from a surfactant on a polymeric fibre specimen. The ion dose was approx. 10^{12} ions cm^{-2}; (a), the positive-ion spectrum showing the PEG ester distribution; (b), the negative-ion spectrum showing the distribution of fatty acids.

dominant mechanism resulting in the observed negative ions is the proton loss from the free fatty acid (eqn (11.10), reaction *c*).

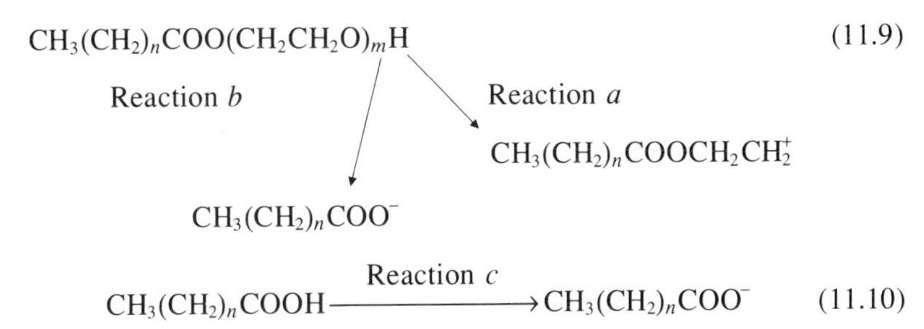

$$CH_3(CH_2)_nCOOH \xrightarrow{\text{Reaction } c} CH_3(CH_2)_nCOO^- \qquad (11.10)$$

It could have been inferred, through the sole consideration of the masses in the positive-ion spectrum, that the PEG ester was being witnessed: fatty acids give a dominant $(M+H)^+$ peak [1] in the SSIMS spectrum, whereas Figure 11.13(a) shows a dominant $(M-H)^+$ peak. However, this would have resulted in the loss of information on fatty acids as well as a possible method for deducing the relative concentration ratio of fatty acid and PEG ester, information that could be vital to the effective working of the surfactant. This particular example shows the importance of considering fragmentation patterns observed in both positive- and negative-ion spectra in order to improve the accuracy of data interpretation. Figure 11.14 is a chemical image showing the surface distribution of one of these esters based on a fatty acid, namely *n*-dodecanoic acid, on the surface of the fibres.

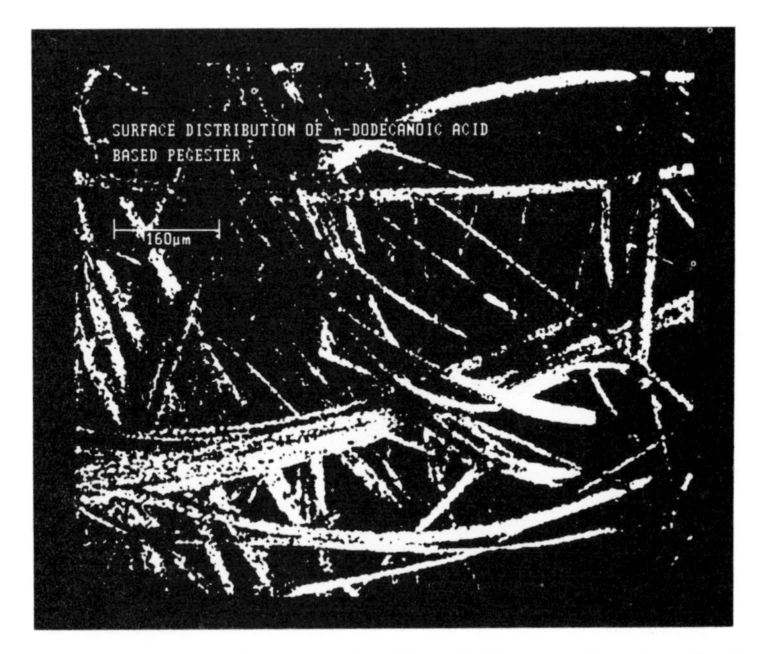

Figure 11.14 The distribution of *n*-dodecanoic acid-based PEG ester on the surface of a polymeric fibrous specimen. The ion dose was approx. 5×10^{12} ions cm^{-2}; the acquisition time was approx. 2 min.

11.3.8 SIMS summary

Since its advent in the early 1970s, SSIMS has found many areas of application where it is routinely used to provide information on the chemistry of surfaces. Typical examples from industry include adhesives, pharmaceuticals, oxides and catalysts. With respect to polymers (and interfaces) the major effort has been into gaining an understanding of the chemical and physical nature of the surface. If effectively used, SSIMS can provide information with a high degree of chemical specificity, with good signal-to-noise ratio and on monolayers of material. However, SSIMS is not the universal tool for surface characterisation. Other techniques are frequently used to confirm, complement and quantify the analysis. The combination, therefore, of SSIMS with other surface spectroscopic techniques will certainly prove to be a powerful methodology.

Acknowledgements

The authors are grateful for the kind assistance of Kratos Analytical plc for providing Figures 11.1 and 11.5, and to Dr Robert Short and Stuart Jackson for providing the PVC–PMMA specimen.

References

1. D. Briggs and M. P. Seah (eds) (1990). *Practical Surface Analysis*, Vol. 1, Wiley.
2. A. Dilks (1981). In *Electron Spectroscopy—Theory, Techniques and Applications*, C. R. Brundle and A. D. Baker (eds), Academic Press, London.
3. A. P. Pijpers and W. A. B. Donners (1985). *J. Polym. Sci., Polym. Chem. Ed.*, **33**, 453.
4. D. Briggs, A. Brown and J. C. Vickerman (1986). *Handbook of Static Secondary Ion Mass Spectrometry*, John Wiley and Sons.
5. A. Benninghoven (1973). *Surf. Sci.*, **35**, 427.
6. J. C. Vickerman, A. Brown and N. M. Reed (eds) (1989). *Secondary Ion Mass Spectrometry, Principles and Applications*, Clarendon Press, Oxford.
7. G. R. Kinsel, J. Lindner and J. Grotemeyer (1992). *J. Phys. Chem.*, **96**(7), 3162–67.
8. C. H. Becker and K. T. Gillen (1984). *Anal. Chem.*, **56**, 1671–74.
9. N. Winograd, J. P. Baxter and F. M. Kimock (1982). *Chem. Phys. Lett.*, **88**, 581.
10. J. J. Thompson (1910). *Phil. Mag.*, **20**, 252.
11. R. F. K. Herzog and F. P. Viehbock (1949). *Phys. Rev.*, **76**, 855L.
12. R. E. Honig (1958). *J. Appl. Phys.*, **29**, 549.
13. A. Benninghoven (1970). *Z. Physik*, **230**, 230.
14. S. J. Pachuta and R. G. Cooks (1985). *Amer. Chem. Soc. Symp. Ser.*, **291**, 1–47.
15. G. J. Legett, J. C. Vickerman and D. Briggs (1991). *Surf. Interface Anal.*, **17**, 737.
16. D. Briggs (1984). Polymer, **25**, 1376.
17. H. Zeininger and R. V Criegern (1990). In *Secondary Ion Mass Spectrometry, SIMSVII*, John Wiley and Sons.
18. E. Zinner (1983). *J. Electrochem. Soc.*, **130**, 199C.
19. E. A. Clark, M. G. Dowsett, H. S. Fox and S. M. Newstead (1990). In *Secondary Ion Mass Spectrometry, SIMSVII*, John Wiley and Sons.
20. J. H. Beynon and A. G. Brenton (1982). *An Introduction to Mass Spectrometry*, University of Wales Press, Cardiff.
21. A. J. Dempster (1918). *Phys. Rev.*, **11**, 316.
22. F. W. Aston (1919). *Phil. Mag.*, **38**, 707.
23. J. J. Thompson (1907). *Phil. Mag.*, **13**, 561.

24. W. H. Paul and H. Steinwedel (1953). *Z. Natur.*, **8a**, 316.
25. J. E. Campana (1980). *Int. J. Mass Spectrom. Ion Phys.*, **33**, 101.
26. G. Lawson and J. F. J. Todd (1974). *Chem. Brit.*, **8**, 373.
27. W. P. Poschenrieder (1972). *Int. J. Mass Spec. Ion Phys.*, **9**, 357.
28. B. A. Mamyrin, V. J. Karataev, D. V. Shmikk and V. A. Zagullin (1973). *Sov. Phys. JETP*, **37**, 45.
29. B. A. Mamyrin and D. V. Shmikk (1979). *Sov. Phys. JETP*, **49**, 762.
30. W. E. Stephens (1946). *Phys. Rev.*, **69**, 691.
31. J. C. Vickerman, A. Brown and N. M. Reed (1989). Chapter 4 in Ref. 6.
32. R. G. Cooks and K. L. Busch (1983). *Int. J. Mass Spectrom. Ion Phys.*, **53**, 111.
33. R. J. Day, S. E. Unger and R. G. Cooks (1980). *Anal. Chem.*, **52**, 557A.
34. F. Honda, G. M. Lancaster, Y. Fakuda and J. W. Rabalais (1978). *J. Chem. Phys.*, **69**, 4931.
35. A. Benninghoven and W. K. Sichtermann (1978). *Anal. Chem.*, **50**, 1180.
36. A. Benninghoven (1983). In *Ion Formation from Organic Solids, Proc. 2nd International Conf.*, Münster, 1982, A. Benninghoven (ed.), Springer-Verlag.
37. R. Short and M. Davies (1989). *Int. J. Mass Spectrom. Ion Proc.*, **89**, 149.
38. D. Briggs, M. J. Hearn, I. W. Fletcher, A. R. Waugh and B. J. McIntosh (1990). *Surf. Interface Anal.*, **15**, 62–65.
39. C. P. Hunt, C. T. Stoddart and M. P. Seah (1981). *Surf. Interface Anal.*, **3**, 157–60.
40. A. Brown and J. C. Vickerman (1986). *Surf. Interface Anal.*, **8**, 75.
41. J. A. Gardella, Jr and D. H. Hercules (1980). *Anal. Chem.*, **52**, 226.
42. J. C. Vickerman, A. Brown and N. M. Reed (1989). Chapter 7 in Ref. 6.
43. D. Briggs and A. B. Wooton (1982). *Surf. Interface Anal.*, **4**, 109.
44. I. M. Deviene (1973). *Vide*, **167**, 193.
45. M. Barber, R. J. Bordoli, R. D. Sedgwick and A. N. Taylor (1981). *Nature*, **293**, 270.
46. D. J. Surman and J. C. Vickerman (1981). *Appl. Surf. Sci.*, **9**, 109.
47. D. J. Surman, J. A. Van den berg and J. C. Vickerman (1982). *Surf. Interface Anal.*, **4**, 160.
48. F. W. McLafferty (1980). In *Interpretation of Mass Spectra*, N. J. Turro (ed.), University Science Books, Cornell University, USA.
49. J. H. Beynon and A. E. Williams. Mass Abundance Tables for Use in Mass Spectrometry, Elsevier, Amsterdam.
50. For example: M. M. Ross, D. A. Kidwell and R. J. Colton (1985). *Int. J. Mass Spectrom. Ion Proc.*, **63**, 141.
51. See Ref. 41.
52. J. A. Gardella, Jr and D. M. Hercules (1981). *Anal. Chem.*, **53**, 1879.
53. J. E. Campana, J. J. Decorpo and R. J. Cotton (1981). *Appl. Surf. Sci.*, **8**, 337.
54. W. L. Baun (1980). *Appl. Surf. Sci.*, **6**, 39.
55. W. L. Baun (1982). *Pure Appl. Chem.*, **54**, 323.
56. D. Briggs (1982). *Surf. Interface Anal.*, **4**, 151.
57. D. Briggs (1983). *Surf. Interface Anal.*, **5**, 113.
58. D. Briggs, M. J. Hearn and B. D. Ratner (1984). *Surf. Interface Anal.*, **6**, 184.
59. D. Briggs and M. J. Hearn (1985). *Int. J. Mass Spectrom Ion Proc.*, **67**, 47.
60. D. Briggs (1986). *Surf. Interface Anal.*, **9**, 391.
61. A. Brown and J. C. Vickerman (1984). *Surf. Interface Anal.*, **6**, 1.
62. A. Brown and J. C. Vickerman (1984). *Analyst*, **109**, 851.
63. D. Briggs, A. Brown, J. A. Van den berg and J. C. Vickerman (1983). In *Ion Formation from Organic Solids*, A. Benninghoven (ed.), Springer-Verlag, p. 162.
64. D. J. Walsh, J. S. Higgins and A. Maconnachie (eds) (1985). *Polymer Blends and Mixtures*, Martinus Nijhoff, Boston.
65. M. B. Clark, Jr, C. A. Burkhardt and J. A. Gardella, Jr (1991). *Macromolecules*, **24**, 799.
66. Q. S. Bhatia, D. H. Pan and J. T. Koberstein (1988). *Macromolecules*, **21**, 2166.
67. R. Chujo, T. Nishi, Y. Sumi, T. Adachi, H. Naito and H. Frenzel (1984). In *Secondary Ion Mass Spectrometry, SIMSIV*, Springer-Verlag, Berlin.
68. D. J. Walsh and J. G. McKeown (1980). *Polymer*, **21**, 1330.
69. J. N. Razinskaya, B. P. Shtarkman, L. L. Butayeva, B. S. Tyves and M. N. Shlykova (1979). *Vysokomol. Soedin.*, **A21**, 1860.
70. A. W. Adamson (1982). *Physical Chemistry of Surfaces*, 4th edn, Wiley-Interscience, New York, Chapter 2.

71. For example: H. Schonhorn (1963). *J. Polym. Sci. A*, **1**, 1860.
72. Th. F. Tadros (ed.) (1984). *Surfactants*, Academic Press, Chapter 8.
73. Th. F. Tadros (ed.) (1984). *Surfactants*, Academic Press, Chapter 9.
74. T. Ogawa, K. Takamura, M. Koishi and T. Kondo (1972). *Bull. Chem. Soc. Japan*, **45**, 2329–31.
75. Th. F. Tadros (ed.) (1984). *Surfactants*, Academic Press, Chapter 10.

Index

WITHDRAWN